Artificial Intelligence and Games

Georgios N. Yannakakis
Julian Togelius

Artificial Intelligence and Games

Georgios N. Yannakakis
Institute of Digital Games
University of Malta
Msida
Malta

Julian Togelius
Department of Computer Science
and Engineering
New York University
Brooklyn, NY
USA

ISBN 978-3-319-87576-7 ISBN 978-3-319-63519-4 (eBook)
https://doi.org/10.1007/978-3-319-63519-4

Softcover re-print of the Hardcover 1st edition 2018

Cover image by Rebecca Portelli

Printed on acid-free paper

This Springer imprint is published by the registered company Springer
International Publishing AG part of Springer Nature
The registered company address is: Gewerbestrasse 11, 6330 Cham, Switzerland

To our families

Foreword

It is my great pleasure to write the foreword for this excellent and timely book. Games have long been seen as the perfect test-bed for artificial intelligence (AI) methods, and are also becoming an increasingly important application area. Game AI is a broad field, covering everything from the challenge of making super-human AI for difficult games such as Go or *StarCraft*, to creative applications such as the automated generation of novel games.

Game AI is as old as AI itself, but over the last decade the field has seen massive expansion and enrichment with the inclusion of video games, which now comprise more than 50% of all published work in the area and enable us to address a broader range of challenges that have great commercial, social, economic and scientific interest. A great surge in research output occurred in 2005, coinciding with both the first IEEE Symposium (Conference) on Computational Intelligence and Games (CIG)—which I co-chaired with Graham Kendall—and the first AAAI AIIDE Conference (Artificial Intelligence in Digital Entertainment). Since then this rich area of research has been more explored and better understood. The Game AI community pioneered much of the research which is now becoming (or about to become) more mainstream AI, such as Monte Carlo Tree Search, procedural content generation, playing games based on screen capture, and automated game design.

Over the last decade, progress in deep learning has had a profound and transformational effect on many difficult problems, including speech recognition, machine translation, natural language understanding and computer vision. As a result, computers can now achieve human-competitive performance in a wide range of perception and recognition tasks. Many of these systems are now available to the programmer via a range of so-called cognitive services. More recently, deep reinforcement learning has achieved ground-breaking success in a number of difficult challenges, including Go and the amazing feat of learning to play games directly from screen capture (playing from pixels). It is fascinating to contemplate what this could mean for games as we stumble towards human-level intelligence in an increasing number of areas. The impacts will be significant for the intelligence of in-game characters, the way in which we interact with them and for the way games are designed and tested.

This book makes an enormous contribution to this captivating, vibrant area of study: an area that is developing rapidly both in breadth and depth as AI is able to cope with a wider range of tasks (and to perform those tasks to increasing levels of excellence). The service to the community will be felt for many years to come: the book provides an easier and more comprehensive entry point for newcomers to the field than previously available, whilst also providing an indispensable reference for existing AI and Games researchers wishing to learn about topics outside their direct field of interest.

Georgios Yannakakis and Julian Togelius have been involved with the field ever since its widespread expansion to video games, and they both presented papers at the first 2005 CIG. Over the years they have made an enormous contribution to the field with a great number of highly cited papers presenting both novel research and comprehensive surveys. It is my opinion that these authors are best qualified to write this book, and they do not disappoint. The book will serve the community very well for many years to come.

London, August 2017 *Simon Lucas*

Preface

Of all the things that wisdom provides for the complete happiness of one's entire life, by far the greatest is friendship.

Epicurus, *Principal Doctrines*, 27

Human beings, viewed as behaving systems, are quite simple. The apparent complexity of our behavior over time is largely a reflection of the complexity of the environment in which we find ourselves.

Herbert A. Simon

It would be an understatement to say that **Artificial Intelligence** (AI) is a popular topic at the moment, and it is unlikely to become any less important in the future. More researchers than ever work on AI in some form, and more non-researchers than ever are interested in the field. It would also be an understatement to say that **games** are a popular application area for AI research. While board games have been central to AI research since the inception of the field, video games have during the last decade increasingly become the domain of choice for testing and showcasing new algorithms. At the same time, video games themselves have become more diverse and sophisticated, and some of them incorporate advances in AI for controlling non-player characters, generating content or adapting to players. Game developers have increasingly realized the power of AI methods to analyze large volumes of player data and optimize game designs. And a small but growing community of researchers and designers experiment with ways of using AI to design and create complete games, automatically or in dialog with humans. It is indeed an exciting time to be working on AI and games!

This is a book about **AI and games**. As far as we know, it is the first *comprehensive* textbook covering the field. With comprehensive, we mean that it features all the major application areas of AI methods within games: game-playing, content generation and player modeling. We also mean that it discusses AI problems in many different types of games, including board games and video games of many genres. The book is also comprehensive in that it takes multiple perspectives of AI and games: how games can be used to test and develop AI, how AI can be used to make games better and easier to develop, and to understand players and design. While this is an academic book which is primarily aimed at students and researchers, we will frequently address problems and methods relevant for game designers and developers.

We wrote this book based on our long experience doing research on AI for games, each on our own and together, and helping lead and shape the research community. We both independently started researching AI methods in games in 2004, and we have been working together since 2009. Together, we played a role in introducing research topics such as procedural content generation and player modeling to the academic research community, and created several of the most widely used game-based AI benchmarks. This book is in a sense a natural outgrowth of the classes on AI and games we have taught at three universities, and the several survey papers of the field and of individual research topics within it that we have published over the years. But the book is also a response to the lack of a good introductory book for the research field. Early discussions on writing such a book date back at least a decade, but no-one actually wrote one, until now.

It could be useful to point out what this book is not. It is not a hands-on book with step-by-step instructions on how to build AI for your game. It does not feature discussions of any particular game engine or software framework, and it does not discuss software engineering aspects or many implementation aspects at all. It is not an introductory book, and it does not give a gentle introduction to basic AI or game design concepts. For all these roles, there are better books available.

Instead, this is a book for readers who already understand AI methods and concepts to the level of having taken an introductory AI course, and the introductory computer science or engineering courses that led up to that course. The book assumes that the reader is comfortable with reading a pseudocode description of an algorithm and implementing it. Chapter 2 is a summary of AI methods used in the book, but is intended more as a reference and refresher than as an introduction. The book also assumes a basic familiarity with games, if not designing them then at least playing them.

The use case for this textbook that we had in mind when writing it is for a one- or a two-semester graduate-level or advanced undergraduate level class. This can take several different shapes to support different pedagogical practices. One way of teaching such a class would be a traditional class, with lectures covering the chapters of the book in order, a conventional pen-and-paper exam at the end, and a small handful of programming exercises. For your convenience, each of the main chapters of the book include suggestions for such exercises. Another way of organizing a class around this book, more in line with how we personally prefer to teach such courses, is to teach the course material during the first half of the semester and spend the second half on a group project.

The material offered by this book can be used in various ways and, thus, support a number of different classes. In our experience, a traditional two-semester class on game artificial intelligence would normally cover Chapter 2 and Chapter 3 in the first semester and then focus on alternative uses of AI in games (Chapters 4 and 5) in the second semester. When teaching the material in compressed (one-semester) fashion instead, it is advisable to skip Chapter 2 (using it as a reference when needed), and focus the majority of the lectures on Chapters 3, 4 and 5. Chapters 6 and 7 can be used as material for inspiring advanced graduate-level projects in the area. Beyond the strict limits of game AI, Chapter 4 (or sections of it) can complement classes with a focus on game design or computational creativity whereas Chapter 5 can complement classes with a focus on affective computing, user experience, and data mining. It is of course also possible to use this book for an introductory undergraduate class for students who have not taken an AI class before, but in that case we advise the instructor to select a small subset of topics to focus on, and to complement the book with online tutorials on specific methods (e.g., best-first search, evolutionary computation) that introduce these topics in a more gentle fashion than this book does.

Chania, Crete, Greece *Georgios N. Yannakakis*
New York, NY, USA *Julian Togelius*

September 2017

Acknowledgments

Writing a book of this size would be impossible without the support and the contributions of a large number of people. First, we would like to thank Ronan Nugent, our editor at Springer, who guided us and helped us from the book proposal phase all the way to its final production.

We also would like to thank the following persons who read all (or parts of a draft of) the book and provided useful feedback: Amy Hoover, Amin Babadi, Sander Bakkes, Vadim Bulitko, Phil Carlisle, Georgios Chalkiadakis, Dave Churchill, Mike Cook, Renato Cunha, Kevin Dill, Nathaniel Du Preez-Wilkinson, Chris Elion, Andy Elmsley, David Fogel, Bernardo Galvão, Kazu-ma Hashimoto, Aaron Isaksen, Emil Johansen, Mark Jones, Niels Justesen, Graham Kendall, Jakub Kowalski, Antonios Liapis, Nir Lipovetzky, Jhovan Mauricio López, Simon Lucas, Jacek Mańdziuk, Luciana Mariñelarena-Dondena, Chris Martens, Sean Mealin, Mark Nelson, Sven Neuhaus, Alexander Osherenko, Santiago Ontañón, Cale Plut, Mike Preuss, Hartmut Procha-ska, Christoffer Holmgård, Florian Richoux, Sebastian Risi, Christoph Salge, Andrea Schiel, Jacob Schrum, Magy Seif El-Nasr, Adam Smith, Gillian Smith, Dennis Soemers, Nathan Sturtevant, Gabriel Synnaeve, Nicolas Szilas, Varunyu Vorachart, James Wen, Marco Wiering, Mark Winands, Junkai Lu, Francesco Calimeri, Diego Pérez Liébana, Corine Jacobs, Junkai Lu, Hana Rudova, and Robert Zubek. Of these, we wish to especially thank Simon Lucas for writing the foreword of the book, Georgios Chalkiadakis for providing substantial input on parts of the book related to game theory, and Mark Nelson, Antonios Liapis, Mike Preuss and Adam Smith for reviewing large parts of the book and providing particularly detailed feedback. We also want to thank all those who granted us their permission to reproduce images and figures from their papers; they are all acknowledged in the figure captions of the book. Special thanks also go to Rebecca Portelli and Daniel Mercieca for the artwork featured on the cover page of the book.

Some chapters of this book build on papers or other book chapters that we have co-authored. In some cases the papers are co-authored by more than the two of us; for those papers our co-authors graciously gave us permission to reuse parts of the material and we wish to thank them for that. In particular,

- Chapter 1: [764, 700].
- Chapter 4: Chapter 2 and Chapter 3 from [616], and [381].
- Chapter 5: [778, 176, 782, 781].
- Chapter 6: [785].
- Chapter 7: [718, 458].

Writing this book has been a very long journey for both of us; a challenging and overwhelming journey at times. There are a number of people who have supported this effort that we would like to thank. Georgios wishes to thank the people at NYU Tandon School of Engineering for hosting him during the early days of book planning and the people at the Technical University of Crete for hosting him during the later stages of this project. Georgios would also like to thank the University of Malta for granting him a sabbatical leave, without which the book would not have been possible. Georgios and Julian also wish to thank each other for putting up with each other despite everything, and announce that they intend to buy each other drinks to celebrate when they meet next.

Last but not least, both of us wish to thank our families and all the people who showed us their support, care, encouragement and love at times when they were absolutely needed. You are too many to list here, but big thanks go to you all! Georgios would not have been able to write this book without the love and support of his family: Amyrsa and Myrto have been his core inspiration, Stavroula has been the main driving force behind the writing at all times; this book is dedicated to you!

Contents

Part III The Road Ahead

Acronyms

A3C	Asynchronous Advantage Actor-Critic
ABL	A Behavior Language
AFC	Alternative Forced Choice
AI	Artificial Intelligence
AIIDE	Artificial Intelligence and Interactive Digital Entertainment
ALE	Arcade Learning Environment
ANN	Artificial Neural Network
ASP	Answer Set Programming
BDI	Belief-Desire-Intention
BT	Behavior Tree
BWAPI	Brood War API
CA	Cellular Automata
CI	Computational Intelligence
CIG	Computational Intelligence and Games
CFR	Counterfactual Regret Minimization
CMA-ES	Covariance Matrix Adaptation Evolution Strategy
CNN	Convolutional Neural Network
CPPN	Compositional Pattern Producing Network
DQN	Deep Q Network
EA	Evolutionary Algorithm
ECG	Electrocardiography
EDPCG	Experience-Driven Procedural Content Generation
EEG	Electroencephalography
EMG	Electromyography
FPS	First-Person Shooter
FSM	Finite State Machine
FSMC	Functional Scaffolding for Musical Composition
GA	Genetic Algorithm
GDC	Game Developers Conference
GGP	General Game Playing
GSP	Generalized Sequential Patterns

GSR	Galvanic Skin Response
GVGAI	General Video Game Artificial Intelligence
HCI	Human-Computer Interaction
ID3	Iterative Dichotomiser 3
JPS	Jump Point Search
LSTM	Long Short-Term Memory
MCTS	Monte Carlo Tree Search
MDP	Markov Decision Process
MLP	Multi-Layer Perceptron
MOBA	Multiplayer Online Battle Arenas
NEAT	NeuroEvolution of Augmenting Topologies
NES	Natural Evolution Strategy
NLP	Natural Language Processing
NPC	Non-Player Character
PC	Player Character
PCG	Procedural Content Generation
PENS	Player Experience of Need Satisfaction
PLT	Preference Learning Toolbox
RBF	Radial Basis Function
ReLU	Rectified Linear Unit
RPG	Role-Playing Game
RTS	Real-Time Strategy
RL	Reinforcement learning
TCIAIG	Transactions on Computational Intelligence and AI in Games
TD	Temporal Difference
ToG	Transactions on Games
TORCS	The Open Racing Car Simulator
TRU	Tomb Raider: Underworld
TSP	Traveling Salesman Problem
SC:BW	StarCraft: Brood War
SOM	Self-Organizing Map
STRIPS	Stanford Research Institute Problem Solver
SVM	Support Vector Machine
UT2k4	Unreal Tournament 2004
VGDL	Video Game Description Language

Website

http://gameaibook.org/

This book is associated with the above website. The website complements the material covered in the book with up-to-date exercises, lecture slides and readings.

Part I
Background

Chapter 1
Introduction

Artificial Intelligence (AI) has seen immense progress in recent years. It is both a thriving research field featuring an increasing number of important research areas and a core technology for an increasing number of application areas. In addition to algorithmic innovations, the rapid progress in AI is often attributed to increasing computational power due to hardware advancements. The success stories of AI can be experienced in our daily lives through its many practical applications. AI advances have enabled better understanding of images and speech, emotion detection, self-driving cars, web searching, AI-assisted creative design, and game-playing, among many other tasks; for some of these tasks machines have reached human-level status or beyond.

There is, however, a difference between what machines can do well and what humans are good at. In the early days of AI, researchers envisaged computational systems that would exhibit aspects of human intelligence and achieve human-level problem solving or decision making skills. These problems were presented to the machines as a set of formal mathematical notions within rather narrow and controlled spaces, which could be solved by some form of symbol manipulation or search in symbolic space. The highly formalized, symbolic representation allowed AI to succeed in many cases. Naturally, games—especially **board games**—have been a popular domain for early AI attempts as they are formal and highly constrained, yet complex, decision making environments.

Over the years the focus of much AI research has shifted to tasks that are relatively simple for humans to do but are hard for us to describe how to do, such as remembering a face or recognizing our friend's voice over the phone. As a result, AI researchers began to ask questions such as: *How can AI detect and express emotion? How can AI educate people, be creative or artistically novel? How can AI play a game it has not seen before? How can AI learn from a minimal number of trials? How can AI feel guilt?* All these questions pose serious challenges to AI and correspond to tasks that are not easy for us to formalize or define objectively. Perhaps surprisingly (or unsurprisingly after the fact), tasks that require relatively low cognitive effort from us often turn out to be much harder for machines to tackle. Again, games have provided a popular domain to investigate such abilities as they feature

G. N. Yannakakis and J. Togelius, *Artificial Intelligence and Games*, https://doi.org/10.1007/978-3-319-63519-4_1

aspects of a subjective nature that cannot be formalized easily. These include, for instance, the experience of play or the creative process of game design [599].

Ever since the birth of the idea of artificial intelligence, **games** have been helping AI research progress. Games not only pose interesting and complex problems for AI to solve—e.g., playing a game well; they also offer a canvas for creativity and expression which is *experienced* by users (people or even machines!). Thus, arguably, games are a rare domain where science (problem solving) meets art and interaction: these ingredients have made games a unique and favorite domain for the study of AI. But it is not only AI that is advanced through games; games have also been advanced through AI research. We argue that AI has been helping games to get better on several fronts: in the way we play them, in the way we understand their inner functionalities, in the way we design them, and in the way we understand play, interaction and creativity. This book is dedicated to all aspects of the intersection of games and AI and the numerous ways both games and AI have been challenged, but nevertheless, advanced through this relationship. It is a book about *AI for games* and *games for AI*.

1.1 This Book

The study of AI **in** and **for** games is what this book defines as the research field of **game artificial intelligence** (in brief **game AI**, also occasionally referred to as **AI and games**). The book offers an academic perspective of game AI and serves as a comprehensive guidebook for this exciting and fast-moving research field. Game AI—in particular video game or computer game AI—has seen major advancements in the (roughly) fifteen years of its existence as a separate research field. During this time, the field has seen the establishment and growth of important yearly meetings—including the IEEE Conference on Computational Intelligence and Games (CIG) and the AAAI Artificial Intelligence and Interactive Digital Entertainment (AIIDE) conference series—as well as the launch of the IEEE TRANSACTIONS ON COMPUTATIONAL INTELLIGENCE AND AI IN GAMES (TCIAIG) journal—which will be renamed IEEE TRANSACTIONS ON GAMES (ToG) from January 2018. Since the early days of game AI we have seen numerous success stories within the several subareas of this growing and thriving research field. We can nowadays use AI to play many games better than any human, we can design AI bots that are more believable and human-like than human players, we can collaborate with AI to design better and unconventional (aspects of) games, we can better understand players and play by modeling the overall game experience, we can better understand game design by modeling it as an algorithm, and we can improve game design and tune our monetization strategy by analyzing massive amounts of player data. This book builds on these success stories and the algorithms that took us there by exploring the different **uses** of AI for games and games for AI.

1.1.1 Why Did We Write This Book?

Both of us have been teaching and researching game artificial intelligence at undergraduate and graduate levels in various research and educational institutions across the globe for over a decade. Both of us have felt, at times, that a comprehensive **textbook** on game AI was necessary for our students and a service to the learning objectives of our programs. Meanwhile, an increasing number of fellow academics felt the same way. Such a book was not available, and given our extensive experience with the field, we felt we were well placed to write the book we needed. Given that we have been collaborating on game AI research since 2009, and known each other since 2005, we knew our perspective was coherent enough to actually agree on what should go into the book without undue bickering. While we have been trying hard to write a book that will appeal to many and be useful for both students and researchers from different backgrounds, it ultimately reflects our perspective of what game AI is and what is important within the field.

Looking at the existing literature to allocate readings for a potential course on game AI one can rely partly on a small number of relevant and recent surveys and vision papers for specific game AI research areas. Examples include papers meant to serve as general introductions to game AI [407, 764, 785], general game AI [718], Monte-Carlo Tree Search [77], procedural content generation [783, 720], player modeling [782], emotion in games [781], computational narrative [562], AI for game production [564], neuroevolution in games [567], and AI for games on mobile devices [265]. There are also some earlier surveys reflecting the state of the art at the beginning of this research field, for example, on evolutionary computation in games [406] and computational intelligence in games [405]. No mere paper can however on its own cover the breadth and depth required by a full course on game AI. For this reason, the courses we have taught have generally been structured around a set of papers, some of them surveys and some of them primary research papers, together with slides and course notes.

The first, recently published, edited volumes on game AI research have been great assets for teaching needs in game AI. These are books with a focus on a particular area of game AI research such as procedural content generation [616], emotion in games [325] and game data mining [186]. Because of their more narrow domains they cannot serve as textbooks for a complete course on game AI but rather as parts of a game AI course or, alternatively, as textbooks for independent courses on procedural content generation, affective computing or game data mining, for instance.

Meanwhile several edited volumes or monographs which have covered aspects of game AI programming are edited or written by game AI experts from the game industry. These include the popular *game AI programming wisdom* series [546, 547, 548, 549] and other game AI programming volumes [604, 8, 552, 553, 80, 81, 425]. These books, however, target primarily professional or indie developers, game AI programmers and practitioners, and do not always fulfill the requirements of an academic textbook. As you will see, this is only one part of the field of game AI as we define it. Further, some of the earlier books are somewhat outdated by now given the fast pace of progress of the game AI research field [109, 62]. Among the industry-focused game AI books there are a few that aim to target educators and students of game AI. They have a rather narrow scope, however, as they are limited to non-player character AI [461], which is arguably the most important topic for game AI practitioners in the gaming industry [764, 425] but only one of several research areas in academic game AI research [785]. In our terminology, the perspective of these industry-focused textbooks tends to be almost exclusively on what we call *playing for experience*, in particular generating interesting non-player character (NPC) behavior that looks lifelike and functions within the confines of a game design. Finally there are game AI books that are tied to a particular language or software suite such as Lua [791] or Unity [31], which also limits their usefulness as general textbooks.

In contrast to the above list of books, edited volumes and papers, this book aims to present the research field as a whole and serve (a) as a comprehensive **textbook** for game artificial intelligence, (b) as a **guidebook** for game AI programming, and (c) as a **field guide** for researchers and graduate students seeking to orient themselves within this multifaceted research field. For this reason, we both detail the state of knowledge of the field and also present research and original scholarship in game AI. Thus the book can be used for both research-based teaching and hands-on applications of game AI. We detail our envisaged target audience in the section below.

1.1.2 Who Should Read This Book?

With this book we hope to reach readers with a general interest in the application of AI to games, who already know at least the basics of artificial intelligence. However, while writing the book we particularly envisioned three groups of people benefiting directly from this book. The first group is university **students**, of graduate or advanced undergraduate level, who wish to learn about AI in games and use it to develop their career in game AI programming or in game AI research. In particular, we see this book being used in advanced courses for students who have already taken an introductory AI course, but with care and some supplementary material it could be used for an introductory course as well. The second group is AI **researchers** and **educators** who want to use this book to inspire their research or, instead, use it as a textbook for a class in artificial intelligence and games. We particularly think of active researchers within some AI-related field wanting to start doing research in game AI, and new Ph.D. students in the area. The last target audience is computer game **programmers** and practitioners who have limited AI or machine learning background and wish to explore the various creative uses of AI in their game or software application. Here we provide a complement to the more industry-focused books listed above by taking a broader view of what AI in and for games could be. For further fostering the learning process and widening the practical application of AI in games the book is accompanied by a website that features lectures, exercises and additional resources such as readings and tools.

This book is written with the assumption that its readers come from a **technical background** such as computer science, software engineering or applied math. We assume that our readers have taken courses on the fundamentals of artificial intelligence (or acquired this knowledge elsewhere) as the book does not cover the algorithms in detail; our focus, instead, is on the *use* of the algorithms in games and their modification for that purpose. To be more specific, we assume that the reader is familiar with core concepts in tree search, optimization, supervised learning, unsupervised learning and reinforcement learning, and has implemented some basic algorithms from these categories. Chapter 2 provides an overview of core methods for game AI and a refresher for the reader whose knowledge is a bit rusty. We also assume familiarity with programming and a basic understanding of algebra and calculus.

1.1.3 A Short Note on Terminology

The term "artificial and computational intelligence in games" is often used to refer to the entire field covered in this book (e.g., see the title of [785]). This reflects the dual roots of the field in artificial intelligence and **computational intelligence** (CI) research, and the use of these terms in the names of the major conferences in the field (AIIDE and CIG) and the flagship journal (IEEE TCIAIG) explicitly targeting both CI and AI research. There is no agreement on the exact meaning of the terms AI and CI. Historically, AI has been associated with logic-based or symbolic methods such as reasoning, knowledge representation and planning, and CI has been associated with biologically-inspired or statistical methods such as neural networks (including what is now known as deep learning) and evolutionary computation. However, there is considerable overlap and strong similarities between these fields. Most of the methods proposed in both fields aim to make computers perform tasks that have at some point been considered to require intelligence to perform, and most of the methods include some form of heuristic search. The field of machine learning intersects with both AI and CI, and many techniques could be said to be part of either field.

In the rest of the book we will use the terms "AI and games", "AI in games" and "game AI" to refer to the whole research field, including those approaches that originally come from the CI and machine learning fields. There are three reasons for this: simplicity, readability, and that we think that the distinction between CI and AI is not useful for the purposes of this book or indeed the research field it describes. Our use of these terms is not intended to express any prejudice towards particular methods or research questions. (For a non-exclusive list of methods we believe are part of "AI" according to this definition, see Chapter 2.)

1.2 A Brief History of Artificial Intelligence and Games

Games and artificial intelligence have a long history together. Much research on AI for games is concerned with constructing agents for playing games, with or without a learning component. Historically, this has been the first and, for a long time, the only approach to using AI in games. Even since before artificial intelligence was recognized as a field, early pioneers of computer science wrote game-playing programs because they wanted to test whether computers could solve tasks that seemed to require "intelligence". Alan Turing, arguably the principal inventor of computer science, (re)invented the Minimax algorithm and used it to play Chess [725]. The first software that managed to master a game was programmed by A. S. Douglas in 1952 on a digital version of the Tic-Tac-Toe game and as part of his doctoral dissertation at Cambridge. A few years later, Arthur Samuel was the first to invent the form of machine learning that is now called **reinforcement learning** using a program that learned to play Checkers by playing against itself [591].

Most early research on game-playing AI was focused on classic board games, such as Checkers and Chess. There was a conception that these games, where great complexity can arise from simple rules and which had challenged the best human minds for hundreds or even thousands of years, somehow captured the essence of thought. After over three decades of research on tree search, in 1994, the Chinook Checkers player managed to beat the World Checkers Champion Marion Tinsley [594]; the game was eventually solved in 2007 [593]. For decades Chess was seen as "the drosophila of AI" in the sense of being the "model organism" that uncountable new AI methods were tested on [194]—at least until we developed software capable of playing better than humans, at which point Chess-playing AI somehow seemed a less urgent problem. The software that first exhibited superhuman Chess capability, IBM's Deep Blue, consisted of a Minimax algorithm with numerous Chess-specific modifications and a very highly tuned board evaluation function running on a custom supercomputer [98, 285]. Deep Blue famously won against the reigning grandmaster of Chess, Garry Kasparov, in a much-publicized event back in 1997. Twenty years later, it is possible to download public domain software that will play better than any human player when running on a regular laptop.

A milestone in AI research in games a few years before the successes of Deep Blue and Chinook is the backgammon software named TD-Gammon which was developed by Gerald Tesauro in 1992. TD-Gammon employs an artificial neural network which is trained via temporal difference learning by playing backgammon against itself a few million times [688, 689]. TD-Gammon managed to play backgammon at a level of a top human backgammon player. After Deep Blue IBM's next success story was Watson, a software system capable of answering questions addressed in natural language. In 2011, Watson competed on the *Jeopardy!* TV game and won $1 million against former winners of the game [201].

Following the successes of AI in traditional board games the latest board game AI milestone was reached in 2016 in the game of Go. Soon after Chinook and Deep Blue, the game of Go became the new benchmark for game playing AI with a branching factor that approximates 250 and a vast search space many times larger than that of Chess'. While human level Go playing had been expected sometime in the far future [368], already in 2016 Lee Sedol—a 9-dan professional Go player—lost a five-game match against Google DeepMind's AlphaGo software which featured a deep reinforcement learning approach [629]. Just a few days before the release of the first draft of this book—between 23 and 27 May 2017—AlphaGo won a three-game Go match against the world's number 1 ranking player Ke Jie, running on a single computer. With this victory, Go was the last great classic board game where computers have attained super-human performance. While it is possible to construct classic-style board games that are harder than Go for computers to play, no such games are popular for human players.

But classic board games, with their discrete turn-based mechanics and where the full state of the game is visible to both players, are not the only games in town, and there is more to intelligence than what classic board games can challenge. In the last decade and a half, a research community has therefore grown up around applying AI to games other than board games, in particular video games. A large part of the research in this community focuses on developing AI for **playing games**—either as effectively as possible, or in the style of humans (or a particular human), or with some other property. A notable milestone in video game AI playing was reached in 2014 when algorithms developed by Google DeepMind learned to play several games from the classic Atari 2600 video game console on a super-human skill level just from the raw pixel inputs [464]. One of the Atari games that proved to be hard to play well with that approach was *Ms Pac-Man* (Namco, 1982). The game was practically solved a few days before the release of the second draft of the book (June 2017) by the Microsoft Maluuba team using a hybrid reward architecture reinforcement learning technique [738].

Other uses of AI in video games (as detailed in this book) have come to be very important as well. One of these is **procedural content generation**. Starting in the early 1980s, certain video games created some of their content algorithmically during runtime, rather than having it designed by humans. Two games that became very influential early on are *Rogue* (Toy and Wichmann, 1980), where dungeons and the placement of creatures and items in them are generated every time a new game starts, and *Elite* (Acornsoft, 1984), which stores a large universe as a set of random seeds and creates star systems as the game is played. The great promise of games that can generate some of their own content is that you can get more—potentially infinite—content without having to design it by hand, but it can also help reduce storage space demands among many other potential benefits. The influence of these games can be seen in recent successes such as *Diablo III* (Blizzard Entertainment, 2012), *No Man's Sky* (Hello Games, 2016) and the Chalice Dungeons of *Bloodborne* (Sony Computer Entertainment, 2015).

Relatively recently, AI has also begun to be used to analyze games, and **model players** of games. This is becoming increasingly important as game developers need to create games that can appeal to diverse audiences, and increasingly relevant as most games now benefit from internet connectivity and can "phone home" to the developer's servers. Facebook games such as *FarmVille* (Zynga, 2009) were among the first to benefit from continuous data collection, AI-supported analysis of the data and semi-automatic adaptation of the game. Nowadays, games such as *Nevermind* (Flying Mollusk, 2016) can track the emotional changes of the player and adapt the game accordingly.

Very recently, research on believable agents in games has opened new horizons in game AI. One way of conceptualizing believability is to make agents that can pass game-based Turing tests. A game Turing test is a variant of the Turing test in which a number of judges must correctly guess whether an observed playing behavior in a game is that of a human or an AI-controlled game bot [263, 619]. Most notably, two AI-controlled bot entries managed to pass the game Turing test in *Unreal Tournament 2004* (Epic Games, 2004) on Turing's centenary in 2012 [603].

In the next section we will outline the parallel developments in both academia and industry and conclude the historical section on game AI with ways the two communities managed to exchange practices and transfer knowledge for a common two-fold goal: the advancement of AI and the improvement of games at large.

1.2.1 Academia

In academic game AI we distinguish two main domains and corresponding research activity: board games and video (or computer) games. Below, we outline the two domains in a chronological sequence, even though game AI research is highly active in both of them.

1.2.1.1 Early Days on the Board

When it comes to game AI research, classic board games such as Chess, Checkers and Go are clearly beneficial to work with as they are very simple to model in code and can be simulated extremely fast—one can easily make millions of moves per second on a modern computer—which is indispensable for many AI techniques. Also, board games seem to require thinking to play well, and have the property that they take "a minute to learn, but a lifetime to master". It is indeed the case that games have a lot to do with learning, and good games are able to constantly teach us more about how to play them. Indeed, to some extent the fun in playing a game consists in learning it and when there is nothing more to learn we largely stop enjoying them [351]. This suggests that better-designed games are also better benchmarks for artificial intelligence. As mentioned above, board games were the dominant domain for AI research from the early 1950s until quite recently. As we will see in the other parts of this book—notably in Chapter 3—board games remain a popular game AI research domain even though the arrival of video and arcade games in the 1980s has shifted a large part of the focus since then.

1.2.1.2 The Digital Era

To the best of our knowledge, the first *video* game conference occurred at Harvard's Graduate School of Education[1] in 1983. The core focus of the conference was on the educational benefits and positive social impact of video game playing.

The birth date of the digital game AI field can be safely considered to be around year 2001. The seminal article by Laird and van Lent [360], emphasizing the role of games as the **killer application** for AI, established the foundations of game AI and inspired early work in the field [696, 235, 476, 292, 211, 694, 439, 766, 707]. In those early days AI in digital games was mainly concerned with playing games, agent architectures for NPC behavior [401, 109], sometimes within interactive drama [438, 399, 412, 483, 107], and pathfinding [664]. Early work in these areas was presented primarily in the AAAI Spring Symposia on AI and Interactive Entertainment preceding the AIIDE (which started 2005) and the IEEE CIG (also started in 2005) conferences. Most of the early work in the game AI field was conducted by researchers with AI, optimization and control background and research experience in adaptive behavior, robotics and multi-agent systems. AI academics used the best of their computational intelligence and AI tools to enhance NPC behavior in generally simple, research-focused, non-scalable projects of low commercial value and perspective.

1.2.2 Industry

The first released video games back in the 1970s included little or nothing that we would call artificial intelligence; NPC behaviors were scripted or relied on simple rules, partly because of the primitive state of AI research, but perhaps even more because of the primitive hardware of the time. However, in parallel to developments in academia, the game industry gradually made steps towards integrating more sophisticated AI in their games during the early days of game AI [109, 758].

A non-exhaustive list of AI methods and game features that advanced the game AI state-of-practice in the industry [546] in chronological order includes the first popular application of neural networks in *Creatures* (Millennium Interactive, 1996) with the aim to model the creatures' behavior; the advanced sensory system of guards in *Thief* (EIDOS, 1998); the team tactics and believable combat scenes in the *Halo* series (Microsoft Studios, 2011–2017)—*Halo 2* in particular popularized the use of behavior trees in games; the behavior-based AI of *Blade Runner* (Virgin Interactive, 1997); the advanced opponent tactics in *Half-Life* (Valve, 1998); the fusion of machine learning techniques such as perceptrons, decision trees and reinforcement learning coupled with the belief-desire-intention cognitive model in *Black and White* (EA, 2000)—see Fig. 1.1; the believable agents of *The Sims* series (Electronic Arts, 2000–2017); the imitation learning *Drivatar* system of *Forza Motorsport* (MS

[1] Fox Butterfield, *Video Game Specialists Come to Harvard to Praise Pac-Man; Not to Bury Him.* New York Times, May 24, 1983

Fig. 1.1 A screenshot from *Black and White* (EA, 2000), a highlight in game artificial intelligence history that successfully integrated several AI methods into its design. The game features a creature that learns through positive rewards and penalties in a reinforcement learning fashion. Further, the creature employs the belief-desire-intention model [224] for its decision making process during the game. The desires of the creature about particular goals are modeled via simple perceptrons. For each desire, the creature selects the belief that it has formed the best opinion about; opinions, in turn, are represented by decision trees. Image obtained from Wikipedia (fair use).

Game Studios, 2005); the generation of context-sensitive behaviors via Goal Oriented Action Planning [506]—a simplified STRIPS-like planning method—which was specifically designed for *F.E.A.R.* (Sierra Entertainment, 2005) [507]; the procedurally generated worlds of the *Civilization* series (MicroProse, Activision, Infogrames Entertainment, SA and 2K Games, 1991–2016) and *Dwarf Fortress* (Bay 12 Games, 2006); the AI director of *Left 4 Dead* (Valve, 2008); the realistic gunfights of *Red Dead Redemption* (Rockstar Games, 2010); the personality-based adaptation in *Silent Hill: Shattered Memories* (Konami, 2010); the affect-based cinematographic representation of multiple cameras in *Heavy Rain* (Quantic Dream, 2010); the neuroevolutionary training of platoons in *Supreme Commander* 2 (Square Enix, 2010); the buddy AI (named Ellie) in *The Last of Us* (Sony Computer Entertainment, 2013); the companion character, Elizabeth, in *BioShock Infinite* (2K Games, 2013); the interactive narratives of *Blood & Laurels* (Emily Short, 2014); the alien's adaptive behavior which adjusts its hunting strategy according to the player in *Alien: Isolation* (Sega, 2014); and the procedurally generated worlds of *Spelunky* (Mossmouth, LLC, 2013) and *No Man's Sky* (Hello Games, 2016).

The key criterion that distinguishes successful AI in commercial-standard games had always been the level of integration and interweaving of AI in the design of the game [599, 546]. An unsuccessful coupling of game design and AI may lead to unjustifiable NPC behaviors, break the suspension of disbelief and immediately reduce player immersion. A typical example of such a mismatch between AI and design is the broken navigation of bots that get stuck in a level's dead end; in such instances either the level design is not (re)considered appropriately to match the design of AI or the AI is not tested sufficiently, or both. On the other hand, the successful integration of AI in the design process is likely to guarantee satisfactory outcomes for the playing experience. The character design process, for instance, may consider the limitations of the AI and, in turn, absorb potential "catastrophic" failures of it. An example of such an interwoven process is the character design in *Façade* [441] that was driven, in part, by the limitations of the natural language processing and the interactive narrative components of the game.

It is important to note that this book is not necessarily about game AI as defined and practiced in the game industry. Instead, it is primarily an **academic textbook** that refers to some of the techniques that have been used in and popularized through the game industry—see for instance the *ad-hoc behavior authoring* section of Chapter 2. The reader with an interest in the AI state of practice in the game industry is referred to the several introductory articles (e.g., [171, 369]) available in books such as the *game AI programming wisdom* series [546, 547, 548, 549]. Another valuable resource is the video recorded talks from top game AI programmers which are hosted at the AI summit[2] of the Game Developers Conference (GDC) and are available at the GDC Vault.[3] Finally, talks and courses mostly relevant for game AI programmers are available through the nucl.ai conference webpage.[4]

1.2.3 The "Gap"

During the first decade of academic game AI research, whenever researchers from academia and developers from industry would meet and discuss their respective work, they would arrive at the conclusion that there exists a *gap* between them; the gap had multiple facets such as differences in background knowledge, practice, trends, and state-of-the-art solutions to important problems. Academics and practitioners would discuss ways to bridge that gap for their mutual benefit on a frequent basis [109] but that debate would persist for many years as developments on both ends were slow. The key message from academic AI was that the game industry should adopt a "high risk-high gain" business model and attempt to use sophisticated AI techniques with high potential in their games. On the other end, the central complaint of industrial game AI regarding game AI academics has been the lack of

---

[2] http://www.gdconf.com/conference/ai.html

[3] http://www.gdcvault.com/

[4] https://nucl.ai/

domain-specific knowledge and practical wisdom when it comes to realistic problems and challenges faced during game production. Perhaps above all there is a difference in what is valued, with academics valuing new algorithms and new uses of algorithms that achieve superior performance or create new phenomena or experiences, and AI developers in industry valuing software architectures and algorithmic modifications that reliably support specific game designs. But what happened since then? Does this gap really exist nowadays or is it merely a ghost from the past?

It is still true that the academic game AI research community and the game industry AI development community largely work on different problems, using different methods. There are also some topics and methods explored by the academic community which are generally very unpopular within the game industry. Real-time adaptation and learning in NPCs is one such example; quite a few academic researchers are excited by the idea of NPCs that can learn and develop from their interactions with the player and other NPCs in the game. However, AI developers in industry point out that it would be very hard to predict what these NPCs will learn, and it is very likely to "break the game" in the sense that it no longer works as designed. Conversely, there are methods and problems explored in industry which most academics do not care about, as they only make sense within the complex software architecture of a complete game.

When thinking about the use of AI within modern video games, it is important to remember that most game genres have developed evolutionarily from earlier game designs. For example, the first platformers were released in the mid-1980s and the first first-person shooters and real-time strategy games in the early 1990s. At that time, the ability to build and deploy advanced AI was much less than it is today, so designers had to design around the lack of AI. These basic design patterns have largely been inherited by today's games. It can therefore be said that many games have been designed not to need AI. For the academic who wants to build interesting AI for an in-game role, the best might therefore be to create new game designs that start from the existence of the AI.

Taking a positive stance on the topic we would argue that any existing gap between academic and industrial NPC AI nowadays can be viewed as a healthy indication of a parallel progress with a certain degree of collaboration. As industry and academia do not necessarily attempt to solve the same problems with the same approaches it may be that NPC AI solutions emerging from industry can inspire new approaches in academia and vice versa. In summary, the NPC AI gap is clearly smaller in tasks that both academia and industry care about. Certain aspects of NPC AI, however, are far from being solved in an ideal fashion and others—such as modeling emotion in role playing games—are still at the beginning stages of investigation. So while we can praise the NPC AI of *The Elder Scrolls V: Skyrim* (Bethesda Softworks, 2011) we cannot be as positive about the companion AI of that game. We can view the very existence of such limitations as an opportunity that can bring industry and academia even closer to work on further improving existing NPC AI in games.

A different take on this discussion—which is supported by some game developers and game AI academics—is that NPC AI is almost solved for most production tasks; some take it one step further and claim that game AI research and development should focus solely on non-traditional uses of AI [477, 671]. The level of AI sophistication in recent games such as *Left 4 Dead* (Valve, 2008) and *The Elder Scrolls V: Skyrim* (Bethesda Softworks, 2011) supports this argument and suggests that advances in NPC AI have reached satisfactory levels for many NPC control challenges faced during game production. Due to the rise of robust and effective industrial game AI solutions, the convergence to satisfying NPC performances, the support of the multidisciplinary nature of game AI and a more pragmatic and holistic view of the game AI problem, recent years have seen a shift of academic and industrial interests with respect to game AI. It seems that we have long reached an era where the primary focus of the application of AI in the domain of games is not on agents and NPC behaviors. The focus has, instead, started to shift towards interweaving game design and game technology by viewing the role of AI holistically and integrating aspects of procedural content generation and player modeling within the very notion of game AI [764].

The view we take in this book is that AI can help us to make better games but that this does not necessarily happen through better, more human-like or believable NPCs [764]. Notable examples of non-NPC AI in games include *No Man's Sky* (Hello Games, 2016) and its procedural generation of a quintillion different planets and *Nevermind* (Flying Mollusk, 2016) with its affective-based game adaptation via a multitude of physiological sensors. But there might be other AI roles with game design and game development that are still to be found by AI. Beyond playing games, modeling players or generating content, AI might be able to play the role of a design assistant, a data analyst, a playtester, a game critic, or even a game director. Finally, AI could potentially play and design games as well as model their players in a general fashion. The final chapter (Chapter 7) of this book is dedicated to these frontier research areas for game AI.

1.3 Why Games for Artificial Intelligence

There are a number of reasons why games offer the ideal domain for the study of artificial intelligence. In this section, we list the most important of them.

1.3.1 Games Are Hard and Interesting Problems

Games are engaging due to the effort and skills required from people to complete them or, in the case of puzzles, solve them. It is that *complexity* and *interestingness* of games as a problem that makes them desirable for AI. Games are **hard** because their finite state spaces, such as the possible strategies for an agent, are often vast. Their complexity as a domain rises as their vast search spaces often feature small feasible spaces (solution spaces). Further, it is often the case that the goodness of any game state is hard (or even impossible) to assess properly.

From a computational complexity perspective, many games are NP-hard (NP refers to "nondeterministic polynomial time"), meaning that the worst-case complexity of "solving" them is very high. In other words, in the general case an algorithm for solving a particular game could run for a very long time. Depending on a game's properties complexity can vary substantially. Nevertheless, the list of games that are NP-hard is rather long and includes games such as the two-player incomplete information *Mastermind* game [733, 660], the *Lemmings* (Psygnosis, 1991) arcade game [334] and the *Minesweeper* game by Microsoft [331]. It should be noted that this computational complexity characterization has little to do with how hard the games are to play for humans, and does not necessarily say much about how well heuristic AI methods can play them. However, it is clear that at least in theory, and for arbitrary-size instances, many games are very hard.

The investigations of AI capacity in playing games that are hard and complex has been benchmarked via a number of milestone games. As mentioned earlier, Chess and (to a lesser degree) Checkers have traditionally been seen as the "drosophila for AI research" even from the early days of AI. After the success of Deep Blue and Chinook in these two games we gradually invented and cited other more complex games as AI "drosophilae", or universal benchmarks. *Lemmings* has been characterized as such; according to McCarthy [446] it "connects logical formalizations with information that is incompletely formalizable in practice". In practice, games for which better APIs have been developed—such as *Super Mario Bros* (Nintendo, 1985)[5] and *StarCraft* (Blizzard Entertainment, 1998)—have become more popular benchmarks.

The game of computer Go has also been another core and traditional game AI benchmark with decades of active research. As a measure of problem complexity a typical game in Go has about 10^{170} states. The first AI feature extraction investigations in Go seem to date back to the 1970s [798]. The game received a lot of research attention during several world computer Go championships up until the recent success of AlphaGo [629]. AlphaGo managed to beat two of the best Go professional human players using a combination of deep learning and Monte Carlo tree search. In March 2016, AlphaGo won against Lee Sedol and in May 2017 it won all three games against the world's number 1 ranked player, Ke Jie.

The *StarCraft* (Blizzard Entertainment, 1998) real-time strategy game can be characterized as perhaps the single hardest game for computers to play well. At the time of writing this book the best *StarCraft* bots only reach the level of amateur players.[6] The complexity of the game derives mainly from the multi-objective task of controlling multiple and dissimilar units in a game environment of partial information. While it is not trivial to approximate the state space of *StarCraft*, according to a recent study [729], a typical game has at least $10^{1,685}$ possible states. In comparison, the number of protons in the observable universe is only about 10^{80} [182].

[5] Note that the original game title contains a dot, i.e., *Super Mario Bros.*; for practical reasons, however, we will omit the dot when referring to the game in the remainder of this book.

[6] http://www.cs.mun.ca/~dchurchill/starcraftaicomp/

The number of *StarCraft*'s possible states sounds huge but, interestingly enough, its search space can be of manageable size if represented by bytes. On that basis, we require about 700 bytes of information to represent the *StarCraft* search space whereas the number of protons in the known universe is equivalent to the number of configurations of about 34 bytes.

One could of course design games that are harder on purpose, but there is no guarantee anyone would want to play those games. When doing AI research, working on games that people care about means you are working on relevant problems. This is because games are designed to challenge the human brain and successful games are typically good at this. *StarCraft* (Blizzard Entertainment, 1998)—and its successor *StarCraft II* (Blizzard Entertainment, 2010)—are played by millions of people all over the world, with a very active competition scene of elite professional players and even dedicated TV channels such as OGN[7]—a South Korean cable television channel—or twitch channels[8] that specialize in broadcasting video game-related content and e-sports events.

Many would claim that *StarCraft* (Blizzard Entertainment, 1998) is the next major target for AI research on playing to win. In academia, there is already a rich body of work on algorithms for playing (parts of) *StarCraft* [504, 569, 505, 124], or generating maps for it [712]. Beyond academia, industrial AI leaders Google DeepMind and Facebook seem to be in agreement and on a similar scientific mission. DeepMind recently announced that *StarCraft II* will be one of their major new testbeds, after their success at training deep networks to play Atari games in the Arcade Learning Environment[9] (ALE) [40] framework. At the time of writing this book, DeepMind in collaboration with Blizzard Entertainment opened up *StarCraft II* to AI researchers for testing their algorithms.[10] Facebook AI Research has led the development of *TorchCraft* [681]—a bridge between the deep learning Torch library and *StarCraft*—and recently published their first paper on using machine learning to learn to play *StarCraft* [729], showing that they take this challenge seriously. Another industrial game AI research lab collaborating with academia on solving *StarCraft* is hosted in Alibaba [523]. Given the game's complexity, it is unlikely we will conquer all of it soon [234] but it is a game through which we expect to see AI advancements in the years to come.

1.3.2 Rich Human-Computer Interaction

Computer games are dynamic media by definition and, arguably, offer one of the **richest** forms of human-computer interaction (HCI); at least at the time of writing this book. The richness of interaction is defined in terms of the available options

[7] http://ch.interest.me/ongamenet/

[8] For instance, https://www.twitch.tv/starcraft

[9] http://www.arcadelearningenvironment.org/

[10] Follow developments at: https://deepmind.com/blog/

a player has at any given moment and the ways (modalities) a player can interact with the medium. The available options for the player are linked to the game action space and the complexity associated with it in games such as *StarCraft II* (Blizzard Entertainment, 2010). Further, the modalities one may use to interact with games nowadays extend beyond the traditional keyboard, mouse and tablet-like haptics, to game controllers, physiology such as heart rate variability, body movement such as body stance and gestures, text, and speech. As a result, many games would easily top the list of information bits exchanged between them and their users per second compared to any other HCI medium; however, such comparative studies are not currently available to further support our claims.

Clearly, as we will see later in this book, games offer one of the best and most meaningful domains for the realization of the **affective loop**, which defines a framework that is able to successfully elicit, detect and respond to the cognitive, behavioral and emotive patterns of its user [670]. The potential that games have to influence players is mainly due to their ability to place the player in a continuous mode of interaction with the game which elicits complex cognitive, affective and behavioral responses to the player. This continuous interaction mode is enriched by fast-paced and multimodal forms of user interactivity that are often possible in games. As every game features a player—or a number of players—the interaction between the player and the game is of key importance for AI research as it gives algorithms access to rich player experience stimuli and player emotional manifestations. Such complex manifestations, however, cannot trivially be captured by standard methods in machine learning and data science. Undoubtedly, the study of game-player interaction via artificial intelligence not only advances our knowledge about human behavior and emotion but also contributes to the design of better human-computer interaction. As a result, it further pushes the boundaries of AI methods in order to address the challenges of game-based interaction.

1.3.3 Games Are Popular

While video games, back in the 1980s, were introduced as a niche activity for those having access to a video arcade or to consoles such as Atari 2600, they gradually turned into a multi-billion industry generating, in the 2010s, a global market revenue higher than any other form of creative industry, including film and music. At the time of writing, games generate a worldwide total of almost $100 billion in revenue which is expected to rise to approximately $120 billion by 2019.[11]

But why did games became so popular? Beyond the obvious argument of games being able to enhance a user's intrinsic motivation and engagement by offering interactivity capacities with a virtual environment, it was the technological advancements over the last 40 years that drastically changed the demographics of players [314]. Back in the early 1980s games used to be played solely in arcade entertain-

[11] See, for instance, the global games market report by Newzoo:
https://newzoo.com/solutions/revenues-projections/global-games-market-report/

ment machines; nowadays, however, they can be played using a multitude of devices including a PC (e.g., multi-player online or casual games), a mobile phone, a tablet, a handheld device, a virtual reality device, or a console (and obviously still an arcade entertainment machine!). Beyond the technological advancements that fostered accessibility and democratized gameplay, it is also the culture that follows a new medium and develops it into a new form of art and expression. Not only the independent scene of game design and development[12] has contributed to this culture, but also the outreach achieved from the multitude of purposes and objectives of games beyond mere entertainment: games for art, games as art, games for a change, physical interactive games, games for education, games for training and health, games for scientific discovery, and games for culture and museums. In brief, not only are games everywhere and present in our daily lives but they also shape our social and cultural values at large—as for instance evidenced by the recent massive success of *Pokémon Go* (Niantic, 2016). As a byproduct of their popularity games offer easy access to people with world-class performance in the domain. Experts (i.e., professional players) for many board and digital games that are world-ranked according to their gameplay performance have participated regularly in competitions against AI algorithms; examples include Garry Kasparov (Chess), and Lee Sedol and Ke Jie (Go).

As games become more popular, grow in quantity, and become more complex, new AI solutions are constantly required to meet the new technological challenges. This is where AI meets a domain with a strong industrial backing and a desire to support sophisticated technology for bettering player experience. Furthermore, very few domains for AI offer the privilege of daily accessibility to new **content** and **data** from their popular use. But let us look at these two aspects in more detail below.

1.3.3.1 Popular Means More Content

The more people play (more) games, the more content is required for games. Content takes effort to create but, over the years, mechanisms have been developed that allow both machines and players to design and create various forms of content in games. Games have gradually developed to be **content-intensive** software applications that demand content that is both of direct use in the game and of sufficient novelty. The overwhelming demand for new and novel gaming experiences from a massive community of users constantly pushes the boundaries of human and computational creativity to new grounds; and naturally AI at large.

Content in games, beyond any other form of multimedia or software application, not only covers all possible forms of digital content such as audio, video, image, and text but it also comes in massive numbers of different resolutions and representations. Any algorithm that attempts to retrieve and process the variety and amount of content within or across games is directly faced with the challenges of interoperability and content convergence as well as scalability caused by such big data sets.

[12] http://www.igf.com/

Compare this with a typical robot simulator, where all the environments would have to be painstakingly hand-crafted or adapted from data collected from the real world. When using games as testbeds, there is no such content shortage.

1.3.3.2 Popular Means More Data

Massive content creation (either by games or players) is one major effect of games' popularity; the other is massive data generation of game playthroughs and player behavior. Since the late 2000s, game companies have had access to accurate game telemetry services that allow them to track and monitor player purchases, churn and re-engagement, or the progress of play for debugging either the game or the players' experience. The algorithmic challenges met here follow the general challenges of big data and big data mining research [445], which include data filtering during data acquisition, metadata generation, information extraction from erroneous and missing data, automatic data analysis across dissimilar datasets, appropriate declarative query and mining interfaces, scalable mining algorithms, and data visualization [359]. Luckily enough some of these datasets are nowadays openly available for game analytics and game data mining research. Indicatively, in March 2017, the OpenDota project[13]—a community-maintained open source Dota 2 data platform—released a sanitized archive of over a **billion matches** (!) of *Dota 2* (Valve Corporation, 2013) that were played between March 2011 and March 2016.[14]

1.3.4 There Are Challenges for All AI Areas

Unlike some more narrow benchmarks, games challenge *all* core areas of AI. This can be seen by taking a number of broadly accepted areas of AI and discussing the challenges available for those areas in games. **Signal processing**, for starters, meets great challenges in games. Data from players, for instance, not only come in different resolutions—in-game events vs. head pose vs. the player's physiology—they also originate from multiple modalities of fast-paced interaction in an environment that elicits complex cognitive and affective patterns to the player. Multi-modal interaction and multi-modal fusion are non-trivial problems when building embodied conversational agents and virtual characters. Further, the complexity of the signal processing task in games is augmented due to the spatio-temporal nature of the signals which is caused by the rich and fast-paced interaction with the game.

As discussed in the introductory section of this chapter Checkers, Chess, Jeopardy!, Go and arcade games mark a historical trace of core major milestones for **machine learning** (Go and arcade games), **tree search** (Checkers and Chess), **knowledge representation and reasoning** (Jeopardy!) and **natural language processing**

[13] https://www.opendota.com/

[14] https://blog.opendota.com/2017/03/24/datadump2/

(Jeopardy!, Kinect games) and, in turn, they resulted in major breakthroughs for AI. This historical association between AI accomplishments and games already provides clear evidence that all the above areas have traditionally been challenged by games. While the full potential of machine learning remains to be discovered in games such as *StarCraft II* (Blizzard Entertainment, 2010), natural language processing (NLP) has been challenged deeply through games involving narrative and natural language input. NLP is challenged even further in game environments that wish to realize forms of interactive storytelling [148].

Finally when it comes to **planning** and **navigation**, games have traditionally offered environments of high and increasing complexity for algorithms to be tested in. While games such as *StarCraft* clearly define major milestones for planning algorithms, navigation and pathfinding have reached a certain degree of maturity through simulated and roborealistic game environments featuring multiple entities (agents). An additional benefit of games as a domain for behavioral planning is that they offer a realistic yet a far more convenient and cheaper testbed compared to robotics. Beyond the extensive testing and advancement of variants of A* through games, popular and highly effective tree search variants such as the Monte Carlo tree search [77] algorithm have been invented in response to problems posed by game-playing.

1.3.5 Games Best Realize Long-Term Goals of AI

One of the long-standing questions that AI is faced with is *what is the ultimate long-term goal for AI*? While numerous debates and books have been dedicated to this topic, the collaborative effort of Wikipedia authors addressing this question[15] reveals the areas of **social intelligence** and **affective interaction**, (computational) **creativity**, and **general intelligence** as the most critical long-term goals of AI. Our reference has been critiqued for systemic bias (in any controversial question such as the one above); we argue, however, that any reference on this topic would be subjective anyhow. Without aiming to be exclusive or biased, we believe that the three aforementioned areas collectively contribute to better AI systems and we discuss why games best realize these three goals below. These three long-term goals define frontier research areas for game AI and are further elaborated on in the last chapter of this book.

1.3.5.1 Social and Emotional Intelligence

Affective computing [530] is the multidisciplinary area of study across computer science, cognitive science and psychology that investigates the design and development of intelligent software that is able to elicit, detect, model, and express emotion

[15] Wikipedia (accessed: May 2017): https://en.wikipedia.org/wiki/Artificial_intelligence

and social intelligence. The ultimate aim of affective computing is the realization of the so-called *affective loop* [670] which, as we covered earlier, defines a system that is able to successfully elicit, detect and respond to the emotions of its user. Naturally, both emotive and social aspects of intelligence are relevant for a system that realizes the affective loop.

Games can offer a highly meaningful realization of the affective loop and affective interaction [781]. Games are by definition both *entertaining* (whether used for pure satisfaction, training or education) and *interactive* activities that are played within fantasy worlds. Thus, any limitations of affective interaction—such as the difficulty to justify affective-based game decisions or alterations of content to the user—are absorbed naturally. For example, an erroneous affective response of a game character can still be justified if the design of the character and the game context do not break the *suspension of disbelief* of the player—during which the player ignores the medium and the interaction for the sake of her enjoyment. Further, games are designed to offer affective experiences which are influenced by player feedback and players are willing to go through, e.g., frustrating, anxious, and fearful episodes of play for experiencing involvement. To that end, a user under gaming conditions—more than any other form of human-computer interaction—is generally open to affective-based alterations of the interaction and influences of his/her emotional state.

1.3.5.2 Computational Creativity

Computational creativity studies the potential of software to autonomously generate outcomes that can be considered creative or algorithmic processes that are deemed to be creative [54, 754]. Computer games can be viewed as the killer application domain for computational creativity [381]. It is not only their unique features that we covered in earlier sections of this chapter—i.e., being highly interactive, dynamic and content-intensive software applications. Most importantly it is their *multifaceted* nature. In particular, it is the fusion of the numerous and highly diverse creative domains—visual art, sound design, graphic design, interaction design, narrative, virtual cinematography, aesthetics and environment beautification—within a single software application that makes games the ideal arena for the study of computational creativity. It is also important to note that each art form (or facet) met in games elicits different experiences to its users; their fusion in the final software targeting a rather large and diverse audience is an additional challenge for computational creativity.

As a result, the study of computational creativity *within* and *for* computer games [381] advances in both the field of AI and the domain of games. Games can, first, be improved as products via computational creations (*for*) and/or, second, be used as the ultimate canvas for the study of computational creativity as a process (*within*). Computer games not only challenge computational creativity but they also provide a creative sandbox for advancing the field. Finally, games can offer an opportunity for computational creativity methods to be extensively assessed via a huge population of users of commercial-standard products of high impact and financial value.

1.3.5.3 General Intelligence

AI has investigated the general intelligence capacity of machines within the domain of games more than any other domain thanks to the ideal properties of games for that purpose: controlled yet interesting and computationally hard problems [598]. In particular, the capacity of AI to play unseen games well—i.e., general game playing—has seen a number of advancements in recent years. Starting with the general game playing competition [223], focusing on board games and similar discrete perfect information games, we now also have the Arcade Learning Environment [40] and the General Video Game AI Competition [528], which offer radically different takes on arcade video games. Advancements vary from the efforts to create game description languages suitable for describing games used for general game playing [533, 223, 400, 691, 354, 596, 181, 429] to the establishment of a set of general video game AI benchmarks [223, 528, 40] to the recent success of deep Q-learning in playing arcade games with human-level performance just by processing the screen's pixels [464].

While general game playing is studied extensively and constitutes one of the key areas of game AI [785], we argue that the focus of generality solely with regard to the performance of game-playing agents is *very narrow* with respect to the spectrum of roles for general intelligence in games. The types of general intelligence required within game development include game and level design as well as player behavior and experience modeling. Such skills touch upon a diverse set of cognitive and affective processes which have until now been ignored by general AI in games. For general game AI to be truly general and advance AI algorithmically, it needs to go beyond game playing while retaining the focus on addressing more than a single game or player [718]. We further argue that the challenge of bringing together different types of skillsets and forms of intelligence within autonomous designers of games cannot only advance our knowledge about human intelligence but also advance the capacity of general artificial intelligence.

1.4 Why Artificial Intelligence for Games

The various uses of AI in games are beneficial for the design of better games for a number of reasons. In this section we focus on the benefits obtained by allowing AI to play a game, to generate content and to analyze player experience and behavior.

1.4.1 AI Plays and Improves Your Game

AI can improve games in several ways by merely playing them. The game industry usually receives praise for the AI of their games—in particular, the non player or opponent AI—when the AI of the game adds to the commercial value of the game, it contributes to better game reviews, and it enhances the experience of the player. Whether the underlying AI is based on a simple behavior tree, a utility-based AI or alternatively on a sophisticated machine learned reactive controller is of limited relevance as long as it serves the aforementioned purposes. An unconventional and effective solution to an NPC task can often be a critical factor that shapes management, marketing and monetization strategies during and after production.

As we will see in Chapter 3, AI plays games with two core objectives in mind: play **well** and/or play **believably** (or human-like, or interestingly). Further AI can control either the **player** character or the **non-player** character of the game. AI that plays well as a player character focuses on optimizing the *performance* of play—performance is measured as the degree to which a player meets the objectives of the game solely. Such AI can be of tremendous importance for automatic game testing and for the evaluation of the game design as a whole. AI that plays well as a non-player character, instead, can empower dynamic difficulty adjustment and automatic game balancing mechanisms that will in turn personalize and enhance the experience for the player (as in [651] among many). If the focus of AI is shifted on controlling player characters that play believably or human-like (as in [96, 719, 264] among many) then AI can serve as means for player experience debugging or as demonstration of realistic play for design purposes. Finally, a game that features a rich interaction with NPCs can only benefit from AI that controls NPCs which are expressive and depict human-like and believable behaviors (as in [563, 683, 762] among many).

1.4.2 More Content, Better Content

There are several reasons for game designers and developers to be interested in AI and, in particular, in content generation as covered in detail in Chapter 4. The first and most historical reason is **memory consumption**. Content can typically be compressed by keeping it "unexpanded" until needed. A good example is the classic space trading and adventure game *Elite* (Acornsoft, 1984), which managed to keep hundreds of star systems in a few tens of kilobytes of memory available on the hardware. Further, content generation might foster or further inspire human **creativity** and allow the emergence of completely new types of games, game genres or entirely new spaces of exploration and artistic expression [381]. Moreover, if new content can be generated with sufficient variety, quality and quantity, then it may become possible to create truly endless games with **ultimate replay** value. Finally, when content generation is associated with aspects of play we can expect personalized and **adaptive play** to emerge via the modification of content.

Unlike other areas of game AI—such as general game playing which might be considered more of an academic pursuit—content generation is a commercial necessity [381]. Prior to the academic interest in content generation—which is rather recent [704, 720, 616]—content generation systems had a long history of supporting commercial standard games for creating engaging yet unpredictable game experiences but, most importantly, lessening the burden of manual game content creation. Naturally, games that feature sophisticated content generation systems can garner praise for their technologies—as in *Diablo III* (Blizzard, 2012)—or even build an entire marketing campaign on content generation—like *No Man's Sky* (Hello Games, 2016).

1.4.3 Player Experience and Behavioral Data Analytics

The use of AI for the understanding of player experience can drive and enhance the design process of games. Game designers usually explore and test a palette of mechanics and game dynamics that yield experience patterns they desire to put the player through. Player states such as engagement, fear and stress, frustration, anticipation, and challenge define critical aspects of the design of player experience, which is dependent on the genre, the narrative and the objectives of the game. As a result, the holy grail of game design—that is player experience—can be improved and tailored to each player but also augmented via richer experience-based interaction. Further, as a direct consequence of better and faster design, the whole game development process is boosted and improved.

Beyond the experience of the player, data derived from games, their use and their players provide a new and complementary way of designing games, of making managerial and marketing decisions about games, of affecting the game production, and of offering a better customer service [178]. Any AI-informed decisions about the future of a game's design or development are based on evidence rather than intuition, which showcases the potential of AI—via game analytics and game data mining—for better design, development and quality assurance procedures. In summary, as we will see in the remaining chapters of this book, AI-enabled and data-driven game design can directly contribute to better games.

1.5 Structure of This Book

We structured this book into three main parts. In the first part (Chapter 2) we outline the core game **AI methods** that are important for the study of AI in and for games. In the second part of the book we ask the question: *how can AI be used in games*? Answers to this question define the main game AI areas identified and covered as corresponding chapters:

- AI can **play games** (Chapter 3).
- AI can **generate content** (Chapter 4).
- AI can **model players** (Chapter 5).

In the final part of the book, we attempt a synthesis of the game AI areas that make up this field and discuss the research trends in what we envisage as the **game**

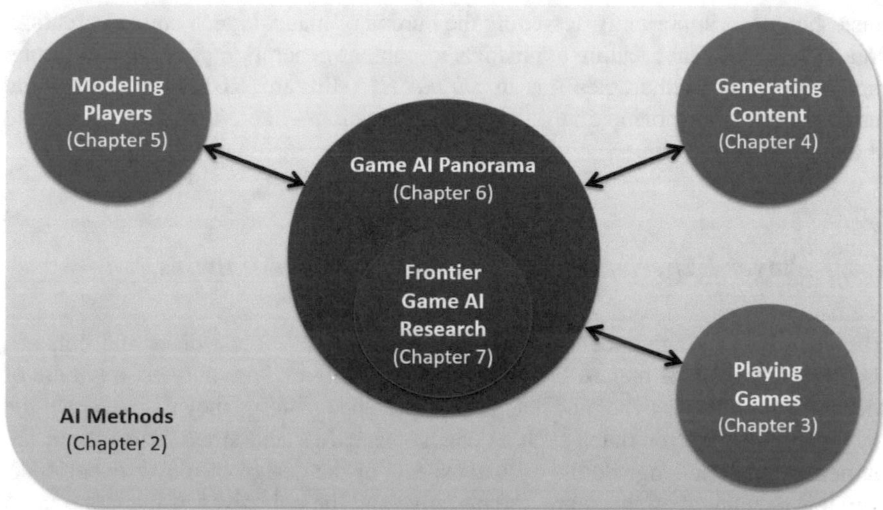

Fig. 1.2 Illustrating the associations among the chapters of this book. Light gray, blue and red areas represent chapters of Part I, Part II, and Part III of the book, respectively.

AI panorama (Chapter 6). Building on this synthesis, we conclude the book with a chapter dedicated to the research areas we view as largely unexplored and important for the study of game AI, namely **frontier game AI research** areas (Chapter 7). An illustration of how the different chapters fit together in this book is presented in Fig. 1.2. The readers of this book may wish to skip parts that are not of interest or not appropriate given their background. For instance, readers with a background in artificial intelligence may wish to skip the first part, while readers who wish to get a rapid overview of the game AI field or a glimpse of frontier research trends in game AI can solely focus on the final part of the book.

1.5.1 What We (Don't) Cover in This Book

The list of core uses of AI in games identified in this book should not be regarded as complete and inclusive of all potential areas of game AI research. It could also be argued that the list of areas we cover is arbitrary. However, this could likely be said of any research field in any discipline. (In software engineering, software design overlaps with software requirements analysis, and in cognitive psychology, memory research overlaps with attention research.) While it might be possible to perform an analysis of this research field so that the individual areas have minimal or no overlap, this would likely lead to a list of artificial areas that do not correspond to the areas game AI students, researchers and practitioners perceive themselves to be working in. It could also be argued that we are omitting certain areas. For example, we only briefly discuss the topic of **pathfinding** in games, whereas some other authors see this as a core concern of game AI. In our view, pathfinding is a relatively isolated area with restricted interaction with the other uses of AI in games. Further, pathfinding has been covered substantially in other game AI textbooks [109, 461, 62] already. Another example is the area of **computational narrative**, which is viewed as a domain for content generation and covered only relatively briefly in Chapter 4. It would certainly be possible to write a whole textbook about computational narrative, but that would be someone else's job. Beyond particular application areas of game AI, in Chapter 2 we cover a number of popular methods used in the field. The list of methods is not inclusive of all AI areas that can find application in games; it is a list, however, we consider sufficient for covering the theoretical foundations of a graduate game AI course. In that regard, we only partially cover **planning** methods and **probabilistic** methods such as Bayesian approaches and methods based on Markov chains, respectively, in Chapter 2 and Chapter 4.

Another important note is the relationship of this book with the general areas of **game theory** [214, 472, 513] and other related research AI areas such as **multiagent systems** [626]. Game theory studies mathematical models of *rational* decision makers within abstract games, for the analysis of economic or social behavior in adversarial [471] or cooperative [108] settings. More specifically, game theory focuses on characterizing or predicting the actions of rational or bounded-rational agents, and studies the related emerging "game solution concepts"—such as the celebrated Nash equilibrium [478]. While the book does not cover these areas in detail, as they are peripheral to the aims of game AI as currently envisioned, we nevertheless believe that it would be fruitful to incorporate foundational ideas and concepts from game theory and multiagent systems research in the game AI field. Particular game AI areas such as game-playing and player (or opponent) modeling [214] could benefit from theoretical models of game-playing [214] and empirical implementations of agent-based systems. Similarly, we believe that game AI research and practice can only help in advancing work on theoretical game theory and multiagent systems. Of course, interweaving these fields properly with the current stream of game AI research is a non-trivial exercise, given the different focus and paths these fields have taken. Further, there are limits to the degrees theoretical models can capture the complexity of games covered in game AI. However, game theory undeniably constitutes a key theoretical pillar for the study of rational decision making, and rational decision making is arguably key for winning in games. Some instances of economic game theory that found successful applications in game AI include the various implementations of theoretical models for playing abstract, card and board games—notably a version of Poker was used as a testbed of game theory by von Neumann and Morgenstern [743] as far back as 1944 [406]. Some of these implementations are discussed in Chapter 3. Another example that bridged rational decision making theory with game AI is the *procedural personas* approach covered in Chapter 5.

Finally, note that this book is the effort of two authors with a high degree of consensus among them and not an edited volume of several authors. As such, it contains **subjective views** on how the game AI field is positioned within the greater

AI research scene and how the game AI areas are synthesized. Your mileage may vary.

1.6 Summary

AI has a long-standing and healthy relationship with games. AI algorithms have been advanced or even invented through games. Games, their design and development, in turn, have benefited largely by the numerous roles AI has taken in games. This book focuses on the main uses of AI in games, namely, for playing games, for generating content, and for modeling players, which are covered extensively in the following chapters. Before delving into the details of these AI uses, in the next chapter we outline the core methods and algorithms used in the field of game artificial intelligence.

Chapter 2
AI Methods

This chapter presents a number of basic AI methods that are commonly used in games, and which will be discussed and referred to in the remainder of this book. These are methods that are frequently covered in introductory AI courses—if you have taken such a course, it should have exposed you to at least half of the methods in this chapter. It should also have prepared you for easily understanding the other methods covered in this chapter.

As noted previously, this book assumes that the reader is already familiar with core AI methods at the level of an introductory university course in AI. Therefore, we recommend you to make sure that you are at least cursorily familiar with the methods presented in this chapter before proceeding to read the rest of the book. The algorithm descriptions in this chapter are **high-level descriptions** meant to refresh your memory if you have learned about the particular algorithm at some previous point, or to explain the general idea of the algorithm if you have never seen it before. Each section comes with pointers to the literature, either research papers or other textbooks, where you can find more details about each method.

In this chapter we divide the relevant parts of AI (for the purposes of the book) into six categories: ad-hoc authoring, tree search, evolutionary computation, supervised learning, reinforcement learning and unsupervised learning. In each section we discuss some of the main algorithms in general terms, and give suggestions for further reading. Throughout the chapter we use the game of *Ms Pac-Man* (Namco, 1982) (or Ms Pac-Man for simplicity) as an overarching testbed for all the algorithms we cover. For the sake of consistency, all the methods we cover are employed to **control** Ms Pac-Man's behavior even though they can find a multitude of other uses in this game (e.g., generating content or analyzing player behavior). While a number of other games could have been used as our testbed in this chapter, we picked Ms Pac-Man for its popularity and its game design simplicity as well as for its high complexity when it comes to playing the game. It is important to remember that Ms Pac-Man is a **non-deterministic** variant of its ancestor *Pac-Man* (Namco, 1980) which implies that the movements of ghosts involve a degree of randomness.

In Section 2.1, we go through a quick overview of two key overarching components of all methods in this book: representation and utility. **Behavior authoring**,

G. N. Yannakakis and J. Togelius, *Artificial Intelligence and Games*, https://doi.org/10.1007/978-3-319-63519-4_2

covered in Section 2.2, refers to methods employing static ad-hoc representations without any form of search or learning such as finite state machines, behavior trees and utility-based AI. **Tree search**, covered in Section 2.3, refers to methods that search the space of future actions and build trees of possible action sequences, often in an adversarial setting; this includes the Minimax algorithm, and Monte Carlo tree search. Covered in Section 2.4, **evolutionary computation** refers to population-based global stochastic optimization algorithms such as genetic algorithms, or evolution strategies. **Supervised learning** (see Section 2.5) refers to learning a model that maps instances of datasets to target values such as classes; target values are necessary for supervised learning. Common algorithms used here are backpropagation (artificial neural networks), support vector machines, and decision tree learning. **Reinforcement learning** is covered in Section 2.6 and refers to methods that solve reinforcement learning problems, where a sequence of actions is associated with positive or negative rewards, but not with a "target value" (the correct action). The paradigmatic algorithm here is temporal difference (TD) learning and its popular instantiation Q-learning. Section 5.6.3 outlines **unsupervised learning** which refers to algorithms that find patterns (e.g., clusters) in datasets that do *not* have target values. This includes clustering methods such as k-means, hierarchical clustering and self-organizing maps as well as frequent pattern mining methods such as Apriori and generalized sequential patterns. The chapter concludes with a number of notable algorithms that combine elements of the algorithms above to yield **hybrid** methods. In particular we cover neuroevolution and TD learning with ANN function approximation as the most popular hybrid algorithms used in the field of game AI.

2.1 General Notes

Before detailing each of the algorithm types we outline two overarching elements that bind together all the AI methods covered in this book. The former is the algorithm's **representation**; the second is its **utility**. On the one hand, any AI algorithm somehow stores and maintains knowledge obtained about a particular task at hand. On the other hand, most AI algorithms seek to find better representations of knowledge. This seeking process is driven by a utility function of some form. We should note that the utility is of no use solely in methods that employ static knowledge representations such as finite state machines or behavior trees.

2.1.1 Representation

Appropriately representing knowledge is a key challenge for artificial intelligence at large and it is motivated by the capacity of the human brain to store and retrieve obtained knowledge about the world. The key questions that drive the design of representations for AI are as follows. How do people represent knowledge and how can AI potentially mimic that capacity? What is the nature of knowledge? How generic can a representation scheme be? General answers to the above questions, however, are far from trivial at this point.

As a response to the open *general* questions regarding knowledge and its representation, AI has identified numerous and very *specific* ways to store and retrieve information which is authored, obtained, or learned. The representation of knowledge about a task or a problem can be viewed as the computational mapping of the task under investigation. On that basis, the representation needs to store knowledge about the task in a format that a machine is able to process, such as a data structure.

To enable any form of artificial intelligence knowledge needs to be represented computationally and the ways this can happen are many. Representation types include **grammars** such as grammatical evolution, **graphs** such as finite state machines or probabilistic models, **trees** such as decision trees, behavior trees and genetic programming, **connectionism** such as artificial neural networks, **genetic** such as genetic algorithms and evolutionary strategies and **tabular** such as temporal difference learning and Q-learning. As we will see in the remainder of this book, all above representation types find dissimilar uses in games and can be associated with various game AI tasks.

One thing is certain for any AI algorithm that is tried on a particular task: the chosen representation has a major impact on the performance of the algorithm. Unfortunately, the type of representation to be chosen for a task follows the *no free lunch theorem* [756], suggesting that there is no single representation type which is ideal for the task at hand. As a general set of guidelines, however, the representation chosen should be as *simple* as possible. Simplicity usually comes as a delicate balance between computational effort and algorithm performance as either being over-detailed or over-simplistic will affect the performance of the algorithm. Furthermore, the representation chosen should be as *small* as possible given the complexity of the task at hand. Neither simplicity nor size are trivial decisions to make with respect to the representation. Good representations come with sufficient practical wisdom and empirical knowledge about the complexity and the qualitative features of the problem the AI is trying to solve.

2.1.2 Utility

Utility in game theory (and economics at large) is a measure of rational choice when playing a game. In general, it can be viewed as a function that is able to assist a search algorithm to decide which path to take. For that purpose, the utility function samples aspects of the search space and gathers information about the "goodness" of areas in the space. In a sense, a utility function is an approximation of the solution we try to find. In other words, it is a **measure of goodness** of the existing representation we search through.

Similar concepts to the utility include the **heuristic** used by computer science and AI as an *approximate* way to solve a problem faster when *exact* methods are too slow to afford, in particular associated with the tree search paradigm. The concept of **fitness** is used similarly as a utility function that measures the degree to which a solution is good, primarily, in the area of evolutionary computation. In mathematical optimization, the **objective**, **loss**, **cost**, or **error** function is the utility function to be minimized (or maximized if that is the objective). In particular, in supervised learning the error function represents how well an approach maps training examples to target (desired) outputs. In the area of reinforcement learning and Markov decision processes instead, the utility is named **reward**, which is a function an agent attempts to maximize by learning to take the right action in a particular state. Finally, in the area of **unsupervised learning** utility is often provided internally and within the representation via e.g., competitive learning or self-organization.

Similarly to selecting an appropriate representation, the selection of a utility function follows the no free lunch theorem. A utility is generally difficult to design and sometimes the design task is basically impossible. The simplicity of its design pays off, but the completeness as well. The quality of a utility function largely depends on thorough empirical research and practical experience, which is gained within the domain under investigation.

2.1.3 Learning = Maximize Utility (Representation)

The utility function is the drive for search and essential for learning. On that basis, the utility function is the *training signal* of any machine learning algorithm as it offers a measure of goodness of the representation we have. Thereby it implicitly provides indications on what to do to further increase the current goodness of the presentation. Systems that do not require learning (such as AI methods that are based on ad-hoc designed representations; or expert-knowledge systems) do not require a utility. In supervised learning the utility is sampled from data—i.e., good input-output patterns. In reinforcement learning and evolutionary computation, instead, the training signal is provided by the environment—i.e., rewards for doing something well and punishments for doing something wrong. Finally, in unsupervised learning the training signal derives from the internal structure of the representation.

2.2 Ad-Hoc Behavior Authoring

In this section we discuss the first, and arguably the most popular, class of AI methods for game development. **Finite state machines**, **behavior trees** and **utility-based AI** are ad-hoc behavior authoring methods that have traditionally dominated the control of non-player characters in games. Their dominance is evident by the fact that the term *game AI* in the game development scene is still nowadays synonymous with the use of these methods.

2.2.1 Finite State Machines

A **Finite State Machine** (FSM) [230]—and FSM variants such as hierarchical FSMs—is the game AI method that dominated the control and decision making processes of non-player characters in games up until the mid-2000s.

FSMs belong to the expert-knowledge systems area and are represented as graphs. An FSM graph is an abstract representation of an interconnected set of objects, symbols, events, actions or properties of the phenomenon that needs to be ad-hoc designed (represented). In particular, the graph contains nodes (states) which embed some mathematical abstraction and edges (transitions) which represent a conditional relationship between the nodes. The FSM can only be in one state at a time; the current state can change to another if the condition in the corresponding transition is fulfilled. In a nutshell, an FSM is defined by three main components:

- A number of **states** which store information about a task—e.g., you are currently on the *explore* state.
- A number of **transitions** between states which indicate a state change and are described by a condition that needs to be fulfilled—e.g., if you hear a fire shot, move to the *alerted* state.
- A set of **actions** that need to be followed within each state—e.g., while in the *explore* state *move randomly* and *seek opponents*.

FSMs are incredibly simple to design, implement, visualize, and debug. Further they have proven they work well with games over the years of their co-existence. However, they can be extremely complex to design on a large scale and are, thereby, computationally limited to certain tasks within game AI. An additional critical limitation of FSMs (and all ad-hoc authoring methods) is that they are not flexible and dynamic (unless purposely designed). After their design is completed, tested and debugged there is limited room for adaptivity and evolution. As a result, FSMs end up depicting very predictable behaviors in games. We can, in part, overcome such a drawback by representing transitions as fuzzy rules [532] or probabilities [109].

2.2.1.1 An FSM for Ms Pac-Man

In this section we showcase FSMs as employed to control the Ms Pac-Man agent. A hypothetical and simplified FSM controller for Ms Pac-Man is illustrated in Fig. 2.1. In this example our FSM has three states (seek pellets, chase ghosts and evade ghosts) and four transitions (ghosts flashing, no visible ghost, ghost in sight, and power pill eaten). While in the *seek pellets* state, Ms Pac-Man moves randomly up until it detects a pellet and then follows a pathfinding algorithm to eat as many pellets as possible and as soon as possible. If a power pill is eaten, then Ms Pac-Man moves to the *chase ghosts* state in which it can use any tree-search algorithm to chase the blue ghosts. When the ghosts start flashing, Ms Pac-Man moves to the *evade ghosts* state in which it uses tree search to evade ghosts so that none is visible

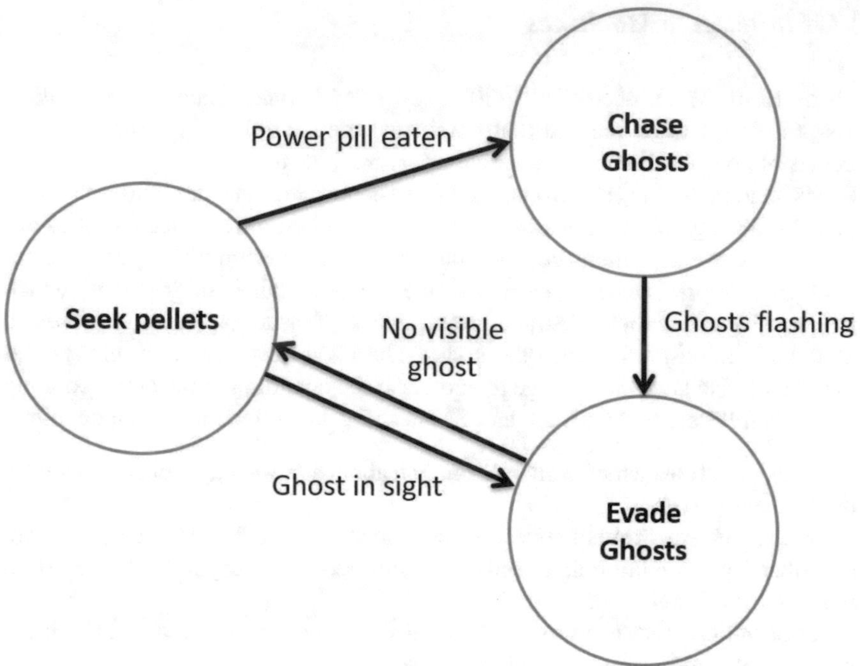

Fig. 2.1 A high-level and simplified FSM example for controlling Ms Pac-Man.

within a distance; when that happens Ms Pac-Man moves back to the *seek pellets* state.

2.2.2 *Behavior Trees*

A **Behavior Tree** (BT) [110, 112, 111] is an expert-knowledge system which, similarly to an FSM, models transitions between a finite set of tasks (or behaviors). The strength of BTs compared to FSMs is their modularity: if designed well, they can yield complex behaviors composed of simple tasks. The main difference between BT and FSMs (or even hierarchical FSMs) is that they are composed of *behaviors* rather than states. As with finite state machines, BTs are easy to design, test and debug, which made them dominant in the game development scene after their successful application in games such as *Halo 2* (Microsoft Game Studios, 2004) [291] and *Bioshock* (2K Games, 2007).

BT employs a tree structure with a root node and a number of parent and corresponding child nodes representing behaviors—see Fig. 2.2 for an example. We traverse a BT starting from the root. We then activate the execution of parent-child pairs as denoted in the tree. A child may return the following values to the parent in predetermined time steps (ticks): *run* if the behavior is still active, *success* if the behavior is completed, *failure* if the behavior failed. BTs are composed of three node types: the **sequence**, the **selector**, and the **decorator** the basic functionality of which is described below:

- **Sequence** (see blue rectangle in Fig. 2.2): if the child behavior succeeds, the sequence continues and eventually the parent node succeeds if all child behaviors succeed; otherwise the sequence fails.
- **Selector** (see red rounded rectangle in Fig. 2.2): there are two main types of selector nodes: the *probability* and the *priority* selectors. When a probability selector is used child behaviors are selected based on parent-child probabilities set by the BT designer. On the other hand if priority selectors are used, child behaviors are ordered in a list and tried one after the other. Regardless of the selector type used, if the child behavior succeeds the selector succeeds. If the child behavior fails, the next child in the order is selected (in priority selectors) or the selector fails (in probability selectors).
- **Decorator** (see purple hexagon in Fig. 2.2): the decorator node adds complexity to and enhances the capacity of a single child behavior. Decorator examples include the number of times a child behavior runs or the time given to a child behavior to complete the task.

Compared to FSM, BTs are more flexible to design and easier to test; they still however suffer from similar drawbacks. In particular, their dynamicity is rather low given that they are static knowledge representations. The probability selector nodes may add to their unpredictability and methods to adapt their tree structures have already shown some promise [385]. There is also a certain degree of similarity between BTs and ABL (A Behavior Language) [440] introduced by Mateas and Stern for story-based believable characters; their dissimilarities have also been reported [749]. Note however that this section barely scratches the surface of what is possible with BT design as there are several extensions to their basic structure that help BTs improve on their modularity and their capacity to deal with more complex behavior designs [170, 627].

2.2.2.1 A BT for Ms Pac-Man

Similarly to the FSM example above we use Ms Pac-Man to demonstrate the use of BTs in a popular game. In Fig. 2.3 we illustrate a simple BT for the *seek pellets* behavior of Ms Pac-Man. While in the *seek pellets* sequence behavior Ms Pac-Man will first *move* (selector), it will then find a pellet and finally it will keep eating pellets until a ghost is found in sight (decorator). While in the *move* behavior—which is a priority selector—Ms Pac-Man will prioritize ghost-free corridors over corridors with pellets and over corridors without pellets.

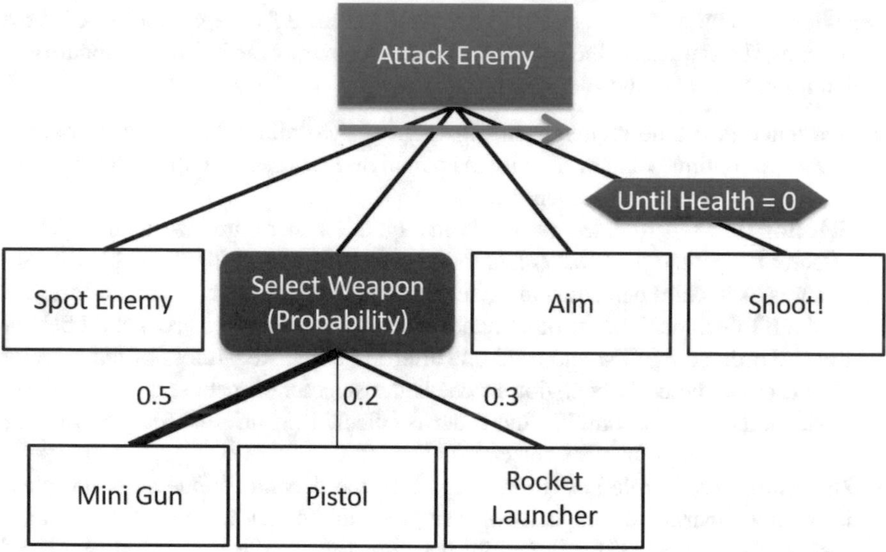

Fig. 2.2 A behavior tree example. The root of the BT is a sequence behavior (attack enemy) which executes the child behaviors *spot enemy*, *select weapon*, *aim* and *shoot* in sequence from left to right. The *select weapon* behavior is a probability selector giving higher probability—denoted by the thickness of the parent-child connecting lines—to the mini gun (0.5) compared to the rocket launcher (0.3) or the pistol (0.2). Once in the *shoot* behavior the decorator *until health = 0* requests the behavior to run until the enemy dies.

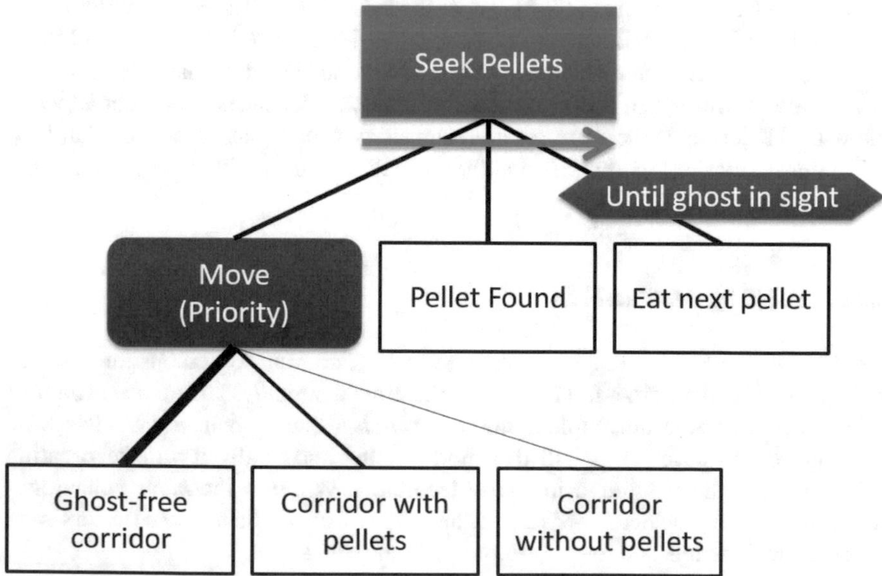

Fig. 2.3 A BT example for the *seek pellets* behavior of Ms Pac-Man.

2.2.3 Utility-Based AI

As has been pointed out by several industrial game AI developers the lack of behavioral modularity across games and in-game tasks is detrimental for the development of high quality AI [605, 171]. An increasingly popular method for ad-hoc behavior authoring that eliminates the modularity limitations of FSMs and BTs is the **utility-based** AI approach which can be used for the design of control and decision making systems in games [425, 557]. Following this approach, instances in the game get assigned a particular utility function that gives a value for the importance of the particular instance [10, 169]. For instance, the importance of an enemy being present at a particular distance or the importance of an agent's health being low in this particular context. Given the set of all utilities available to an agent and all the options it has, utility-based AI decides which is the most important option it should consider at this moment [426]. The utility-based approach is grounded in the utility theory of economics and is based on utility function design. The approach is similar to the design of membership functions in a fuzzy set.

A utility can measure anything from observable objective data (e.g., enemy health) to subjective notions such as emotions, mood and threat. The various utilities about possible actions or decisions can be aggregated into linear or non-linear formulas and guide the agent to take decisions based on the aggregated utility. The utility values can be checked every n frames of the game. So while FSMs and BTs would examine one decision at a time, utility-based AI architectures examine all available options, assign a utility to them and select the option that is most appropriate (highest utility).

As an example of utility-based AI we will build on the one appearing in [426] for weapon selection. For selecting a weapon an agent needs to consider the following aspects: range, inertia, random noise, ammo and indoors. The *range* utility function adds value to the utility of a weapon depending on the distance—for instance, if the distance is short, pistols are assigned higher utility. *Inertia* assigns higher utility value to the current weapon so that changes of weapons are not very frequent. *Random noise* adds non-determinism to the selection so that the agent does not always pick the same weapon given the same game situation. *Ammo* returns a utility about the current level of ammunition and *indoors* penalizes the use of particular weapons indoors such as a grenade through a boolean utility function (e.g., 0 utility value if the grenade is used indoors; 1 otherwise). Our agent makes a regular check of the available weapons, assigns utility scores to all of them and selects the weapon with the best total utility.

Utility-based AI has certain advantages compared to other ad-hoc authoring techniques. It is *modular* as the decision of the game agent is dependent on a number of different factors (or considerations); this list of factors can be dynamic. Utility-based AI is also *extensible* as we can easily author new types of considerations as we see them fit. Finally, the method is *reusable* as utility components can be transfered from one decision to another and from a game to another game. As a result of these advantages utility-based AI is gradually getting traction in the game industry scene [557, 171]. Utility-based AI has seen a widespread use across game genres

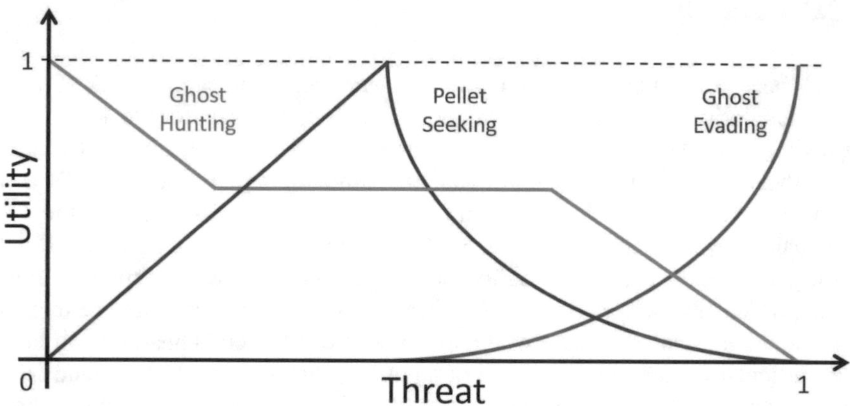

Fig. 2.4 A utility-based approach for controlling Ms Pac-Man behavior. The threat level (x-axis) is a function that lies between 0 and 1 which is based on the current position of ghosts. Ms Pac-Man considers the current level of threat, assigns utility values (through the three different curves) and decides to follow the behavior with the highest utility value. In this example the utility of *ghost evading* rises exponentially with the level of threat. The utility of *ghost hunting* decreases linearly with respect to threat up to a point where it stabilizes; it then decreases linearly as the threat level increases above a threshold value. Finally, the utility of *pellet seeking* increases linearly up to considerable threat level from which point it decreases exponentially.

and has been featured, among others, in *Kohan 2: Kings of War* (Take Two Interactive and Global Star Software, 2004), in *Iron Man* (Sega, 2008) for controlling the boss, in *Red Dead Redemption* (Rockstar Games, 2010) for weapon and dialog selection, and in *Killzone 2* (Sony Computer Entertainment, 2009) and in *F.E.A.R.* (Sierra Entertainment, 2005) for dynamic tactical decision making [426].

2.2.3.1 Utility-Based AI for Ms Pac-Man

Once again we use Ms Pac-Man to demonstrate the use of utility-based AI. Figure 2.4 illustrates an example of three simple utility functions that could be considered by Ms Pac-Man during play. Each function corresponds to a different behavior that is dependent on the current threat level of the game; threat is, in turn, a function of current ghost positions. At any point in the game Ms Pac-Man selects the behavior with the highest utility value.

2.2.3.2 A Short Note on Ad-Hoc Behavior Authoring

It is important to remember that all three methods covered in this section (and, in general, the methods covered in this chapter) represent the very **basic variants** of the algorithms. As a result, the algorithms we covered appear as static representations of states, behaviors or utility functions. It is possible, however, to create dynamic variants of those by adding non-deterministic or fuzzy elements; for instance, one may employ fuzzy transitions in an FSM or evolve behaviors in a BT. Further, it is important to note that these ad-hoc designed architectures can feature any of the methods this book covers in the remainder of this chapter. Basic processing elements such as an FSM state, a BT behavior or a utility function or even more complex hierarchies of nodes, trees or functions can be replaced by any other AI method yielding hybrid algorithms and agent architectures. Note that possible extensions of the algorithms can be found in the work we cite in the corresponding section of each algorithm but also in the reading list we provide next.

2.2.4 Further Reading

Further details on how to build and test FSMs and hierarchical FSMs can be found in [367]. For behavior trees we recommend the online tutorials and blogposts of A. Champandard found at the http://aigamedev.com/ portal [110, 111] and recent adaptations of the basic behavior tree structure as in [627]. Finally, the book of Dave Mark [425] is a good starting point for the study of utility-based AI and its application to control and decision making in games.

When it comes to software, a BT tool has been integrated within the Unreal Engine[1] while several other BT Unity tools[2] are available for the interested reader. Further, the Behave system[3] streamlines the iterative process of designing, integrating and debugging behavior trees and utility-based AI.

2.3 Tree Search

It has been largely claimed that most, if not all, of artificial intelligence is really just search. Almost every AI problem can be cast as a search problem, which can be solved by finding the best (according to some measure) plan, path, model, function, etc. Search algorithms are therefore often seen as being at the core of AI, to the point that many textbooks (such as Russell and Norvig's famous textbook [582]) start with a treatment of search algorithms.

The algorithms presented below can all be characterized as **tree search algorithms** as they can be seen as building a **search tree** where the root is the node representing the state where the search starts. Edges in this tree represent actions the agent takes to get from one state to another, and nodes represent states. Because there are typically several different actions that can be taken in a given state, the tree

---

1 https://docs.unrealengine.com/latest/INT/Engine/

2 For instance, see http://nodecanvas.paradoxnotion.com/ or http://www.opsive.com/.

3 http://eej.dk/community/documentation/behave/0-Introduction.html

branches. Tree search algorithms mainly differ in which branches are explored and in what order.

2.3.1 Uninformed Search

Uninformed search algorithms are algorithms which search a state space without any further information about the goal. The basic uninformed search algorithms are commonly seen as fundamental computer science algorithms, and are sometimes not even seen as AI.

Depth-first search is a search algorithm which explores each branch as far as possible before backtracking and trying another branch. At every iteration of its main loop, depth-first search selects a branch and then moves on to explore the resulting node in the next iteration. When a terminal node is reached—one from which it is not possible to advance further—depth-first search advances up the list of visited nodes until it finds one which has unexplored actions. When used for playing a game, depth-first search explores the consequences of a single move until the game is won or lost, and then goes on to explore the consequences of taking a different move close to the end states.

Breadth-first search does the opposite of depth-first search. Instead of exploring all the consequences of a single action, breadth-first search explores all the actions from a single node before exploring any of the nodes resulting from taking those actions. So, all nodes at depth one are explored before all nodes at depth two, then all nodes at depth three, etc.

While the aforementioned are fundamental uninformed search algorithms, there are many variations and combinations of these algorithms, and new uninformed search algorithms are being developed. More information about uninformed search algorithms can be found in Chapter 4 of [582].

It is rare to see uninformed search algorithms used effectively in games, but there are exceptions such as iterative width search [58], which does surprisingly well in general video game playing, and the use of breadth-first search to evaluate aspects of strategy game maps in *Sentient Sketchbook* [379]. Also, it is often illuminating to compare the performance of state-of-the-art algorithms with a simple uninformed search algorithm.

2.3.1.1 Uninformed Search for Ms Pac-Man

A depth-first approach in Ms Pac-Man would normally consider the branches of the game tree until Ms Pac-Man either completes the level or loses. The outcome of this search for each possible action would determine which action to take at a given moment. Breadth-first instead would first explore all possible actions of Ms Pac-Man at the current state of the game (e.g., going left, up, down or right) and would then explore all their resulting nodes (children) and so on. The game tree of either method is too big and complex to visualize within a Ms Pac-Man example.

2.3.2 Best-First Search

In **best-first search**, the expansion of nodes in the search tree is informed by some knowledge about the goal state. In general, the node that is closest to the goal state by some criterion is expanded first. The most well-known best-first search algorithm is A* (pronounced A star). The A* algorithm keeps a list of "open" nodes, which are next to an explored node but which have not themselves been explored. For each open node, an estimate of its distance from the goal is made. New nodes are chosen to explore based on a lowest cost basis, where the cost is the distance from the origin node plus the estimate of the distance to the goal.

A* can easily be understood as navigation in two- or three-dimensional space. Variants of this algorithm are therefore commonly used for **pathfinding** in games. In many games, the "AI" essentially amounts to non-player characters using A* pathfinding to traverse scripted points. In order to cope with large, deceptive spaces numerous modifications of this basic algorithm have been proposed, including hierarchical versions of A* [61, 661], real-time heuristic search [82], **jump point search** for uniform-cost grids [246], 3D pathfinding algorithms [68], planning algorithms for dynamic game worlds [495] that enable the animation of crowds in collision-free paths [631] and approaches for pathfinding in navigation meshes [68, 722]. The work of Steve Rabin and Nathan Sturtevant on grid-based pathfinding [551, 662] and pathfinding architectures [550] are notable examples. Sturtevant and colleagues have also been running a dedicated competition to grid-based path-planning [665] since 2012.[4] For the interested reader Sturtevant [663] has released a list of benchmarks for grid-based pathfinding in games[5] including *Dragon Age: Origins* (Electronic Arts, 2009), *StarCraft* (Blizzard Entertainment, 1998) and *Warcraft III: Reign of Chaos* (Blizzard Entertainment, 2002).

However, A* can also be used to search in the space of game states, as opposed to simply searching physical locations. This way, best-first search can be used for **planning** rather than just navigation. The difference is in taking the changing state of the world (rather than just the changing state of a single agent) into account. Planning with A* can be surprisingly effective, as evidenced by the winner of the 2009 Mario AI Competition—where competitors submitted agents playing *Super Mario Bros* (Nintendo, 1985)—being based on a simple A* planner that simply tried to get to the right end of the screen at all times [717, 705] (see also Fig. 2.5).

[4] http://movingai.com/GPPC/

[5] http://movingai.com/benchmarks/

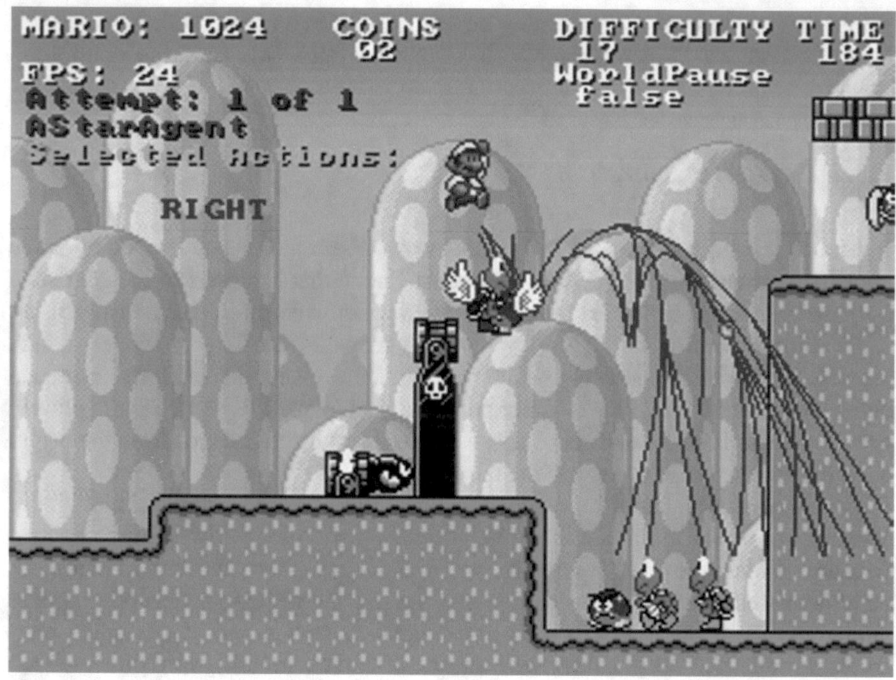

Fig. 2.5 The A* controller of the 2009 Mario AI Competition champion by R. Baumgarten [705]. The red lines illustrate possible future trajectories considered by the A* controller of Mario, taking the dynamic nature of the game into account.

2.3.2.1 Best-First Search for Ms Pac-Man

Best-first search can be applicable in Pac-Man in the form of A*. Following the paradigm of the 2009 Mario AI competition champion, Ms Pac-Man can be controlled by an A* algorithm that searches through possible game states within a short time frame and takes a decision on where to move next (up, down, left or right). The game state can be represented in various ways: from a very direct, yet costly, representation that takes ghost and pellet coordinates into account to an indirect representation that considers the distance to the closest ghost or pellet. Regardless of the representation chosen, A* requires the design of a cost function that will drive the search. Relevant cost functions for Ms Pac-Man would normally reward moves to areas containing pellets and penalizing areas containing ghosts.

2.3.3 Minimax

For single-player games, simple uninformed or informed search algorithms can be used to find a path to the optimal game state. However, for two-player adversarial games, there is another player that tries to win as well, and the actions of each player depend very much on the actions of the other player. For such games we need adversarial search, which includes the actions of two (or more) adversarial players. The basic adversarial search algorithm is called **Minimax**. This algorithm has been used very successfully for playing classic perfect-information two-player board games such as Checkers and Chess, and was in fact (re)invented specifically for the purpose of building a Chess-playing program [725].

The core loop of the Minimax algorithm alternates between player 1 and player 2—such as the white and black player in Chess—named the *min* and the *max* player. For each player, all possible moves are explored. For each of the resulting states, all possible moves by the other player are also explored, and so on until all the possible combinations of moves have been explored to the point where the game ends (e.g., with a win, a loss or a draw). The result of this process is the generation of the whole game tree from the root node down to the leaves. The outcome of the game informs the utility function which is applied onto the leaf nodes. The utility function estimates how good the current game configuration is for a player. Then, the algorithm traverses up the search tree to determine what action each player would have taken at any given state by backing-up values from leaves through the branch nodes. In doing so, it assumes that each player tries to play optimally. Thus, from the standpoint of the *max* player, it tries to maximize its score, whereas *min* tries to minimize the score of *max*; hence, the name *Minimax*. In other words, a *max* node of the tree computes the max of its child values whereas a *min* node computes the min of its child values. The optimal winning strategy is then obtained for *max* if, on *min*'s turn, a win is obtainable for *max* for *all moves* that *min* can make. The corresponding optimal strategy for *min* is when a win is possible independently of what move *max* will take. To obtain a winning strategy for max, for instance, we start at the root of the tree and we iteratively choose the moves leading to child nodes of highest value (on min's turn the child nodes with the lowest value are selected instead). Figure 2.6 illustrates the basic steps of Minimax through a simple example.

Of course, exploring all possible moves and countermoves is infeasible for any game of interesting complexity, as the size of the search tree increases exponentially with the depth of the game or the number of moves that are simulated. Indicatively, tic-tac-toe has a game tree size of $9! = 362,880$ states which is feasible to traverse through; however, the Chess game tree has approximately 10^{154} nodes which is infeasible to search through with modern computers. Therefore, almost all actual applications of the Minimax algorithm cut off search at a given depth, and use a **state evaluation** function to evaluate the desirability of each game state at that depth. For example, in Chess a simple state evaluation function would be to merely sum the number of white pieces on the board and subtract the number of black pieces; the higher this number is, the better the situation is for the white player. (Of course, much more sophisticated board evaluation functions are commonly used.) Together with improvements to the basic Minimax algorithm such as **α-β pruning** and the use of non-deterministic state evaluation functions, some very competent programs emerged for many classic games (e.g., IBM's Deep Blue). More information about Minimax and other adversarial search algorithms can be found in Chapter 6 of [582].

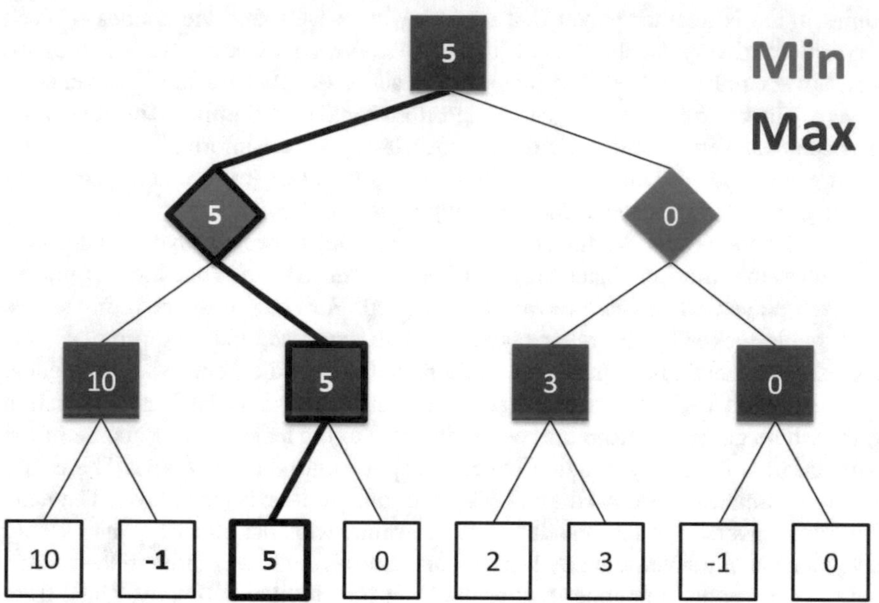

Fig. 2.6 An abstract game tree illustrating the Minimax algorithm. In this hypothetical game of two options for each player max (represented as red squares) plays first, min (represented as blue diamonds) plays second and then max plays one last time. White squares denote terminal nodes containing a winning (positive), a losing (negative) or a draw (zero) score for the max player. Following the Minimax strategy, the scores (utility) are traversed up to the root of the game tree. The optimal play for max and min is illustrated in bold. In this simple example if both players play optimally, max wins a score of 5.

2.3.3.1 Minimax for Ms Pac-Man

Strictly speaking, Minimax is not applicable to Ms Pac-Man as the game is non-deterministic and, thus, the Minimax tree is formally unknown. (Of course Minimax variants with heuristic evaluation functions can be eventually applicable.) Minimax is however applicable to Ms Pac-Man's deterministic ancestor, *Pac-Man* (Namco, 1980). Again strictly speaking, Pac-Man is a single-player adversarial game. As such Minimax is applicable only if we assume that Pac-Man plays against adversaries (ghosts) who make optimal decisions. It is important to note that ghosts' movements are not represented by tree nodes; instead, they are simulated based on their assumed optimal play. Game tree nodes in Pac-Man may represent the game state including the position of Pac-Man, the ghosts, and the current pellets and power pills available. The branches of the Minimax tree are the available moves of the Pac-Man in each game state. The terminal nodes can, for instance, feature either a binary utility (1 if Pac-Man completes the level; 0 if Pac-Man was killed by a ghost) or the final score of the game.

2.3.4 Monte Carlo Tree Search

There are many games which Minimax will not play well. In particular, games with a high **branching factor** (where there are many potential actions to take at any given point in time) lead to Minimax that will only ever search a very shallow tree. Another aspect of games which frequently throws spanners in the works of Minimax is when it is hard to construct a good state evaluation function. The board game Go is a deterministic, perfect information game that is a good example of both of these phenomena. Go has a branching factor of approximately 300, whereas Chess typically has around 30 actions to choose from. The positional nature of the Go game, which is all about surrounding the adversary, makes it very hard to correctly estimate the value of a given board state. For a long time, the best Go-playing programs in the world, most of which were based on Minimax, could barely exceed the playing strength of a human beginner. In 2007, **Monte Carlo Tree Search** (MCTS) was invented and the playing strength of the best Go programs increased drastically.

Beyond complex perfect information, deterministic games such as Go, Chess and Checkers, **imperfect information** games such a Battleship, Poker, Bridge and/or **non-deterministic** games such as backgammon and monopoly cannot be solved via Minimax due to the very nature of the algorithm. In such games, MCTS not only overcomes the tree size limitation of Minimax but, given sufficient computation, it approximates the Minimax tree of the game.

So how does MCTS handle high branching factors, lack of good state evaluation functions, and lack of perfect information and determinism? To begin with, it does not search all branches of the search tree to an even depth, instead it concentrates on the more promising branches. This makes it possible to search certain branches to a considerable depth even though the branching factor is high. Further, to get around the lack of good evaluation functions, determinism and imperfect information, the standard formulation of MCTS uses **rollouts** to estimate the quality of the game state, randomly playing from a game state until the end of the game to see the expected win (or loss) outcome. The utility values obtained via the random simulations may be used efficiently to adjust the policy towards a best-first strategy (a Minimax tree approximation).

At the start of a run of the MCTS algorithm, the tree consists of a single node representing the current state of the game. The algorithm then iteratively builds a search tree by adding and evaluating new nodes representing game states. This process can be interrupted at any time, rendering MCTS an **anytime** algorithm. MCTS requires only two pieces of information to operate: the *game rules* that would, in turn, yield the available moves in the game and the *terminal state evaluation*—whether that is win, a loss, a draw, or a game score. The vanilla version of MCTS does not require a heuristic function, which is, in turn, a key advantage over Minimax.

The core loop of the MCTS algorithm can be divided into four steps: **Selection**, **Expansion** (the first two steps are also known as *tree policy*), **Simulation** and **Backpropagation**. The steps are also depicted in Fig. 2.7.

Selection: In this phase, it is decided which node should be expanded. The process starts at the root of the tree, and continues until a node is selected which has unexpanded children. Every time a node (action) is to be selected within the existing tree a child node j is selected to maximise the UCB1 formula:

$$\text{UCB1} = \overline{X}_j + 2C_p\sqrt{\frac{2\ln n}{n_j}} \tag{2.1}$$

where $\overline{X}_j$ is the average reward of all nodes beneath this node, C_p is an exploration constant (often set to $1/\sqrt{2}$), n is the number of times the parent node has been visited, and n_j is the number of times the child node j has been visited. It is important to note that while UCB1 is the most popular formula used for action selection it is certainly not the only one available. Beyond equation (2.1) other options include epsilon-greedy, Thompson sampling, and Bayesian bandits. For instance, Thompson sampling selects actions stochastically based on their posterior probabilities of being optimal [692].

Expansion: When a node is selected that has unexpanded children—i.e., that represents a state from which actions can be taken that have not been attempted yet—one of these children is chosen for *expansion*, meaning that a simulation is done starting in that state. Selecting which child to expand is often done at random.

Simulation (Default Policy): After a node is expanded, a simulation (or *rollout*) is done starting from the non-terminal node that was just expanded until the end of game to produce a value estimate. Usually, this is performed by taking random actions until a termination state is reached, i.e., until the game is either won or lost. The state at the end of the game (e.g., -1 if losing, $+1$ if winning, but could be more nuanced) is used as the reward (Δ) for this simulation, and propagated up the search tree.

Backpropagation: The reward (the outcome of the simulation) is added to the total reward X of the new node. It is also "backed up": added to the total reward of its parent node, its parent's parent and so on until the root of the tree.

The simulation step might appear counter-intuitive—taking random actions seems like no good way to play a game—but it provides a relatively unbiased estimate of the quality of a game state. Essentially, the better a game state is, the more simulations are likely to end up winning the game. At least, this is true for games like Go where a game will always reach a terminal state within a certain relatively small number of moves (400 for Go). For other games like Chess, it is theoretically possible to play an arbitrary number of moves without winning or losing the game. For many video games, it is *probable* that any random sequence of actions will not end the game unless some timer runs out, meaning that most simulations will be

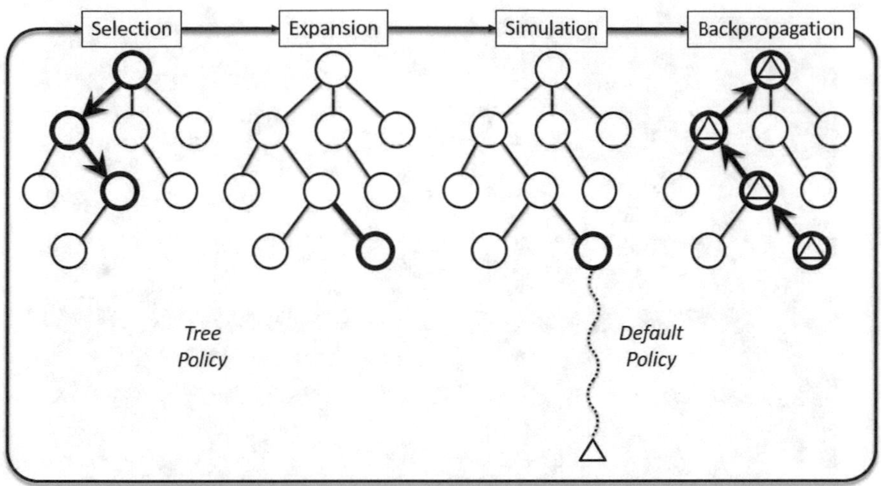

Fig. 2.7 The four basic steps of MCTS exemplified through one iteration of the algorithm. The figure is a recreation of the corresponding MCTS outline figure by Chaslot et al. [118].

very long (tens or hundreds of thousands of steps) and not yield useful information. For example, in *Super Mario Bros* (Nintendo,1985), the application of random actions would most likely make Mario dance around his starting point until his time is up [294]. In many cases it is therefore useful to complement the simulation step with a state evaluation function (as commonly used in Minimax), so that a simulation is performed for a set number of steps and if a terminal state is not reached a state evaluation is performed in lieu of a win-lose evaluation. In some cases it might even be beneficial to replace the simulation step entirely with a state evaluation function.

It is worth noting that there are many variations of the basic MCTS algorithm—it may in fact be more useful to see MCTS as an algorithm family or framework rather than a single algorithm.

2.3.4.1 MCTS for Ms Pac-Man

MCTS can be applicable to the real-time control of the Ms Pac-Man agent. There are obviously numerous ways to represent a game state (and thereby a game tree node) and design a reward function for the game, which we will not discuss in detail here. In this section, instead, we will outline the approach followed by Pepels et al. [524] given its success in obtaining high scores for Ms Pac-Man. Their agent, named Maastricht, managed to obtain over 87,000 points and was ranked first (among 36 agents) in the Ms Pac-Man competition of the IEEE Computational Intelligence and Games conference in 2012.

When MCTS is used for real-time decision making a number of challenges become critical. First, the algorithm has limited rollout computational budget which increases the importance of heuristic knowledge. Second, the action space can be

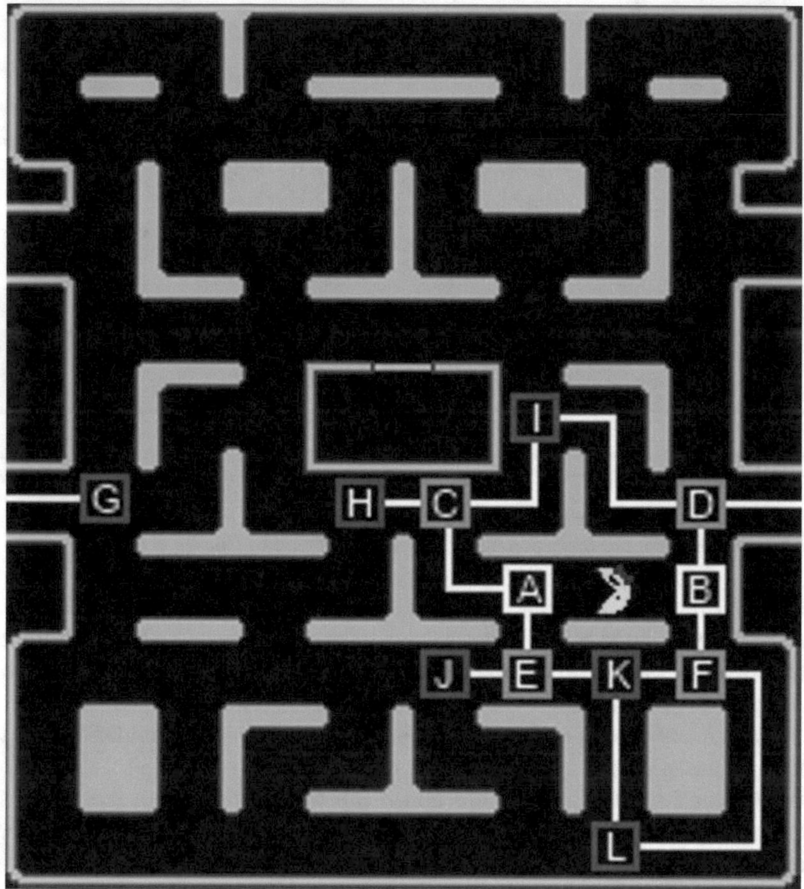

Fig. 2.8 The junction-based representation of a game state for the Maastricht MCTS controller [524]. All letter nodes refer to game tree nodes (decisions) for Ms Pac-Man. Imaged adapted from [524] with permission from authors.

particularly fine-grained which suggests that macro-actions are a more powerful way to model the game tree; otherwise the agent's planning will be very short-term. Third, there might be no terminal node in sight which calls for good heuristics and possibly restricting the simulation depth. The MCTS agent of Pepels et al. [524] managed to cope with all the above challenges of using MCTS for real-time control by using a restricted game tree and a junction-based game state representation (see Fig. 2.8).

2.3.5 *Further Reading*

The basic search algorithms are well covered in Russell and Norvig's classic AI textbook [582]. The A* algorithm was invented in 1972 for robot navigation [247]; a good description of the algorithm can be found in Chapter 4 of [582]. There is plenty of more advanced material on tailoring and optimizing this algorithm for specific game problems in dedicated game AI books such as [546]. The different components of Monte Carlo tree search [141] were invented in 2006 and 2007 in the context of playing Go [142]; a good overview of and introduction to MCTS and some of its variants is given in a survey paper by Browne et al. [77].

2.4 Evolutionary Computation

While tree search algorithms start from the root node representing an origin state, and build a search tree based on the available actions, *optimization* algorithms do not build a search tree; they only consider complete solutions, and not the path taken to get there. As mentioned earlier in Section 2.1, all optimization algorithms assume that there is something to optimize solutions for; there must be an **objective**, alternatively called **utility function**, **evaluation function** or **fitness function**, which can assign a numerical value (the **fitness**) to a solution, which can be maximized (or minimized). Given a utility function, an optimization algorithm can be seen as an algorithm that seeks in a search space solutions that have the highest (or lowest) value of that utility.

A broad family of optimization algorithms is based on randomized variation of solutions, where one or multiple solutions are kept at any given time, and new solutions (or candidates, or search points; different terminology is used by different authors) are created through randomly changing some of the existing solutions, or maybe combining some of them. Randomized optimization algorithms which keep multiple solutions are called **evolutionary algorithms**, by analogy with natural evolution.

Another important concept when talking about optimization algorithms (and AI at large as covered in Section 2.1) is their **representation**. All solutions are represented in some way, for example, as fixed-size vectors of real numbers, or variable-length strings of characters. Generally, the same artifact can be represented in many different ways; for example, when searching for a sequence of actions that solves a maze, the action sequence can be represented in several different ways. In the most direct representation, the character at step t determines what action to take at time step $t+1$. A somewhat more indirect representation for a sequence of actions would be a sequence of tuples, where the character at time step t decides what action to take and the number $t+n$ determines for how many time steps n to take that action. The choice of representation has a big impact on the efficiency and efficacy of the search algorithm, and there are several tradeoffs at play when making these choices.

Optimization is an extremely general concept, and optimization algorithms are useful for a wide variety of tasks in AI as well as in computing more generally. Within AI and games, optimization algorithms such as evolutionary algorithms have been used in many roles as well. In Chapter 3 we explain how optimization algorithms can be used for searching for game-playing agents, and also for searching for action sequences (these are two very different uses of optimization that are both in the context of game-playing); in Chapter 4 we explain how we can use optimization to create game content such as levels; and in Chapter 5 we discuss how to use optimization to find player models.

2.4.1 Local Search

The simplest optimization algorithms are the local optimization algorithms. These are so called because they only search "locally", in a small part of the search space, at any given time. A local optimization algorithm generally just keeps a single solution candidate at any given time, and explores variations of that solution.

The arguably simplest possible optimization algorithm is the **hill climber**. In its most common formulation, which we can call the deterministic formulation, it works as follows:

1. *Initialization:* Create a solution s by choosing a random point in search space. Evaluate its fitness.
2. Generate all possible neighbors of s. A neighbor is any solution that differs from s by at most a certain given distance (for example, a change in a single position).
3. Evaluate all the neighbors with the fitness function.
4. If none of the neighbors has a better fitness score than s, exit the algorithm and return s.
5. Otherwise, replace s with the neighbor that has the highest fitness value and go to step 2.

The deterministic hill climber is only practicable when the representation is such that each solution has a small number of neighbors. In many representations there are an astronomically high number of neighbors. It is therefore preferable to use variants of hill climbers that may guide the search effectively. One approach is the **gradient-based hill climber** that follows the gradient towards minimizing a cost function. That algorithmic approach trains artificial neural networks for instance (see Section 2.5). Another approach that we cover here is the **randomized hill climber**. This instead relies on the concept of **mutation**: a small, random change to a solution. For example, a string of letters can be mutated by randomly flipping one or two characters to some other character (see Fig. 2.9), and a vector of real

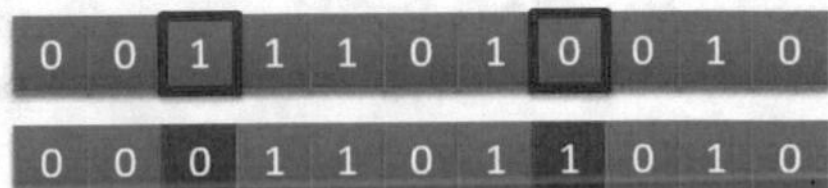

(a) **Mutation**: A number of genes is selected to be mutated with a small probability e.g., less than 1%. The selected genes are highlighted with a red outline at the top chromosome and are mutated by flipping their binary value (red genes) at the bottom chromosome.

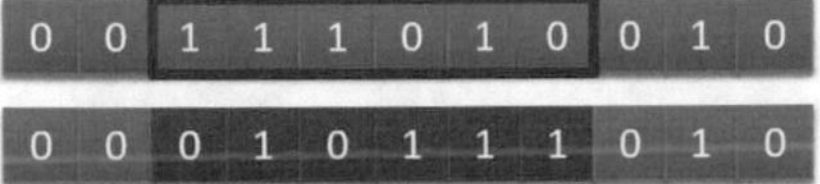

(b) **Inversion**: Two positions in the offspring are randomly chosen and the positions between them—the gene sequence highlighted by a red outline at the top chromosome—are inversed (red genes) at the bottom chromosome.

Fig. 2.9 Two ways of mutating a binary chromosome. In this example we use a chromosome of eleven genes. A chromosome is selected (top bit-string) and mutated (bottom bit-string).

numbers can be mutated by adding another vector to it drawn from a random distribution around zero, and with a very small standard deviation. Macro-mutations such as gene **inversion** can also be applied as visualized in Fig. 2.9. Given a representation, fitness function and mutation operator, the randomized hill climber works as follows:

1. *Initialization*: Create a solution s by choosing a random point in the search space. Evaluate its fitness.
2. *Mutation*: Generate an offspring s' by mutating s.
3. *Evaluation*: Evaluate the fitness of s'.
4. *Replacement*: If s' has higher fitness than s, replace s with s'.
5. Go to step 2.

While very simple, the randomized hill climber can be surprisingly effective. Its main limitation is that it is liable to get stuck in local optima. A **local optimum** is sort of a "dead end" in search space from which there is "no way out"; a point from which there are no better (higher-fit) points within the immediate vicinity. There are many ways of dealing with this problem. One is to simply restart the hill climber at a new randomly chosen point in the search space whenever it gets stuck. Another is **simulated annealing**, to accept moving to solutions with *lower* fitness with a given probability; this probability gradually diminishes during the search. A far more popular response to the problem of local optima is to keep not just a single solution at any time, but a **population** of solutions.

2.4.1.1 Local Search for Ms Pac-Man

While we can think of a few ways one can apply local search in Ms Pac-Man we outline an example of its use for controlling path-plans. Local search could, for instance, evolve short local plans (action sequences) of Ms Pac-Man. A solution could be represented as a set of actions that need to be taken and its fitness could be determined by the score obtained after following this sequence of actions.

2.4.2 Evolutionary Algorithms

Evolutionary algorithms are randomized **global** optimization algorithms; they are called global rather than local because they search many points in the search space simultaneously, and these points can be far apart. They accomplish this by keeping a population of solutions in memory at any given time. The general idea of evolutionary computation is to optimize by "breeding" solutions: generate many solutions, throw away the bad ones and keep the good (or at least less bad) ones, and create new solutions from the good ones.

The idea of keeping a population is taken from Darwinian evolution by natural selection, from which evolutionary algorithms also get their name. The size of the population is one of the key parameters of an evolutionary algorithm; a population size of 1 yields something like a randomized hill climber, whereas populations of several thousand solutions are not unheard of.

Another idea which is taken from evolution in nature is **crossover**, also called **recombination**. This is the equivalent of sexual reproduction in the natural world; two or more solutions (called **parents**) produce an offspring by combining elements of themselves. The idea is that if we take two good solutions, a solution that is a combination of these two—or intermediate between them—ought to be good as well, maybe even better than the parents. The offspring operator is highly dependent on the solution representation. When the solution is represented as a string or a vector, operators such as uniform crossover (which flips a fair coin and randomly picks values from each parent for each position in the offspring) or one-point crossover (where a position p in the offspring is randomly chosen, and values of positions before p are taken from parent 1 and values of positions after p are taken from parent 2) can be used. Crossover can be applied to any chromosome representation varying from a bit-string to a real-valued vector. Figure 2.10 illustrates these two crossover operators. It is in no way guaranteed, however, that the crossover operator generates an offspring that is anything as highly fit as the parents. In many cases, crossover can be highly destructive. If crossover is used, it is therefore important that the offspring operator is chosen with care for each problem. Figure 2.11 illustrates this possibility through a simple two-dimensional example.

The basic template for an evolutionary algorithm is as follows:

1. *Initialization:* The population is filled with N solutions created randomly, i.e., random points in search space. Known highly-fit solutions can also be added to this initial population.

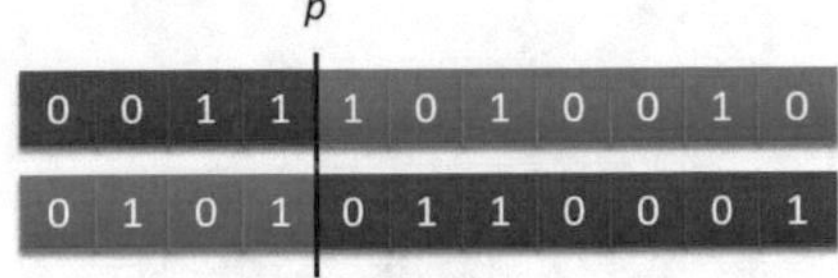

(a) **1-point crossover**: The vertical line across the two parents denotes the crossover point at position p.

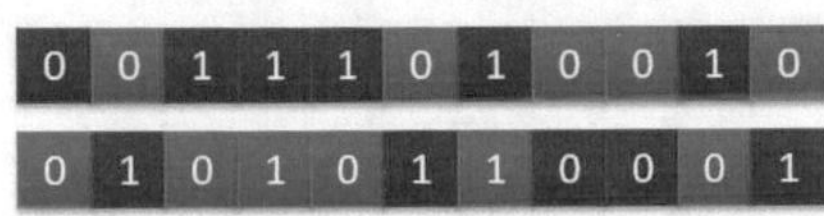

(b) **Uniform crossover**: To select genes from each parent to form offspring the operator flips a fair coin at each position of the chromosome.

Fig. 2.10 Two popular types of crossover used in evolutionary algorithms. In this example we use a **binary** representation and a chromosome size of eleven genes. The two bit-strings used in both crossover operators represent the two parents selected for recombination. Red and blue genes represent the two different offspring emerged from each crossover operator. Note that the operators are directly applicable to **real-valued** (floating point) representations too.

2. *Evaluation:* The fitness function is used to evaluate all solutions in the population and assign fitness values to them.
3. *Parent selection:* Based on fitness and possibly other criteria, such as distance between solutions, those population members that will be used for reproduction are selected. Selection strategies include methods directly or indirectly dependent on the fitness of the solutions, including roulette-wheel (proportionally to fitness), ranking (proportionally to rank in population) and tournament.
4. *Reproduction:* Offspring are generated through crossover from parents, or through simply copying parent solutions, or some combination of these.
5. *Variation:* Mutation is applied to some or all of the parents and/or offspring.
6. *Replacement:* In this step, we select which of the parents and/or offspring will make it to the next generation. Popular replacement strategies of the current population include the **generational** (parents die; offspring replace them), **steady state** (offspring replaces worst parent if and only if offspring is better) and **elitism** (generational, but best $x\%$ of parents survive) approaches.
7. *Termination:* Are we done yet? Decide based on how many generations or evaluations have elapsed (**exhaustion**), the highest fitness attained by any solution (**success**), and/or some other termination condition.
8. Go to step 2.

Every iteration of the main loop (i.e., every time we reach step 2) is called a **generation**, keeping with the nature-inspired terminology. The total number of fitness evaluations performed is typically proportional to the size of the population times the number of generations.

This high-level template can be implemented and expanded in a myriad different ways; there are thousands of evolutionary or evolution-like algorithms out there, and

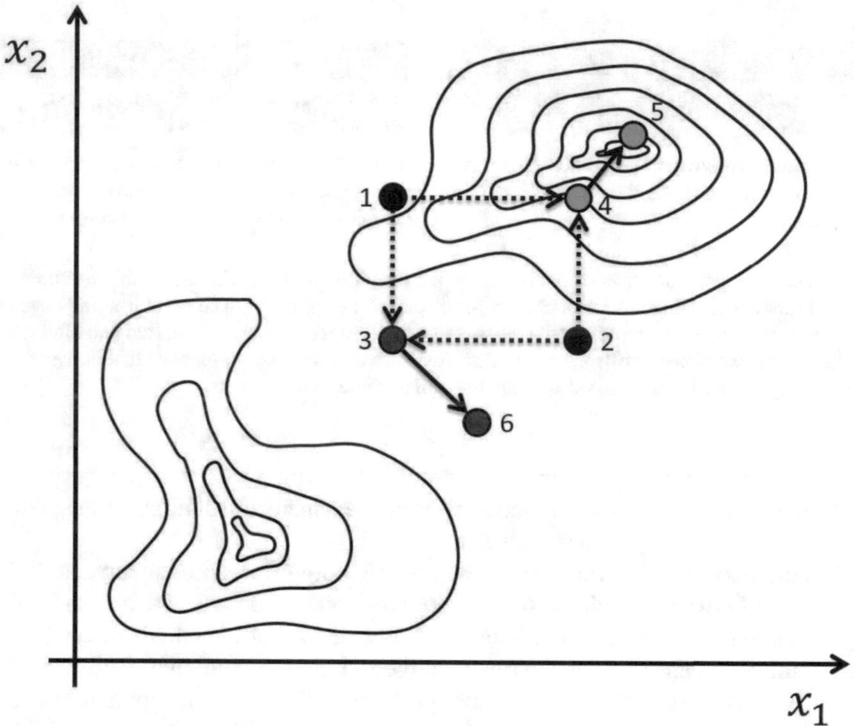

Fig. 2.11 An illustration of the mutation and crossover operators in a simplified two-dimensional fitness landscape. The problem is represented by two real-valued variables (x_1 and x_2) that define the two genes of the vector chromosome. The fitness landscape is represented by the contour lines on the 2D plane. Chromosomes 1 and 2 are selected to be parents. They are recombined via 1-point crossover (dotted arrows) which yields offspring 3 and 4. Both offspring are mutated (solid arrows) to yield solutions 5 and 6. Operators that lead to poorer-fit or higher-fit solutions are, respectively, depicted with green and red color.

many of them rearrange the overall flow, add new steps and remove existing steps. In order to make this template a bit more concrete, we will give a simple example of a working evolutionary algorithm below. This is a form of **evolution strategy**, one of the main families of evolutionary algorithms. While the $\mu + \lambda$ evolution strategy is a simple algorithm that can be implemented in 10 to 20 lines of code, it is a fully functional global optimizer and quite useful. The two main parameters are μ, which signifies the "elite" or the size of the part of the population that is kept every generation, and λ, the size of the part of the population that is re-generated every generation.

1. Fill the population with $\mu + \lambda$ randomly generated solutions.
2. Evaluate the fitness of all solutions.

3. Sort the population by decreasing fitness, so that the lowest-numbered solutions have highest fitness.
4. Remove the least fit λ individuals.
5. Replace the removed individuals with copies of the μ best individuals.
6. Mutate the offspring.
7. Stop if success or exhaustion. Otherwise go to step 2.

Evolution strategies, the type of algorithms which the $\mu + \lambda$ evolution strategy above is a simple example of, are characterized by a reliance on mutation rather than crossover to create variation, and by the use of self-adaptation to adjust mutation parameters (though that is not part of the simple algorithm above). They are also generally well suited to optimize artifacts represented as vectors of real numbers, so-called continuous optimization. Some of the very best algorithms for continuous optimization, such as the covariance matrix adaptation evolution strategy (CMA-ES) [245] and the natural evolution strategy (NES) [753], are conceptual descendants of this family of algorithms.

Another prominent family of evolutionary algorithms is **genetic algorithms** (GAs). These are characterized by a reliance on crossover rather than mutation for variation (some genetic algorithms have no mutation at all), fitness-proportional selection and solutions being often represented as bit-strings or other discrete strings. It should be noted, however, that the distinctions between different types of evolutionary algorithms are mainly based on their historical origins. These days, there are so many variations and such extensive hybridization that it often makes little sense to categorize a particular algorithm as belonging to one or the other family.

A variant of evolutionary algorithms emerges from the need of satisfying particular constraints within which a solution is not only fit but also **feasible**. When evolutionary algorithms are used for constrained optimization we are faced with a number of challenges such as that mutation and crossover cannot preserve or guarantee the feasibility of a solution. It may very well be that a mutation or a recombination between two parents may yield an infeasible offspring. One approach to deal with constraint handling is **repair**, which could be any process that turns infeasible individuals into feasible ones. A second approach is to modify the genetic operators so that the probability of an infeasible individual to appear becomes smaller. A popular approach is to merely penalize the existence of infeasible solutions by assigning them low fitness values or, alternatively, in proportion to the number of constraint violations. This strategy however may over-penalize the actual fitness of a solution which in turn will result in its rapid elimination from the population. Such a property might be undesirable and is often accused for the weak performance of evolutionary algorithms on handling constraints [456]. As a response to this limitation the **feasible-infeasible 2-population** (FI-2pop) algorithm [341] evolves two populations, one with feasible and one with infeasible solutions. The infeasible population optimizes its members towards minimizing the distance from feasibility. As the infeasible population converges to the border of feasibility, the likelihood of discovering new feasible individuals increases. Feasible offspring of infeasible parents are transferred to the feasible population, boosting its diversity (and vice versa for infeasible offspring). FI-2pop has been used in games on instances where we require fit and feasible solutions such as well-designed and playable game levels [649, 379].

Finally, another blend of evolutionary algorithms considers more than one objective when attempting to find a solution to a problem. For many problems it is hard to combine all requirements and specifications into a single objective measure. It is also often true that these objectives are conflicting; for instance, if our objectives are to buy the fastest and cheapest possible laptop we will soon realize the two objectives are partially conflicting. The intuitive solution is to merely add the different objective values—as a weighted sum—and use this as your fitness under optimization. Doing so, however, has several drawbacks such as the non-trivial ad-hoc design of the weighting among the objectives, the lack of insight on the interactions between the objectives (e.g., what is the price threshold above which faster laptops are not more expensive?) and the fact that a weighted-sum single-objective approach cannot reach solutions that achieve an optimal compromise among their weighted objectives. The response to these limitations is the family of algorithms known as **multiobjective evolutionary algorithms**. A multiobjective evolutionary algorithm considers at least two objective functions—that are partially conflicting—and searches for a **Pareto front** of these objectives. The Pareto front contains solutions that cannot be improved in one objective without worsening in another. Further details about multiobjective optimization by means of evolutionary algorithms can be found in [126]. The approach is applicable in game AI on instances where more than one objective is relevant for the problem we attempt to solve: for instance, we might wish to optimize both the balance and the asymmetry of a strategy game map [712, 713], or design non-player characters that are interestingly diverse in their behavioral space [5].

2.4.2.1 Evolutionary Algorithms for Ms Pac-Man

A simple way to employ evolutionary algorithms (EAs) in Ms Pac-Man is as follows. You could design a utility function based on a number of important parameters Ms Pac-Man must consider for taking the right decision on where to move next. These parameters, for instance, could be the current placement of ghosts, the presence of power pills, the number of pellets available on the level and so on. The next step would be to design a utility function as the weighted sum of these parameters. At each junction, Ms Pac-Man would need to consult its utility function for all its possible moves and pick the move with the highest utility. The weights of the utility function are unknown of course and this is where an EA can be of help by evolving the weights of the utility so that they optimize the score for Ms Pac-Man. In other words, the fitness of each chromosome (weight vector of utility) is determined by the score obtained from Ms Pac-Man within a number of simulation steps, or game levels played.

2.4.3 Further Reading

We recommend three books for further reading on evolutionary computation: Eiben and Smith's *Introduction to Evolutionary Computing* [184], Ashlock's *Evolutionary Computation for Modeling and Optimization* [21] and finally, the genetic programming field guide by Poli et al. [536].

2.5 Supervised Learning

Supervised learning is the algorithmic process of approximating the underlying function between labeled data and their corresponding attributes or features [49]. A popular example of supervised learning is that of a machine that is asked to distinguish between apples and pears (**labeled data**) given a set of **features** or **data attributes** such as the fruits' color and size. Initially, the machine learns to classify between apples and pears by *seeing* a number of available fruit examples—which contain the color and size of each fruit, on one hand, and their corresponding label (apple or pear) on the other. After learning is complete, the machine should ideally be able to tell whether a new and *unseen* fruit is a pear or an apple based solely on its color and size. Beyond distinguishing between apples and pears supervised learning nowadays is used in a plethora of applications including financial services, medical diagnosis, fraud detection, web page categorization, image and speech recognition, and user modeling (among many).

Evidently, supervised learning requires a set of labeled training examples; hence supervised. More specifically, the training signal comes as a set of supervised labels on the data (e.g., this is an apple whereas that one is a pear) which acts upon a set of characterizations of these labels (e.g., this apple has red color and medium size). Consequently, each data example comes as a pair of a set of labels (or outputs) and features that correspond to these labels (or inputs). The ultimate goal of supervised learning is not to merely learn from the input-output pairs but to derive a function that approximates (better, imitates) their relationship. The derived function should be able to map well to new and *unseen* instances of input and output pairs (e.g., unseen apples and pears in our example), a property that is called **generalization**. Here are some examples of input-output pairs one can meet in games and make supervised learning relevant: $\{$player health, own health, distance to player$\} \rightarrow \{$action (shoot, flee, idle)$\}$; $\{$player's previous position, player's current position$\} \rightarrow \{$player's next position$\}$; $\{$number of kills and headshots, ammo spent$\} \rightarrow \{$skill rating$\}$; $\{$score, map explored, average heart rate$\} \rightarrow \{$level of player frustration$\}$; $\{$Ms Pac-Man and ghosts position, pellets available$\} \rightarrow \{$Ms Pac-Man direction$\}$.

Formally, supervised learning attempts to derive a function $f : X \rightarrow Y$, given a set of N training examples $\{(\mathbf{x}_1, \mathbf{y}_1), \ldots, (\mathbf{x}_N, \mathbf{y}_N)\}$; where X and Y is the input and output space, respectively; $\mathbf{x}_i$ is the feature (input) vector of the i-th example and $\mathbf{y}_i$ is its corresponding set of labels. A supervised learning task has two core steps. In the first **training** step, the training samples—attributes and corresponding labels—

are presented and the function f between attributes and labels is derived. As we will see in the list of algorithms below f can be represented as a number of classification rules, decision trees, or mathematical formulae. In the second **testing** step f can be used to predict the labels of unknown data given their attributes. To validate the generalizability of f and to avoid overfitting to the data [49], it is common practice that f is evaluated on a new independent (test) dataset using a performance measure such as accuracy, which is the percentage of test samples that are correctly predicted by our trained function. If the accuracy is acceptable, we can use f to predict new data samples.

But how do we derive this f function? In general, an algorithmic process modifies the parameters of this function so that we achieve a good match between the given labels of our training samples and the function we attempt to approximate. There are numerous ways to find and represent that function, each one corresponding to a different supervised learning algorithm. These include artificial neural networks, case-based reasoning, decision tree learning, random forests, Gaussian regression, naive Bayes classifiers, k-nearest neighbors, and support vector machines [49]. The variety of supervised learning algorithms available is, in part, explained by the fact that there is no single learning algorithm that works best on all supervised learning problems out there. This is widely known as the *no free lunch theorem* [756].

Before covering the details of particular algorithms we should stress that the data type of the label determines the output type and, in turn, the type of the supervised learning approach that can be applied. We can identify three main types of supervised learning algorithms depending on the data type of the labels (outputs). First, we meet **classification** [49] algorithms which attempt to predict categorical class labels (discrete or nominal) such as the apples and pears of the previous example or the level in which a player will achieve her maximum score. Second, if the output data comes as an interval—such as the completion time of a game level or retention time—the supervised learning task is metric **regression** [49]. Finally, **preference learning** [215] predicts ordinal outputs such as ranks and preferences and attempts to derive the underlying global order that characterizes those ordinal labels. Examples of ordinal outputs include the ranked preferences of variant camera viewpoints, or a preference of a particular sound effect over others. The training signal in the preference learning paradigm provides information about the *relative* relation between instances of the phenomenon we attempt to approximate, whereas regression and classification provide information, respectively, about the *intensity* and the *classes* of the phenomenon.

In this book, we focus on a subset of the most promising and popular supervised learning algorithms for game AI tasks such as game playing (see Chapter 3), player behavior imitation or player preference prediction (see Chapter 5). The three algorithms outlined in the remainder of this section are artificial neural networks, support vector machines and decision tree learning. All three supervised learning algorithms covered can be used for either classification, prediction or preference learning tasks.

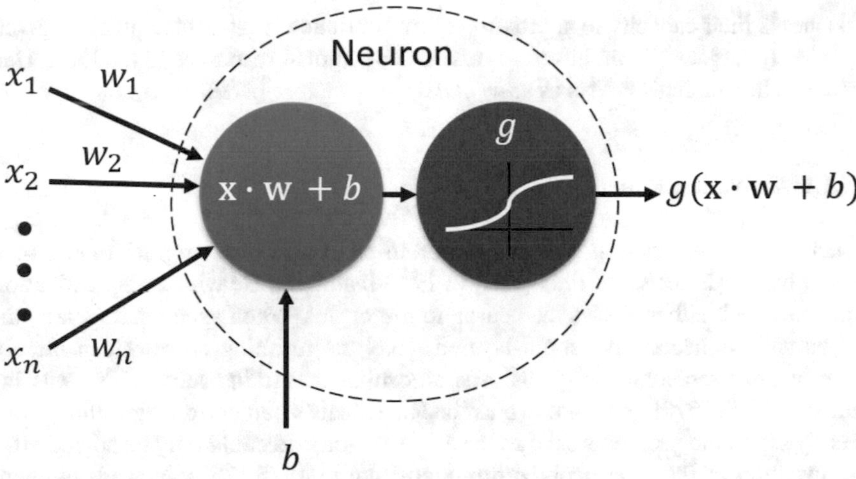

Fig. 2.12 An illustration of an artificial neuron. The neuron is fed with the input vector $\mathbf{x}$ through n connections with corresponding weight values $\mathbf{w}$. The neuron processes the input by calculating the weighted sum of inputs and corresponding connection weights and adding a bias weight (b): $\mathbf{x} \cdot \mathbf{w} + b$. The resulting formula feeds an activation function (g), the value of which defines the output of the neuron.

2.5.1 Artificial Neural Networks

Artificial Neural Networks (ANNs) are a bio-inspired approach for computational intelligence and machine learning. An ANN is a set of interconnected processing units (named **neurons**) which was originally designed to model the way a biological brain—containing over 10^{11} neurons—processes information, operates, learns and performs in several tasks. Biological neurons have a cell body, a number of dendrites which bring information into the neuron and an axon which transmits electrochemical information outside the neuron. The artificial neuron (see Fig. 2.12) resembles the biological neuron as it has a number of **inputs** $\mathbf{x}$ (corresponding to the neuron dendrites) each with an associated **weight** parameter $\mathbf{w}$ (corresponding to the synaptic strength). It also has a processing unit that combines inputs with their corresponding weights via an inner product (weighted sum) and adds a **bias** (or threshold) weight b to the weighted sum as follows: $\mathbf{x} \cdot \mathbf{w} + b$. This value is then fed to an **activation function** g (cell body) that yields the **output** of the neuron (corresponding to an axon terminal). ANNs are essentially simple mathematical models defining a function $f : \mathbf{x} \rightarrow \mathbf{y}$.

Various forms of ANNs are applicable for regression analysis, classification, and preference learning, and even unsupervised learning (via e.g., Hebbian learning [256] and self-organizing maps [347]). Core application areas include pattern recognition, robot and agent control, game-playing, decision making, gesture, speech and text recognition, medical and financial applications, affective modeling, and image recognition. The benefits of ANNs compared to other supervised learning approaches is their capacity to approximate any continuous real-valued function given sufficiently large ANN architectures and computational resources [348, 152]. This capacity characterizes ANNs as *universal approximators* [279].

2.5.1.1 Activation Functions

Which activation function should one use in an ANN? The original model of a neuron by McCulloch and Pitts [450] in 1943 featured a Heaviside step activation function which either allows the neuron to *fire* or not. When such neurons are employed and connected to a multi-layered ANN the resulting network can merely solve linearly separable problems. The algorithm that trains such ANNs was invented in 1958 [576] and is known as the Rosenblatt's perceptron algorithm. Non-linearly separable problems such as the exclusive-or gate could only be solved after the invention of the **backpropagation** algorithm in 1975 [752]. Nowadays, there are several activation functions used in conjunction with ANNs and their training. The use of the activation function, in turn, yields different types of ANNs. Examples include Gaussian activation function that is used in radial basis function (RBF) networks [71] and the numerous types of activation functions that can be used in the compositional pattern producing networks (CPPNs) [653]. The most common function used for ANN training is the sigmoid-shaped logistic function ($g(x) = 1/(1+e^{-x})$) for the following properties: 1) it is bounded, monotonic and non-linear; 2) it is continuous and smooth and 3) its derivative is calculated trivially as $g'(x) = g(x)(1-g(x))$. Given the properties above the logistic function can be used in conjunction with gradient-based optimization algorithms such as backpropagation which is described below. Other popular activation functions for training **deep** architectures of neural networks include the **rectifier**—named **rectified linear unit** (ReLU) when employed to a neuron—and its smooth approximation, the **softplus** function [231]. Compared to sigmoid-shaped activation functions, ReLUs allow for faster and (empirically) more effective training of deep ANNs, which are generally trained on large datasets (see more in Section 2.5.1.6).

2.5.1.2 From a Neuron to a Network

To form an ANN a number of neurons need to be structured and connected. While numerous ways have been proposed in the literature the most common of them all is to structure neurons in layers. In its simplest form, known as the **multi-layer perceptron** (MLP), neurons in an ANN are layered across one or more layers but not connected to other neurons in the same layer (see Fig. 2.13 for a typical MLP structure). The output of each neuron in each layer is connected to all the neurons in the next layer. Note that a neuron's output value feeds merely the neurons of the next layer and, thereby, becomes their input. Consequently, the outputs of the neurons in the last layer are the outputs of the ANN. The last layer of the ANN is also known as the **output layer** whereas all intermediate layers between the output

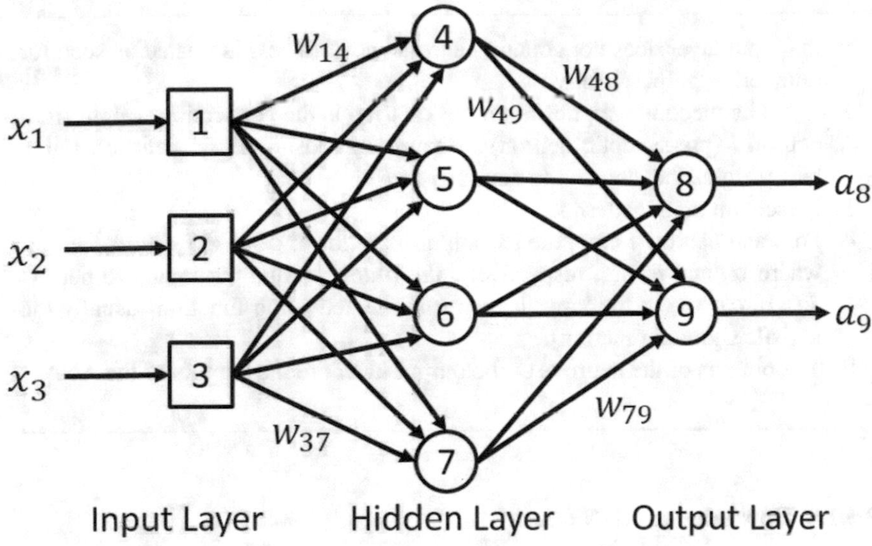

Fig. 2.13 An MLP example with three inputs, one hidden layer containing four hidden neurons and two outputs. The ANN has labeled and ordered neurons and example connection weight labels. Bias weights b_j are not illustrated in this example but are connected to each neuron j of the ANN.

and the input are the **hidden layers**. It is important to note that the inputs of the ANN, **x**, are connected to all the neurons of the first hidden layer. We illustrate this with an additional layer we call the **input layer**. The input layer does not contain neurons as it only distributes the inputs to the first layer of neurons. In summary, MLPs are 1) *layered* because they are grouped in layers; 2) *feed-forward* because their connections are unidirectional and always forward (from a previous layer to the next); and 3) *fully connected* because every neuron is connected to all neurons of the next layer.

2.5.1.3 Forward Operation

In the previous section we defined the core components of an ANN whereas in this section we will see how we compute the output of the ANN when an input pattern is presented. The process is called **forward operation** and propagates the inputs of the ANN throughout its consecutive layers to yield the outputs. The basic steps of the forward operation are as follows:

1. Label and order neurons. We typically start numbering at the input layer and increment the numbers towards the output layer (see Fig. 2.13). Note that

the input layer does not contain neurons, nevertheless is treated as such for numbering purposes only.

2. Label connection weights assuming that w_{ij} is the connection weight from neuron i (pre-synaptic neuron) to neuron j (post-synaptic neuron). Label bias weights that connect to neuron j as b_j.
3. Present an input pattern **x**.
4. For each neuron j compute its output as follows: $\alpha_j = g(\sum_i \{w_{ij}\alpha_i\} + b_j)$, where α_j and α_i are, respectively, the output of and the inputs to neuron j (n.b. $\alpha_i = x_i$ in the input layer); g is the activation function (usually the logistic sigmoid function).
5. The outputs of the neurons of the output layer are the outputs of the ANN.

2.5.1.4 How Does an ANN Learn?

How do we approximate $f(\mathbf{x};\mathbf{w},\mathbf{b})$ so that the outputs of the ANN match the desired outputs (labels) of our dataset, $\mathbf{y}$? We will need a training algorithm that adjusts the weights ($\mathbf{w}$ and $\mathbf{b}$) so that $f : \mathbf{x} \rightarrow \mathbf{y}$. A training algorithm as such requires two components. First, it requires a cost function to evaluate the quality of any set of weights. Second, it requires a search strategy within the space of possible solutions (i.e., the weight space). We outline these aspects in the following two subsections.

Cost (Error) Function

Before we attempt to adjust the weights to approximate f, we need some measure of MLP performance. The most common performance measure for training ANNs in a supervised manner is the squared Euclidean distance (error) between the vectors of the actual output of the ANN (α) and the desired labeled output y (see equation 2.2).

$$E = \frac{1}{2}\sum_j (y_j - \alpha_j)^2 \tag{2.2}$$

where the sum is taken over all the output neurons (the neurons in the final layer). Note that the y_j labels are constant values and more importantly, also note that E is a function of all the weights of the ANN since the actual outputs depend on them. As we will see below, ANN training algorithms build strongly upon this relationship between error and weights.

Backpropagation

The **backpropagation** (or backprop) [579] algorithm is based on gradient descent optimization and is arguably the most common algorithm for training ANNs. Backpropagation stands for *backward propagation of errors* as it calculates weight updates that minimize the error function—that we defined earlier (2.2)—from the output to the input layer. In a nutshell, backpropagation computes the partial derivative (gradient) of the error function E with respect to each weight of the ANN and adjusts the weights of the ANN following the (opposite direction of the) gradient that minimizes E.

As mentioned earlier, the squared Euclidean error of (2.2) depends on the weights as the ANN output which is essentially the $f(\mathbf{x};\mathbf{w},\mathbf{b})$ function. As such we can calculate the gradient of E with respect to any weight ($\frac{\theta E}{\theta w_{ij}}$) and any bias weight ($\frac{\theta E}{\theta b_j}$) in the ANN, which in turn will determine the degree to which the error will change if we change the weight values. We can then determine how much of such change we desire through a parameter $\eta \in [0,1]$ called **learning rate**. In the absence of any information about the general shape of the function between the error and the weights but the existence of information about its gradient it appears that a **gradient descent** approach would seem to be a good fit for attempting to find the global minimum of the E function. Given the lack of information about the E function, the search can start from some random point in the weight space (i.e., random initial weight values) and follow the gradient towards lower E values. This process is repeated iteratively until we reach E values we are happy with or we run out of computational resources.

More formally, the basic steps of the **backpropagation** algorithm are as follows:

1. Initialize $\mathbf{w}$ and $\mathbf{b}$ to random (commonly small) values.
2. For each training pattern (input-output pair):
   (a) Present input pattern $\mathbf{x}$, ideally normalized to a range (e.g., $[0,1]$).
   (b) Compute ANN actual outputs α_j using the forward operation.
   (c) Compute E according to (2.2).
   (d) Compute error derivatives with respect to each weight $\frac{\theta E}{\theta w_{ij}}$ and bias weight $\frac{\theta E}{\theta b_j}$ of the ANN from the output all the way to the input layer.
   (e) Update weights and bias weights as $\Delta w_{ij} = -\eta \frac{\theta E}{\theta w_{ij}}$ and $\Delta b_j - \eta \frac{\theta E}{\theta b_j}$, respectively.
3. If E is *small* or you are out of *computational budget*, stop! Otherwise go to step 2.

Note that we do not wish to detail the derivate calculations of step 2(d) as doing so would be out of scope for this book. We instead refer the interested reader to the original backpropagation paper [579] for the exact formulas and to the reading list at the end of this section.

Limitations and Solutions

It is worth noting that backpropagation is not guaranteed to find the global minimum of E given its local search (hill-climbing) property. Further, given its gradient-based (local) search nature, the algorithm fails to overcome potential plateaux areas in the error function landscape. As these are areas with near-zero gradient, crossing them results in near-zero weight updates and further in **premature convergence** of the algorithm. Typical solutions and enhancements of the algorithm to overcome convergence to local minima include:

- **Random restarts**: One can rerun the algorithm with new random connection weight values in the hope that the ANN is not too dependent on luck. No ANN model is good if it depends too much on luck—for instance, if it performs well only in one or two out of ten runs.
- **Dynamic learning rate**: One can either modify the learning rate parameter and observe changes in the performance of the ANN or introduce a dynamic learning rate parameter that increases when convergence is slow whereas it decreases when convergence to lower E values is fast.
- **Momentum**: Alternatively, one may add a momentum amount to the weight update rule as follows:

$$\Delta w_{ij}^{(t)} = m\Delta w_{ij}^{(t-1)} - \eta \frac{\theta E}{\theta w_{ij}} \tag{2.3}$$

  where $m \in [0,1]$ is the momentum parameter and t is the iteration step of the weight update. The addition of a momentum value of the previous weight update ($a\Delta w_{ij}^{(t-1)}$) attempts to help backpropagation to overcome a potential local minimum.

While the above solutions are directly applicable to ANNs of small size, practical wisdom and empirical evidence with modern (deep) ANN architectures, however, suggests that the above drawbacks are largely eliminated [366].

Batch vs. Non-batch Training

Backpropagation can be employed following a batch or a non-batch learning mode. In **non-batch** mode, weights are updated every time a training sample is presented to the ANN. In **batch mode**, weights are updated after all training samples are presented to the ANN. In that case, errors are accumulated over the samples of the batch prior to the weight update. The non-batch mode is more unstable as it iteratively relies on a single data point; however, this might be beneficial for avoiding a convergence to a local minimum. The batch mode, on the other hand, is naturally a more stable gradient descent approach as weight updates are driven by the average error of all training samples in the batch. To best utilize the advantages of both approaches it is common to apply batch learning of randomly selected samples in small batch sizes.

2.5.1.5 Types of ANNs

Beyond the standard feedforward MLP there are numerous other types of ANN used for classification, regression, preference learning, data processing and filtering, and clustering tasks. Notably, **recurrent** neural networks (such as Hopfield networks [278], Boltzmann machines [4] and Long Short-Term Memory [266]) allow connections between neurons to form directed cycles, thus enabling an ANN to capture dynamic and temporal phenomena (e.g., time-series processing and prediction). Further, there are ANN types mostly used for clustering and data dimensionality reduction such as Kohonen self-organizing maps [347] and Autoencoders [41].

2.5.1.6 From Shallow to Deep

A critical parameter for ANN training is the size of the ANN. So, how *wide* and *deep* should my ANN architecture be to perform well on this particular task? While there is no formal and definite answer to this question, there is a generally accepted rule-of-thumb suggesting that the size of the network should match the complexity of the problem. According to Goodfellow et al. in their deep learning book [231] an MLP is essentially a deep (feedforward) neural network. Its *depth* is determined by the number of hidden layers it contains. Goodfellow et al. state that "It is from this terminology that the name **deep learning** arises". On that basis, training of ANN architectures containing (at least) a hidden layer can be viewed as a deep learning task whereas single output-layered architectures can be viewed as *shallow*. Various methods have been introduced in recent years to enable training of deep architectures containing several layers. The methods largely rely on gradient search and are covered in detail in [231] for the interested reader.

2.5.1.7 ANNs for Ms Pac-Man

As with every other method in this chapter we will attempt to employ ANNs in the Ms Pac-Man game. One straightforward way to use ANNs in Ms Pac-Man is to attempt to imitate expert players of the game. Thus, one can ask experts to play the game and record their playthroughs, through which a number of features can be extracted and used as the input of the ANN. The resolution of the ANN input may vary from simple statistics of the game—such as the average distance between ghosts and Ms Pac-Man—to detailed pixel-to-pixel RGB values of the game level image. The output data, on the other hand, may contain the actions selected by Ms Pac-Man in each frame of the game. Given the input and desired output pairs, the ANN is trained via backpropagation to predict the action performed by expert players (ANN output) given the current game state (ANN input). The size (width and depth) of the ANN depends on both the amount of data available from the expert Ms Pac-Man players and the size of the input vector considered.

2.5.2 Support Vector Machines

Support vector machines (SVMs) [139] are an alternative and very popular set of supervised learning algorithms that can be used for classification, regression [179] and preference learning [302] tasks. A support vector machine is a binary linear classifier that is trained so as to maximize the margin between the training examples of the separate classes in the data (e.g., apples and pears). As with every other supervised learning algorithm, the attributes of new and unseen examples are seeding the SVM which predicts the class they belong to. SVMs have been used widely for text categorization, speech recognition, image classification, and hand-written character recognition among many other areas.

Similarly to ANNs, SVMs construct a hyperplane that divides the input space and represents the function f that maps between the input and the target outputs. Instead of implicitly attempting to minimize the difference between the model's actual output and the target output following the gradient of the error (as backpropagation does), SVMs construct a hyperplane that maintains the largest distance to the nearest training-data point of any other class. That distance is called a **maximum-margin** and its corresponding hyperplane divides the points ($\mathbf{x_i}$) of class with label (y_i) 1 from those with label -1 in a dataset of n samples in total. In other words, the distance between the derived hyperplane and the nearest point $\mathbf{x}_i$ from either class is maximized. Given the input attributes of a training dataset, $\mathbf{x}$, the general form of a hyperplane can be defined as: $\mathbf{w} \cdot \mathbf{x} - b = 0$ where, as in backpropagation training, $\mathbf{w}$ is the weight (normal) vector of the hyperplane and $\frac{b}{||\mathbf{w}||}$ determines the offset (or weight threshold/bias) of the hyperplane from the origin (see Fig. 2.14). Thus, formally put, an SVM is a function $f(\mathbf{x};\mathbf{w},b)$ that predicts target outputs ($\mathbf{y}$) and attempts to

$$\textbf{minimize } ||\mathbf{w}||, \tag{2.4}$$

$$\textbf{subject to } y_i(\mathbf{w} \cdot \mathbf{x}_i - b) \geq 1, \text{for } i = 1, \ldots, n \tag{2.5}$$

The weights $\mathbf{w}$ and b determine the SVM classifier. The $\mathbf{x}_i$ vectors that lie nearest to the derived hyperplane are called **support vectors**. The above problem is solvable if the training data is linearly separable (also known as a **hard-margin** classification task; see Fig. 2.14). If the data is not linearly separable (**soft-margin**) the SVM instead attempts to

$$\textbf{minimize } \left[\frac{1}{n}\sum_{i=1}^{n} \max\left(0, 1 - y_i(\mathbf{w} \cdot \mathbf{x}_i - b)\right)\right] + \lambda||\mathbf{w}||^2 \tag{2.6}$$

which equals $\lambda||\mathbf{w}||^2$ if the hard constraints of equation 2.5 are satisfied—i.e., if all data points are correctly classified on the right side of the margin. The value of equation (2.6) is proportional to the distance from the margin for misclassified data and λ is designed so as to qualitatively determine the degree to which the margin-size should be increased versus ensuring that the $\mathbf{x}_i$ will lie on the correct side of the

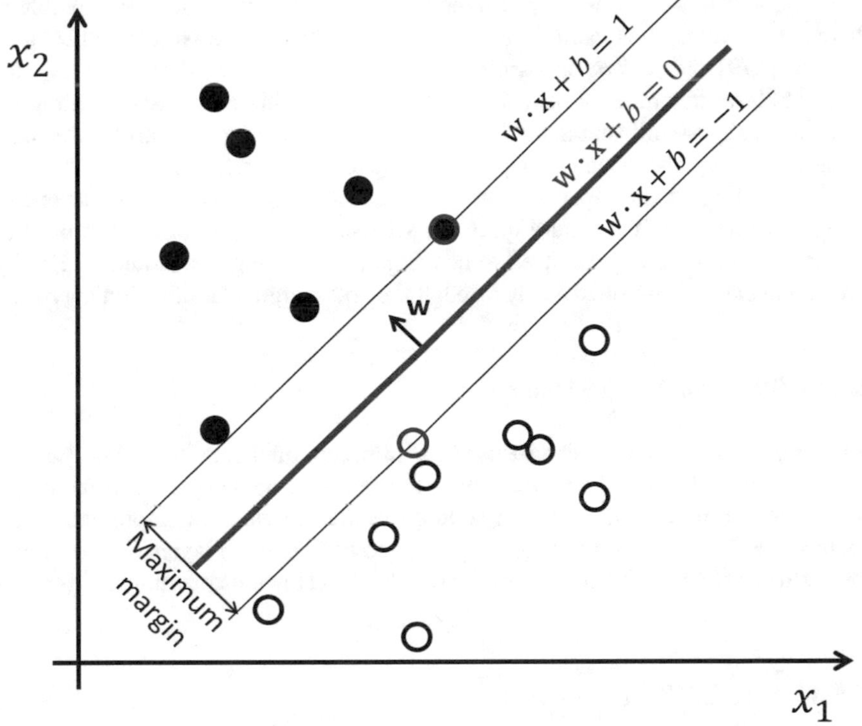

Fig. 2.14 An example of a maximum-margin hyperplane (red thick line) and margins (black lines) for an SVM which is trained on data samples from two classes. Solid and empty circles correspond to data with labels 1 and -1, respectively. The classification is mapped onto a two-dimensional input vector (x_1, x_2) in this example. The two data samples on the margin—the circles depicted with red outline—are the support vectors.

margin. Evidently, if we choose a small value for λ we approximate the hard-margin classifier for linearly separable data.

The standard approach for training soft-margin classifiers is to treat the learning task as a quadratic programming problem and search the space of $\mathbf{w}$ and b to find the widest possible margin that matches all data points. Other approaches include sub-gradient descent and coordinate descent.

In addition to linear classification tasks, SVMs can support non-linear classification by employing a number of different non-linear **kernels** which map the input space onto higher-dimensional feature spaces. The SVM task remains similar, except that every dot product is replaced by a nonlinear kernel function. This allows the algorithm to fit the maximum-margin hyperplane in a transformed feature space. Popular kernels used in conjunction with SVMs include polynomial functions, Gaussian radial basis functions or hyperbolic tangent functions.

While SVMs were originally designed to tackle **binary** classification problems there exist several SVM variants that can tackle multi-class classification [284], regression [179] and preference learning [302] that the interested reader can refer to.

SVMs have a number of advantages compared to other supervised learning approaches. They are efficient in finding solutions when dealing with large, yet sparse, datasets as they only depend on support vectors to construct hyperplanes. They also handle well large feature spaces as the learning task complexity does not depend on the dimensionality of the feature space. SVMs feature a simple convex optimization problem which can be guaranteed to converge to a single global solution. Finally, overfitting can be controlled easily through the soft margin classification approach.

2.5.2.1 SVMs for Ms Pac-Man

Similarly to ANNs, SVMs can be used for imitating the behavior of Ms Pac-Man expert players. The considerations about the feature (input) space and the action (output) space remain the same. In addition to the design of the input and output vectors, the size and quality of the data obtained from expert players will determine the performance of the SVM controlling Ms Pac-Man towards maximizing its score.

2.5.3 Decision Tree Learning

In **decision tree learning** [67], the function f we attempt to derive uses a decision tree representation which maps attributes of data observations to their target values. The former (inputs) are represented as the nodes and the latter (outputs) are represented as the leaves of the tree. The possible values of each node (input) are represented by the various branches of that node. As with the other supervised learning algorithms, decision trees can be classified depending on the output data type they attempt to learn. In particular, decision trees can be distinguished into classification, regression and rank trees if, respectively, the target output is a finite set of values, a set of continuous (interval) values, or a set of ordinal relations among observations.

An example of a decision tree is illustrated in Fig. 2.15. Tree nodes correspond to input attributes; there are branches to children for each of the possible values of each input attribute. Further leaves represent values of the output—car type in this example—given the values of the input attributes as determined by the path from the root to the leaf.

The goal of decision tree learning is to construct a mapping (a tree model) that predicts the value of target outputs based on a number of input attributes. The basic and most common approach for learning decision trees from data follows a top-down **recursive** tree induction strategy which has the characteristics of a greedy process. The algorithm assumes that both the input attributes and the target outputs have finite discrete domains and are of categorical nature. If inputs or outputs are continuous values, they can be discretized prior to constructing the tree. A tree is

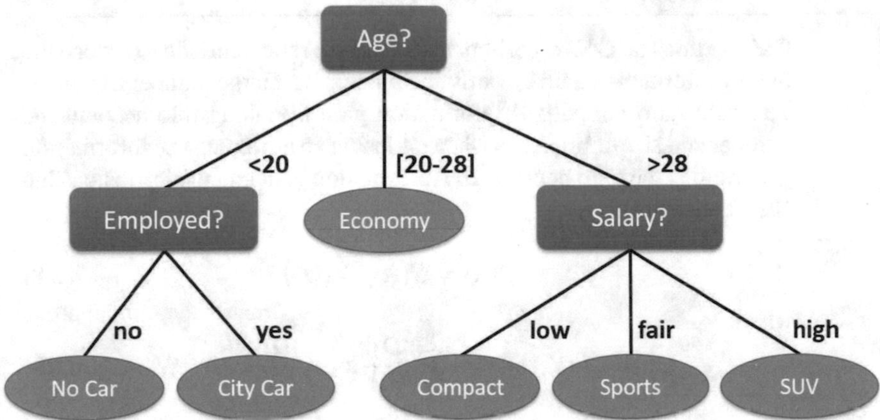

Fig. 2.15 A decision tree example: Given *age*, *employment* status and *salary* (data attributes) the tree predicts the *type of car* (target value) a person owns. Tree nodes (blue rounded rectangles) represent data attributes, or inputs, whereas leaves (gray ovals) represent target values, or outputs. Tree branches represent possible values of the corresponding parent node of the tree.

gradually constructed by splitting the available training dataset into subsets based on selections made for the attributes of the dataset. This process is repeated on a attribute-per-attribute basis in a recursive manner.

There are several variants of the above process that lead to dissimilar decision tree algorithms. The two most notable variants of decision tree learning, however, are the **Iterative Dichotomiser 3 (ID3)** [544] and its successor **C4.5** [545]. The basic tree learning algorithm has the following general steps:

1. At start, all the training examples are at the root of the tree.
2. Select an attribute on the basis of a heuristic and pick the attribute with the maximum heuristic value. The two most popular heuristics are as follows:

   - **Information gain**: This heuristic is used by both the ID3 and the C4.5 tree-generation algorithms. Information gain $G(A)$ is based on the concept of entropy from information theory and measures the difference in entropy H from before to after the dataset D is split on an attribute A.

$$G(A) = H(D) - H_A(D) \tag{2.7}$$

   where $H(D)$ is the entropy of D ($H(D) = -\sum_i^m p_i \log_2(p_i)$); p_i is the probability that an arbitrary sample in D belongs to class i; m is the total number of classes; $H_A(D)$ is the information needed (after using attribute A to split D into v partitions) to classify D and is calculated as $H_A(D) = -\sum_j^v(|D_j|/|D|)H(D_j)$ with $|x|$ being the size of x.

- **Gain ratio**: The C4.5 algorithm uses the gain ratio heuristic to reduce the bias of information gain towards attributes with a large number of values. The gain ratio normalizes information gain by taking into account the number and size of branches when choosing an attribute. The information gain ratio is the ratio between the information gain and the intrinsic value IV_A of attribute A:

$$GR(A) = G(A)/IV_A(D) \tag{2.8}$$

where

$$IV_A(D) = -\sum_{j}^{v} \frac{|D_j|}{|D|} \log_2\left(\frac{|D_j|}{|D|}\right) \tag{2.9}$$

3. Based on the selected attribute from step 2, construct a new node of the tree and split the dataset into subsets according to the possible values of the selected attribute. The possible values of the attribute become the branches of the node.
4. Repeat steps 2 and 3 until one of the following occurs:
    - All samples for a given node belong to the same class.
    - There are no remaining attributes for further partitioning.
    - There are no data samples left.

2.5.3.1 Decision Trees for Ms Pac-Man

As with ANNs and SVMs, decision tree learning requires data to be trained on. Presuming that data from expert Ms Pac-Man players would be of good quality and quantity, decision trees can be constructed to predict the strategy of Ms Pac-Man based on a number of ad-hoc designed attributes of the game state. Figure 2.16 illustrates a simplified hypothetical decision tree for controlling Ms Pac-Man. According to that example if a ghost is nearby then Ms Pac-Man checks if power pills are available in a close distance and aims for those; otherwise it takes actions so that it evades the ghost. If alternatively, ghosts are not visible Ms Pac-Man checks for pellets. If those are nearby or in a fair distance then it aims for them; otherwise it aims for the fruit, if that is available on the level. It is important to note that the leaves of the tree in our example represent control strategies (macro-actions) rather than actual actions (up, down, left, right) for Ms Pac-Man.

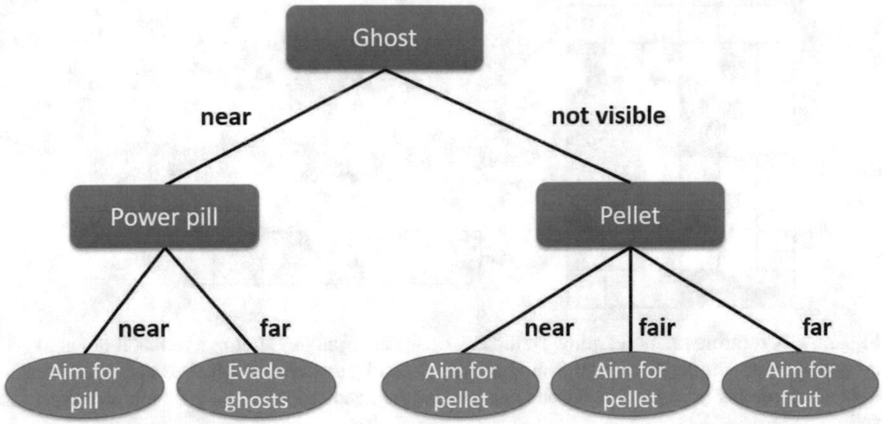

Fig. 2.16 A decision tree example for controlling Ms Pac-Man. The tree is trained on data from expert Ms Pac-Man players. Given the distance from the nearest *ghost*, *power pill* and *pellet* (data attributes) the tree predicts the *strategy* Ms Pac-Man needs to follow.

2.5.4 Further Reading

The core supervised learning algorithms are covered in detail in the Russell and Norvig classic AI textbook [582] including decision tree learning (Chapter 18) and artificial neural networks (Chapter 19). Detailed descriptions of artificial neural networks and backpropagation can also be found in the book of Haykin [253]. Deep architectures of ANNs are covered in great detail in the deep learning book by Goodfellow et al. [231]. Finally, support vector machines are covered in the tutorial paper of Burges [86].

The preference learning version of backpropagation in shallow and deep architectures can be found in [430, 436] whereas RankSVM is covered in the original paper of Joachims [303].

2.6 Reinforcement Learning

Reinforcement Learning (RL) [672] is a machine learning approach inspired by behaviorist psychology and, in particular, the way humans and animals learn to take decisions via (positive or negative) rewards received by their environment. In reinforcement learning, samples of good behavior are usually not available (as in supervised learning); instead, similarly to evolutionary (reinforcement) learning, the training signal of the algorithm is provided by the environment based on how an agent is interacting with it. At a particular point in time t, the agent is on a particular **state** s and decides to take an **action** a from all the available actions in its current state. As a response the environment delivers an immediate **reward**, r. Through

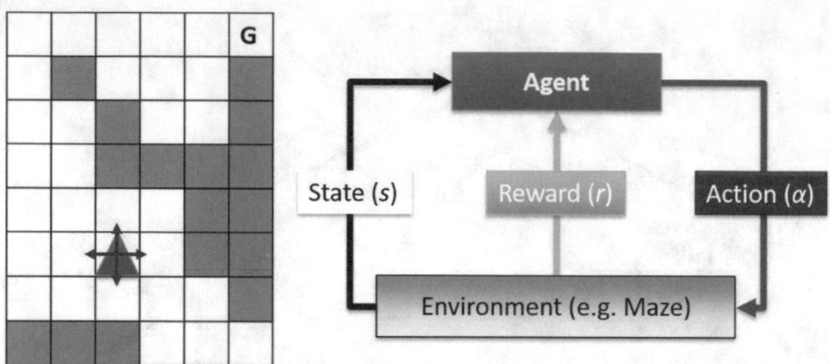

Fig. 2.17 A reinforcement learning example. The agent (triangle) attempts to reach the goal (G) by taking an action (a) among all available actions in its current state (s). The agent receives an immediate reward (r) and the environment notifies the agent about its new state after taking the action.

the continuous interaction between the agent and its environment, the agent gradually learns to select actions that maximize its sum of rewards. RL has been studied from a variety of disciplinary perspectives including operations research, game theory, information theory, and genetic algorithms and has been successfully applied in problems which involve a balance between long-term and short-term rewards such as robot control and games [464, 629]. An example of the reinforcement problem is illustrated through a maze navigation task in Fig. 2.17.

More formally, the aim of the agent is to discover a **policy** (π) for selecting actions that maximize a measure of a long-term reward such as the expected cumulative reward. A policy is a strategy that the agent follows in selecting actions, given the state it is in. If the function that characterizes the value of each action either exists or is learned, the optimal policy (π^*) can be derived by selecting the action with the highest value. The interactions with the environment occur in discrete time steps ($t = \{0, 1, 2, \ldots\}$) and are modeled as a **Markov decision process** (MDP). The MDP is defined by

- S: A set of states $\{s_1, ..., s_n\} \in S$. The environment states are a function of the agent's information about the environment (i.e., the agent's inputs).
- A: A set of actions $\{a_1, ..., a_m\} \in A$ possible in each state s. The actions represent the different ways the agent can act in the environment.
- $P(s, s', a)$: The probability of transition from s to s' given a. P gives the probability of ending in state s' after picking action a in state s and it follows the **Markov property** implying that future states of the process depend only upon the present state, not on the sequence of events that preceded it. As a result, the Markov property of P makes predictions of 1-step dynamics possible.
- $R(s, s', a)$: The reward function on transition from s to s' given a. When the agent in state s picks an action a and moves to state s', it receives an immediate reward r from the environment.

P and R define the **world model** and represent, respectively, the environment's dynamics (P) and the long-term reward (R) for each policy. If the world model is *known* there is no need to learn to estimate the transition probability and reward function and we thus directly calculate the optimal strategy (policy) using **model-based** approaches such as dynamic programming [44]. If, instead, the world model is *unknown* we approximate the transition and the reward functions by learning estimates of future rewards given by picking action a in state s. We then calculate our policy based on these estimates. Learning occurs via **model-free** methods such as Monte Carlo search and **temporal difference learning** [672]. In this section we put an emphasis on the latter set of algorithms and in particular, we focus on the most popular algorithm of TD learning: Q-learning. Before delving into the details of the Q-learning algorithm, we first discuss a few core RL concepts and provide a high-level taxonomy of RL algorithms according to RL problems and tools used for tackling them. We will use this taxonomy to place Q-learning with respect to RL as a whole.

2.6.1 Core Concepts and a High-Level Taxonomy

A central question in RL problems is the right balance between the **exploitation** of current learned knowledge versus the **exploration** of new unseen territories in the search space. Both randomly selecting actions (no exploitation) and always greedily selecting the best action according to a measure of performance or reward (no exploration) are strategies that generally yield poor results in stochastic environments. While several approaches have been proposed in the literature to address the exploration-exploitation balance issue, a popular and rather efficient mechanism for RL action selection is called ε-greedy, determined by the $\varepsilon \in [0,1]$ parameter. According to ε-greedy the RL agent chooses the action it believes will return the highest future reward with probability $1-\varepsilon$; otherwise, it chooses an action uniformly at random.

RL problems can be classified into **episodic** versus **incremental**. In the former class, algorithm training occurs offline and within a finite horizon of multiple training instances. The finite sequence of states, actions and reward signals received within that horizon is called an **episode**. Monte Carlo methods that rely on repeated random sampling, for instance, are a typical example of episodic RL. In the latter class of algorithms, instead, learning occurs online and it is not bounded by an horizon. We meet TD learning under incremental RL algorithms.

Another distinction is between **off-policy** and **on-policy** RL algorithms. An off-policy learner approximates the optimal policy independently of the agent's actions. As we will see below, Q-learning is an off-policy learner since it estimates the return for state-action pairs assuming that a greedy policy is followed. An on-policy RL algorithm instead approximates the policy as a process being tied to the agent's actions including the exploration steps.

Bootstrapping is a central notion within RL that classifies algorithms based on the way they optimize state values. Bootstrapping estimates how good a state is based on how good we think the next state is. In other words, with bootstrapping we update an estimate based on another estimate. Both TD learning and dynamic programming use bootstrapping to learn from the experience of visiting states and updating their values. Monte Carlo search methods instead do not use bootstrapping and thus learn each state value separately.

Finally, the notion of **backup** is central in RL and acts as a distinctive feature among RL algorithms. With backup we go backwards from a state in the future, s_{t+h}, to the (current) state we want to evaluate, s_t, and consider the in-between state values in our estimates. The backup operation has two main properties: its **depth**—which varies from one step backwards to a full backup—and its **breadth**—which varies from a (randomly) selected number of sample states within each time step to a full-breadth backup.

Based on the above criteria we can identify three major RL algorithm types:

1. **Dynamic programming**. In dynamic programming knowledge of the world model (P and R) is required and the optimal policy is calculated via bootstrapping.
2. **Monte Carlo methods**. Knowledge of the world model is not required for Monte Carlo methods. Algorithms of this class (e.g., MCTS) are ideal for off-line (episodic) training and they learn via sample-breadth and full-depth backup. Monte Carlo methods do not use bootstrapping, however.
3. **TD learning**. As with Monte Carlo methods knowledge of the world model is not required and it is thus estimated. Algorithms of this type (e.g., Q-learning) learn from experience via bootstrapping and variants of backup.

In the following section we cover the most popular TD learning algorithm in the RL literature with the widest use in game AI research.

2.6.2 Q-Learning

Q-learning [748] is a model-free, off-policy, TD learning algorithm that relies on a tabular representation of $Q(s,a)$ values (hence its name). Informally, $Q(s,a)$ represents how good it is to pick action a in state s. Formally, $Q(s,a)$ is the expected discounted reinforcement of taking action a in state s. The Q-learning agent learns from experience by picking actions and receiving rewards via bootstrapping.

The goal of the Q-learning agent is to maximize its expected reward by picking the right action at each state. The reward, in particular, is a weighted sum of the expected values of the discounted future rewards. The Q-learning algorithm is a simple update on the Q values in an iterative fashion. Initially, the Q table has arbitrary values as set by the designer. Then each time the agent selects an action

a from state s, it visits state s', it receives an immediate reward r, and updates its $Q(s,a)$ value as follows:

$$Q(s,a) \leftarrow Q(s,a) + \alpha\{r + \gamma \max_{a'} Q(s',a') - Q(s,a)\} \tag{2.10}$$

where $\alpha \in [0,1]$ is the **learning rate** and $\gamma \in [0,1]$ is the **discount factor**. The learning rate determines the extent to which the new estimate for Q will override the old estimate. The discount factor weights the importance of earlier versus later rewards; the closer γ is to 1, the greater the weight is given to future reinforcements. As seen from equation (2.10), the algorithm uses bootstrapping since it maintains estimates of how good a state-action pair is (i.e., $Q(s,a)$) based on how good it thinks the next state is (i.e., $Q(s',a')$). It also uses a one-step-depth, full-breadth backup to estimate Q by taking into consideration all Q values of all possible actions a' of the newly visited state s'. It is proven that by using the learning rule of equation (2.10) the $Q(s,a)$ values converge to the expected future discounted reward [748]. The optimal policy can then be calculated based on the Q-values; the agent in state s selects the action a with the highest $Q(s,a)$ value. In summary, the basic steps of the algorithm are as follows:

Given an immediate reward function r and a table of $Q(s,a)$ values for all possible actions in each state:

1. Initialize the table with arbitrary Q values; e.g., $Q(s,a) = 0$.
2. $s \leftarrow$ Start state.
3. While not finished* do:

   (a) Choose an action a based on policy derived from Q (e.g., ε-greedy).
   (b) Apply the action, transit to state s', and receive an immediate reward r.
   (c) Update the value of $Q(s,a)$ as per (2.10).
   (d) $s \leftarrow s'$.

*The most commonly used termination conditions are the algorithm's *speed*—i.e., stop within a number of iterations—or the *quality* of convergence—i.e., stop if you are satisfied with the obtained policy.

2.6.2.1 Limitations of Q-Learning

Q-learning has a number of limitations associated primarily with its tabular representation. First of all, depending on the chosen state-action representation the size of the state-action space might be computationally very expensive to handle. As the Q table size grows our computational needs for memory allocation and information retrieval increase. Further, we may experience very long convergence since learning time is exponential to the size of the state-action space. To overcome these obstacles and get decent performance from RL learners we need to devise a way of reducing the state-action space. Section 2.8 outlines the approach of using artificial neural networks as Q-value function approximators, directly bypassing the Q-table limitation and yielding compressed representations for our RL learner.

2.6.2.2 Q-Learning for Ms Pac-Man

Q-learning is applicable for controlling Ms Pac-Man as long as we define a suitable state-action space and we design an appropriate reward function. A state in Ms Pac-Man could be represented directly as the current snapsnot of the game—i.e., where Ms Pac-Man and ghosts are and which pellets and power pills are still available. That representation, however, yields a prohibitive number of game states for a Q-table to be constructed and processed. Instead, it might be preferred to choose a more indirect representation such as whether ghosts and pellets are nearby or not. Possible actions for Ms Pac-Man could be that it either keeps its current direction, it turns backward, it turns left, or it turns right. Finally, the reward function can be designed to reward Ms Pac-Man positively when it eats a pellet, a ghost or a power pill, whereas it could penalize Ms Pac-Man when it dies.

It is important to note that both Pac-Man and Ms Pac-Man follow the Markov property in the sense that any future game states may depend only upon the present game state. There is one core difference however: while the transition probability in Pac-Man is known given its deterministic nature, it is largely unknown in Ms Pac-Man given the stochastic behavior of the ghosts in that game. Thereby, Pac-Man can theoretically be solved via model-based approaches (e.g., dynamic programming) whereas the world model of Ms Pac-Man can only be approximated via model-free methods such as temporal difference learning.

2.6.3 Further Reading

The RL book of Sutton and Barto [672] is highly recommended for a thorough presentation of RL including Q-learning (Chapter 6). The book is freely available online.[6] A draft version of the latest (2017) version of the book is also available.[7] The survey paper of Kaelbling et al. [316] is another recommended reading of the approaches covered. Finally, for an in-depth analysis of model-based RL approaches you are referred to the dynamic programming book of Bertsekas [44].

[6] http://incompleteideas.net/sutton/book/ebook/the-book.html

[7] http://incompleteideas.net/sutton/book/the-book-2nd.html

2.7 Unsupervised Learning

As stated earlier, the utility type (or training signal) determines the class of the AI algorithm. In supervised learning the training signal is provided as data labels (target outputs) and in reinforcement learning it is derived as a reward from the environment. Unsupervised learning instead attempts to discover associations of the input by searching for patterns among all input data attributes and without having access to a target output—a machine learning process that is usually inspired by Hebbian learning [256] and the principles of self-organization [20]. With unsupervised learning we focus on the intrinsic structure of and associations in the data instead of attempting to imitate or predict target values. We cover two unsupervised learning tasks with corresponding algorithms: **clustering** and **frequent pattern mining**.

2.7.1 Clustering

Clustering is the unsupervised learning task of finding unknown groups of a number of data points so that data within a group (or else, **cluster**) is similar to each other and dissimilar to data from other clusters. Clustering has found applications in detecting groups of data across multiple attributes and in data reduction tasks such as data compression, noise smoothing, outlier detection and dataset partition. Clustering is of key importance for games with applications in player modeling, game playing and content generation.

As with classification, clustering places data into classes; the labels of the classes, however, are unknown a priori and clustering algorithms aim to discover them by assessing their quality iteratively. Since the correct clusters are unknown, similarity (and dissimilarity) depends only on the data attributes used. Good clusters are characterized by two core properties: 1) high *intra*-cluster similarity, or else, high compactness and 2) low *inter*-cluster similarity, or else, good separation. A popular measure of compactness is the average distance between every sample in the cluster and the closest representative point—e.g., centroid—as used in the k-means algorithm. Examples of separation measures include the **single link** and the **complete link**: the former is the *smallest* distance between any sample in one cluster and any sample in the other cluster; the latter is the *largest* distance between any sample in one cluster and any sample in the other cluster. While compactness and separation are objective measures of cluster validity, it is important to note that they are not indicators of cluster meaningfulness.

Beyond the validity metrics described above, clustering algorithms are defined by a **membership function** and a **search procedure**. The membership function defines the structure of the clusters in relation to the data samples. The search procedure is a strategy we follow to cluster our data given a membership function and a validity metric. Examples of such strategies include splitting all data points into clusters at once (as in k-means), or recursively merging (or splitting) clusters (as in hierarchical clustering).

Clustering can be realized via a plethora of algorithms including hierarchical clustering, k-means [411], k-medoids [329], DBSCAN [196] and self-organizing maps [347]. The algorithms are dissimilar in the way they define what a cluster is and how they form it. Selecting an appropriate clustering algorithm and its corresponding parameters, such as which distance function to use or the number of clusters to expect, depends on the aims of the study and the data available. In the remainder of the section we outline the clustering algorithms we find to be the most useful for the study of AI in games.

2.7.1.1 K-Means Clustering

K-means [411] is a vector quantization method that is considered the most popular clustering algorithm as it offers a good balance between simplicity and effectiveness. It follows a simple data partitioning approach according to which it partitions a database of objects into a set of k clusters, such that the sum of squared Euclidean distances between data points and their corresponding cluster center (centroid) is minimized—this distance is also known as the **quantization error**.

In k-means each cluster is defined by one point, that is the centroid of the cluster, and each data sample is assigned to the closest centroid. The centroid is the mean of the data samples in the cluster. The intra-cluster validity metric used by k-means is the average distance to the centroid. Initially, the data samples are randomly assigned to a cluster and then the algorithm proceeds by alternating between the re-assignment of data into clusters and the update of the resulting centroids. The basic steps of the algorithm are as follows:

Given k

1. Randomly partition the data points into k nonempty clusters.
2. Compute the position of the centroids of the clusters of the current partitioning. Centroids are the centers (mean points) of the clusters.
3. Assign each data point to the cluster with the nearest centroid.
4. Stop when the assignment does not change; otherwise go to step 2.

While k-means is very popular due to its simplicity it has a number of considerable weaknesses. First, it is applicable only to data objects in a continuous space. Second, one needs to specify the number of clusters, k, in advance. Third, it is not suitable to discover clusters with non-convex shapes as it can only find hyper-spherical clusters. Finally, k-means is sensitive to outliers as data points with extremely large (or small) values may substantially distort the distribution of the data and affect the performance of the algorithm. As we will see below, hierarchical clustering manages to overcome some of the above drawbacks, suggesting a useful alternative approach to data clustering.

2.7.1.2 Hierarchical Clustering

Clustering methods that attempt to build a hierarchy of clusters fall under the **hierarchical clustering** approach. Generally speaking there are two main strategies available: the *agglomerative* and the *divisive*. The former constructs hierarchies in a bottom-up fashion by gradually merging data points together, whereas the latter constructs hierarchies of clusters by gradually splitting the dataset in a top-down fashion. Both clustering strategies are greedy. Hierarchical clustering uses a distance matrix as the clustering strategy (whether agglomerative or divisive). This method does not require the number of clusters k as an input, but needs a termination condition.

Indicatively, we present the basic steps of the agglomerative clustering algorithm which are as follows:

Given k

1. Create one cluster per data sample.
2. Find the two closest data samples—i.e., find the shortest Euclidean distance between two points (single link)—which are not in the same cluster.
3. Merge the clusters containing these two samples.
4. Stop if there are k clusters; otherwise go to step 2.

In divisive hierarchical clustering instead, all data are initially in the same cluster which is split until every data point is on its own cluster following a split strategy—e.g., DIvisive ANAlysis Clustering (DIANA) [330]—or employing another clustering algorithm to split the data in two clusters—e.g., 2-means.

Once clusters of data are iteratively merged (or split), one can visualize the clusters by decomposing the data into several levels of nested partitioning. In other words, one can observe a tree representation of clusters which is also known as a **dendrogram**. The clustering of data is obtained by cutting the dendrogram at the desired level of squared Euclidean distance. For the interested reader, a dendrogram example is illustrated in Chapter 5.

Hierarchical clustering represents clusters as the set of data samples contained in them and, as a result, a data sample belongs to the same cluster as its closest sample. In k-means instead, each cluster is represented by a centroid and thus a data sample belongs to the cluster represented by the closest centroid. Further, when it comes to cluster validity metrics, agglomerative clustering uses the shortest distance between any sample in one cluster and a sample in another whereas k-means uses the average distance to the centroid. Due to these different algorithmic properties hierarchical clustering has the capacity to cluster data that come in any form of a connected shape; k-means, on the other hand, is only limited to hyper-spherical clusters.

2.7.1.3 Clustering for Ms Pac-Man

One potential application of clustering for controlling Ms Pac-Man would be to model ghost behaviors and use that information as an input to the controller of Ms Pac-Man. Whether it is k-means or hierarchical clustering, the algorithm would consider different attributes of ghost behavior—such as level exploration, behavior divergence, distance between ghosts, etc.—and cluster the ghosts into behavioral patterns or profiles. The controller of Ms Pac-Man would then consider the ghost profile met in a particular level as an additional input for guiding the agent better.

Arguably, beyond agent control, we can think of better uses of clustering for this game such as profiling Ms Pac-Man players and generating appropriate levels or challenges for them so that the game is balanced. As mentioned earlier, however, the focus of the Ms Pac-Man examples is on the control of the playing agent for the purpose of maintaining a consistent paradigm throughout this chapter.

2.7.2 Frequent Pattern Mining

Frequent pattern mining is a set of techniques that attempt to derive frequent patterns and structures in data. Patterns include sequences and itemsets. Frequent pattern mining was first proposed for mining association rules [6], which aims to identify a number of data attributes that frequently associate to each other, thereby forming conditional rules among them. There are two types of frequent pattern mining that are of particular interest for game AI: **frequent itemset mining** and **frequent sequence mining**. The former aims to find structure among data attributes that have no particular internal order whereas the latter aims to find structure among data attributes based on an inherent temporal order. While associated with the unsupervised learning paradigm, frequent pattern mining is dissimilar in both the aims and the algorithmic procedures it follows.

Popular and scalable frequent pattern mining methods include the **Apriori** algorithm [6] for itemset mining, and SPADE [793] and **GSP** [652, 434, 621] for sequence mining. In the remainder of this section we outline Apriori and GSP as representative algorithms for frequent itemset and frequent sequence mining, respectively.

2.7.2.1 Apriori

Apriori [7] is an algorithm for frequent itemset mining. The algorithm is appropriate for mining datasets that contain sets of instances (also named transactions) that each feature a set of items, or an **itemset**. Examples of transactions include books bought by an Amazon customer or apps bought by a smartphone user. The algorithm is very simple and can be described as follows: given a predetermined threshold named **support** (T), Apriori detects the itemsets which are subsets of at least T transactions in the database. In other words, Apriori will attempt to identify all itemsets that have at least a minimum support which is the minimum number of times an itemset exists in the dataset.

To demonstrate Apriori in a game example, below we indicatively list events from four players of an online role playing game:

- <Completed more than 10 levels; Most achievements unlocked; Bought the shield of the magi>
- <Completed more than 10 levels; Bought the shield of the magi>
- <Most achievements unlocked; Bought the shield of the magi; Found the Wizard's purple hat>
- <Most achievements unlocked; Found the Wizard's purple hat; Completed more than 10 levels; Bought the shield of the magi>

If in the example dataset above we assume that the support is 3, the following 1-itemsets (sets of only one item) can be found: <Completed more than 10 levels>, <Most achievements unlocked> and <Bought the shield of the magi>. If instead, we seek 2-itemsets with a support threshold of 3 we can find <Completed more than 10 levels, Bought the shield of the magi>, as three of the transactions above contain both of these items. Longer itemsets are not available (not frequent) for support count 3. The process can be repeated for any support threshold we wish to detect frequent itemsets for.

2.7.2.2 Generalized Sequential Patterns

Frequent itemset mining algorithms are not adequate if the **sequence** of events is the critical information we wish to mine from a dataset. The dataset may contain events in an ordered set of sequences such as temporal sequence data or time series. Instead, we need to opt for a frequent sequence mining approach. The sequence mining problem can be simply described as the process of finding frequently occurring subsequences given a sequence or a set of sequences.

More formally, given a dataset in which each sample is a sequence of events, namely a **data sequence**, a sequential pattern defined as a subsequence of events is a **frequent sequence** if it occurs in the samples of the dataset regularly. A frequent sequence can be defined as a sequential pattern that is supported by, at least, a minimum amount of data-sequences. This amount is determined by a threshold named minimum support value. A data sequence supports a sequential pattern if and only if it contains all the events present in the pattern in the same order. For example, the data-sequence $< x_0, x_1, x_2, x_3, x_4, x_5 >$ supports the pattern $< x_0, x_5 >$. As with frequent itemset mining, the amount of data sequences that support a sequential pattern is referred as the **support count**.

The Generalized Sequential Patterns (GSP) algorithm [652] is a popular method for mining frequent sequences in data. GSP starts by extracting the frequent sequences with a single event, namely 1-sequences. That set of sequences is self-joined to generate all 2-sequence candidates for which we calculate their support count. Those sequences that are frequent (i.e., their support count is greater than a threshold value) are then self-joined to generate the set of 3-sequence candidates. The algorithm is gradually increasing the length of the sequences in each algorithmic step until the next set of candidates is empty. The basic principle of the algorithm is that if a sequential pattern is frequent, then its *contiguous* subsequences are also frequent.

2.7.2.3 Frequent Pattern Mining for Ms Pac-Man

Patterns of events of sequences can be extracted to assist the control of Ms Pac-Man. Itemsets may be identified across successful events of expert Ms Pac-Man players given a particular support count. For instance, an Apriori algorithm running on events across several different expert players might reveal that a frequent 2-itemset is the following: <player went for the upper left corner first, player ate the bottom right power pill first>. Such information can be useful explicitly for designing rules for controlling Ms Pac-Man.

Beyond itemsets, frequencies of ghost events can be considered for playing Ms Pac-Man. For example, by running GSP on extracted attributes of ghosts it might turn out that when Ms Pac-Man eats a power pill it is very likely that the Blinky ghost moves left (<power pill, Blinky left>). Such frequent sequences can form additional inputs of any Ms Pac-Man controller—e.g., an ANN. Chapter 5 details an example on this frequent sequence mining approach in a 3D prey-predator game.

2.7.3 Further Reading

A general introduction to frequent pattern mining is offered in [6]. The Apriori algorithm is detailed in the original article of Agrawal and Srikant [7] whereas GSP is covered throughly in [652].

2.8 Notable Hybrid Algorithms

AI methods can be interwoven in numerous ways to yield new sophisticated algorithms that aggregate the strengths of their combined parts, often with an occurring *gestalt* effect. You can, for instance, let GAs evolve your behavior trees or FSMs; you can instead empower MCTS with ANN estimators for tree pruning; or you can add a component of local search in every search algorithm covered earlier. We name the resulting combinations of AI methods as **hybrid** algorithms and in this section we cover the two most influential, in our opinion, hybrid game AI algorithms: neuroevolution and temporal difference learning with ANN function approximators.

2.8.1 Neuroevolution

The evolution of artificial neural networks, or else **neuroevolution**, refers to the design of artificial neural networks—their connection weights, their topology, or both—using evolutionary algorithms [786]. Neuroevolution has been successfully applied in the domains of artificial life, robot control, generative systems and computer games. The algorithm's wide applicability is primarily due to two main reasons. First, many AI problems can be viewed as function optimization problems whose underlying general function can be approximated via an ANN. Second, neuroevolution is a method grounded in biological metaphors and evolutionary theory and inspired by the way brains evolve [567].

This evolutionary (reinforcement) learning approach is applicable either when the error function available is not differentiable or when target outputs are not available. The former may occur, for instance, when the activation functions employed in the ANN are not continuous and, thus, not differentiable. (This is a prominent phenomenon, for instance, in the compositional pattern producing networks [653].) The latter may occur in a domain for which we have no samples of good (or bad) behavior or it is impossible to define objectively what a good behavior might be. Instead of backpropagating the error and adjusting the ANN based on gradient search, neuroevolution designs ANNs via metaheuristic (evolutionary) search. In contrast to supervised learning, neuroevolution does not require a dataset of input-output pairs to train ANNs. Rather, it requires only a measure of a ANN's performance on the problem under investigation, for instance, the score of a game playing agent that is controlled by an ANN.

The core algorithmic steps of neuroevolution are as follows:

1. A population of chromosomes that represent ANNs is evolved to optimize a fitness function that characterizes the utility (quality) of the ANN representation. The population of chromosomes (ANNs) is typically initialized randomly.
2. Each chromosome is encoded into an ANN which is, in turn, tested on the task under optimization.
3. The testing procedure assigns a fitness value for each ANN of the population. The fitness of an ANN defines its measure of performance on the task.
4. Once the fitness values for all genotypes in the current population are determined, a selection strategy (e.g., roulette-wheel, tournament) is applied to pick the parents for the next generation.
5. A new population of offspring is generated by applying genetic operators on the selected ANN-encoded chromosomes. Mutation and/or crossover are applied on the chromosomes in the same way as in any evolutionary algorithm.
6. A replacement strategy (e.g., steady-state, elitism, generational) is applied to determine the final members of the new population.

7. Similarly to a typical evolutionary algorithm, the generational loop (steps 2 to 6) is repeated until we exhaust our computational budget or we are happy with the obtained fitness of the current population.

Typically there are two types of neuroevolution approaches: those that consider the evolution of a network's connection weights only and those that evolve both the connection weights and the topology of the network (including connection types and activation functions). In the former type of neuroevolution, the weight vector is encoded and represented genetically as a chromosome; in the latter type, the genetic representation includes an encoding of the ANN topology. Beyond simple MLPs, the ANN types that have been considered for evolution include the NeuroEvolution of Augmenting Topologies (NEAT) [655] and the compositional pattern producing networks [653].

Neuroevolution has found extensive use in the games domain in roles such as those of evaluating the state-action space of a game, selecting an appropriate action, selecting among possible strategies, modeling opponent strategies, generating content, and modeling player experience [567]. The algorithm's efficiency, scalability, broad applicability, and open-ended learning are a few of the reasons that make neuroevolution a good general method for many game AI tasks [567].

2.8.1.1 Neuroevolution for Ms Pac-Man

One simple way to implement neuroevolution in Ms Pac-Man is to first design an ANN that considers the game state as input and output actions for Ms Pac-Man. The weights of the ANN can be evolved using a typical evolutionary algorithm and following the steps of neuroevolution as described above. The fitness of each ANN in the population is obtained by equipping Ms Pac-Man with each ANN in the population and letting her play the game for a while. The performance of the agent within that simulation time (e.g., the score) can determine the fitness value of the ANN. Figure 2.18 illustrates the steps of ANN encoding and fitness assignment in this hypothetical implementation of neuroevolution in Ms Pac-Man.

2.8.2 TD Learning with ANN Function Approximators

Reinforcement learning typically uses tabular representations to store knowledge. As mentioned earlier in the RL section, representing knowledge this way may drain our available computational resources since the size of the look-up table increases exponentially with respect to the action-state space. The most popular way of addressing this challenge is to use an ANN as a value (or Q value) approximator, thereby replacing the table. Doing so makes it possible to apply the algorithm to

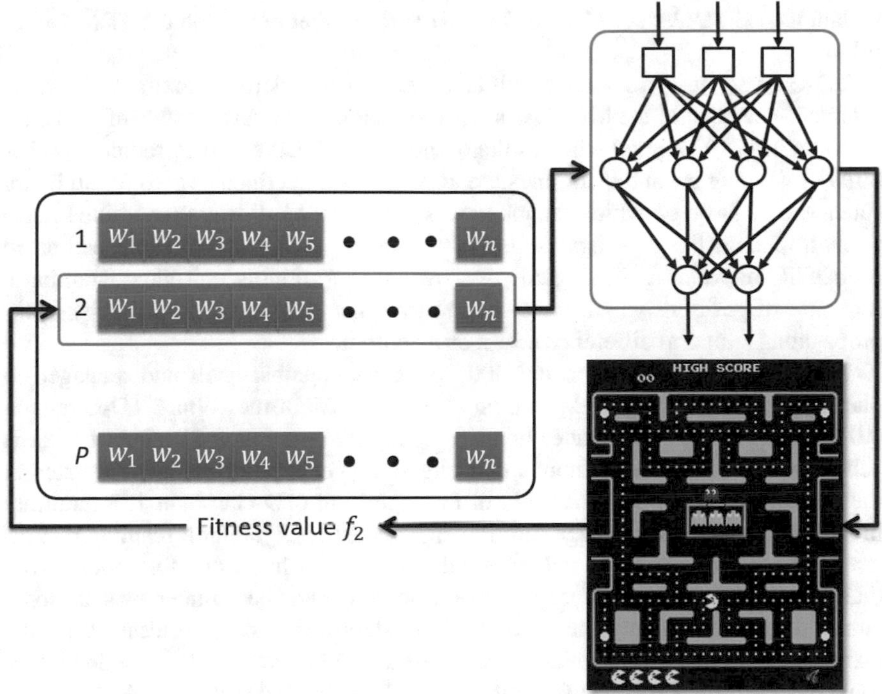

Fig. 2.18 Neuroevolution in Ms Pac-Man. The figure visualizes step 2 (ANN encoding) and step 3 (fitness assignment) of the algorithm for assigning a fitness value to chromosome 2 in the population (of size *P*). In this example, only the weights of the ANN are evolved. The *n* weights of the chromosome are first encoded in the ANN and then the ANN is tested in Ms Pac-Man for a number of simulation steps (or game levels). The result of the game simulation determines the fitness value (f_2) of the ANN.

larger spaces of action-state representations. Further, an ANN as a function approximator of Q, for instance, can handle problems with continuous state spaces which are infinitely large.

In this section, we outline two milestone examples of algorithms that utilize the ANN universal approximation capacity for temporal difference learning. The algorithms of TD-Gammon and deep Q network have been applied, respectively, to master the game of backgammon and play Atari 2600 arcade games at super-human level. Both algorithms are applicable to any RL task beyond these particular games, but the games that made them popular are used to describe the algorithms below.

2.8.2.1 TD-Gammon

Arguably one of the most popular success stories of AI in games is that of Tesauro's TD-Gammon software that plays backgammon on the grandmaster-level [689]. The learning algorithm was a hybrid combination of an MLP and a temporal difference variant named TD(λ); see Chapter 7 of [672] for further details on the TD(λ) algorithm.

TD-Gammon used a standard multilayer neural network to approximate the value function. The input of the MLP was a representation of the current state of the board (Tesauro used 192 inputs) whereas the output of the MLP was the predicted probability of winning given the current state. Rewards were defined as zero for all board states except those on which the game was won. The MLP was then trained iteratively by playing the game against itself and selecting actions based on the estimated probability of winning. Each game was treated as a training episode containing a sequence of positions which were used to train the weights of the MLP by backpropagating temporal difference errors of its output.

TD-Gammon 0.0 played about 300,000 games against itself and managed to play as well as the best backgammon computer of its time. While TD-Gammon 0.0 did not win the performance horse race, it gave us a first indication of what is achievable with RL even without any backgammon expert knowledge integrated in the AI algorithm. The next iteration of the algorithm (TD-Gammon 1.0) naturally incorporated expert knowledge through specialized backgammon features that altered the input of the MLP and achieved substantially higher performance. From that point onwards the number of hidden neurons and the number of self-payed games determined greatly the version of the algorithm and its resulting capacity. From TD-Gammon 2.0 (40 hidden neurons) to TD-Gammon 2.1 (80 hidden neurons) the performance of TD-Gammon gradually increased and, with TD Gammon 3.0 (160 hidden neurons), it reached the playing strength of the best human player in backgammon [689].

2.8.2.2 Deep Q Network

While the combination of RL and ANNs results in very powerful hybrid algorithms, the performance of the algorithm traditionally depended on the design of the input space for the ANN. As we saw earlier, even the most successful applications of RL such as the TD-Gammon agent managed to reach human-level playing performance by integrating game specific features in the input space, thereby adding expert knowledge about the game. It was up until very recently that the combination of RL and ANNs managed to reach human-level performance in a game without considering ad-hoc designed features but rather discovering them merely through learning. A team from Google's DeepMind [464] developed a reinforcement learning agent called *deep Q network* (DQN) that trains a deep convolutional ANN via Q-learning. DQN managed to reach or exceed human-level playing performance in 29 out of 46 arcade (Atari 2600) games of the Arcade Learning Environment [40] it was trained on [464].

DQN is inspired by and based upon TD-Gammon since it uses an ANN as the function approximator for TD learning via gradient descent. As in TD-Gammon, the gradient is calculated by backpropagating the temporal difference errors. However, instead of using TD(λ) as the underlying RL algorithm, DQN uses Q-learning. Further, the ANN is not a simple MLP but rather a deep convolutional neural network. DQN played each game of ALE for a large amount of frames (50 million frames). This amounts to about 38 days of playing time for each game [464].

The DQN analyses a sequence of four game screens simultaneously and approximates the future game score per each possible action given its current state. In particular, the DQN uses the pixels from the four most recent game screens as its inputs, resulting in ANN input size of 84×84 (screen size in pixels) $\times 4$. No other game-specific knowledge was given to the DQN beyond the screen pixel information. The architecture used for the convolutional ANN has three hidden layers that yield 32 20×20, 64 9×9 and 64 7×7 feature maps, respectively. The first (low-level) layers of the DQN process the pixels of the game screen and extract specialized visual features. The convolutional layers are followed by a fully connected hidden layer and an output layer. Each hidden layer is followed by a rectifier nolinearity. Given a game state represented by the network's input, the outputs of the DQN are the estimated optimal action values (optimal Q-values) of the corresponding state-action pairs. The DQN is trained to approximate the Q-values (the actual score of the game) by receiving immediate rewards from the game environment. In particular, the reward is $+1$ if the score increases in between two successive time steps (frames), it is -1 if the score decreases, and 0 otherwise. DQN uses an ε-greedy policy for its action-selection strategy. It is worth mentioning that, at the time of writing, there are newer and more efficient implementations of the deep reinforcement learning concept such as the Asynchronous Advantage Actor-Critic (A3C) algorithm [463].

2.8.2.3 TD Learning with ANN Function Appoximator for Ms Pac-Man

We can envisage a DQN approach for controlling Ms Pac-Man in a similar fashion to that with which ALE agents were trained [464]. A deep convolutional neural network scans the level image on a pixel-to-pixel basis (see Fig. 2.19). The image goes through a number of convolution and fully connected layers which eventually feed the input of an MLP that outputs the four possible actions for Ms Pac-Man (keep direction, move backwards, turn left, turn right). Once an action is applied, the score of the game is used as the immediate reward for updating the weights of the deep network (the convolutional ANN and the MLP). By playing for a sufficient time period the controller gathers experience (image snapshots, actions, and corresponding rewards) which trains the deep ANN to approximate a policy that maximizes the score for Ms Pac-Man.

2.8.3 *Further Reading*

For a recent thorough survey on the application of neuroevolution in games the reader may refer to [567]. For a complete review of neuroevolution please refer to Floreano et al. [205]. CPPNs and NEAT are covered in detail in [653] and [655]

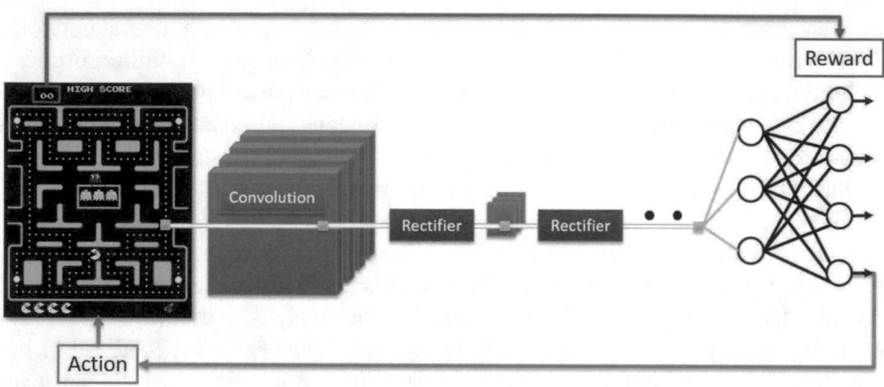

Fig. 2.19 A deep Q-learning approach for Ms Pac-Man. Following [464], the network's first part contains a set of convolution layers which are followed by rectifier nonlinearities. The final layers of the DQN we present in this example are fully connected employing ReLUs, as in [464].

respectively. TD-Gammon and DQN are covered in detail in [689] and [464], respectively. Both are also placed within the greater RL field in the upcoming second edition of [672]. Details about the A3C algorithm can be found in [463] and implementations of the algorithm can be found directly as part of Tensorflow.

2.9 Summary

This chapter covered the AI methods we feel the reader of this book needs to be familiar with. We expect, however, that our readers have a basic background in AI or have completed a course in fundamentals of AI prior to reading this book. Hence, the algorithms were not covered in detail since the emphasis of this book is on the application of AI within the domain of games and not on AI per se. On that basis, we used the game of Ms Pac-Man as the overarching application testbed of all algorithms throughout this chapter.

The families of algorithms we discussed include traditional ad-hoc behavior authoring methods (such as finite state machines and behavior trees), tree search (such as best-first, Minimax and Monte Carlo tree search), evolutionary computation (such as local search and evolutionary algorithms), supervised learning (e.g., neural networks, support vector machines and decision trees), reinforcement learning (e.g., Q-learning), unsupervised learning (such as clustering and frequent pattern mining), and hybrid algorithms such as evolving artificial neural networks and artificial neural networks as approximators of expected rewards.

With this chapter we reached the end of the first, introductory, part of the book. The next part begins with a chapter on the most traditional and widely explored task of AI in games: playing!

Part II
Ways of Using AI in Games

Chapter 3
Playing Games

When most people think of AI in games they think of an AI playing the game, or controlling the non-player characters you meet in the game. This might be because of the association between AI and the idea of autonomous action, or the association between game characters and robots. While playing games is far from the only interesting application for AI in games, it is a very important one and the one with the longest history. Many methods for content generation (Chapter 4) and player modeling (Chapter 5) are also dependent on methods for playing games, and therefore it makes sense to discuss playing games before content generation and player modeling.

This chapter is devoted to AI methods for playing games, including methods for creating interesting non-player characters in games. While winning a game, appearing human-like, and providing entertainment are very different objectives, they face many of the same challenges. In fact, there are many different reasons why one might want to use AI methods to play a game. We start the chapter with discussing these various motivations (Section 3.1). Regardless of why you want to use AI to play a game, which methods you can effectively use to play the game is determined by the various characteristics of the game that, in turn, affect the choice and design of the AI method. So the next section in this chapter (Section 3.2) is devoted to characterizing games and AI algorithms according to several criteria. Once you have understood your game sufficiently, you can make an informed choice of which algorithm to play it. The following section (Section 3.3) is devoted to discussing the various methods that can be used to play games, and how the right choice of method depends on the characteristics of the game. Most of the methods discussed here will have been briefly and somewhat abstractly discussed in Chapter 2, but this chapter will go into some depth about the application of these methods to playing games.

Next, a long section (Section 3.4) divides up the space of games by game genre, and discusses how AI methods can be applied in various types of games. This section will contain plenty of examples from the literature, and some from published games. This section also introduces several commonly used game-based frameworks and competitions for testing AI game-playing algorithms. Throughout the

G. N. Yannakakis and J. Togelius, *Artificial Intelligence and Games*, https://doi.org/10.1007/978-3-319-63519-4_3

chapter we will mostly discuss the use of AI methods to play to **win**, but also make numerous references to the **experience**-creation aspect of game-playing.

3.1 Why Use AI to Play Games?

The question of why you might want to deploy some kind or artificial intelligence to play a game can be reduced to two more specific questions:

Is the AI playing to win?

The question here is whether achieving as high a performance as possible in the game is the overarching goal of the AI method. High performance here means getting a high score, winning over the opponent, surviving for a long time or similar. It is not always possible to define what high performance and "playing to win" means—for example, *The Sims* (Electronic Arts, 2000) has no clear winning state and the winning condition in *Minecraft* (Mojang, 2011) is not strongly related to playing the game well—but in a very large number of games, from *Tetris* (Alexey Pajitnov and Vladimir Pokhilko, 1984) to Go to the *Halo* (Microsoft Studios, 2001–2015) series, it is straightforward to define what playing better means. However, not all players play to win, and few players play to win in every game all the time. Players play to pass time, relax, test new strategies, explore the game, role-play, keep their friends company and so on (see a more detailed discussion on this topic in Chapter 5). An AI algorithm might likewise be used in a number of roles beyond simply playing as well as possible. For example, the agent might play in a human-like manner, play in an entertaining manner, or behave predictably. It is important to note that optimizing an agent for playing a game to win might be at odds with some of the other ways of playing: many high-performing AI agents play in distinctly non-human, boring and/or unpredictable ways, as we will see in some case studies.

Is the AI taking the role of a human player?

Some games are single-player, and some games are multi-player where all players are human. This is particularly true for classic board games. But many, probably most, video games include various non-player characters. These are controlled by the computer software in some way—in fact, for many game developers "game AI" refers to the program code that controls the NPCs, regardless of how simple or sophisticated that code is. Obviously, the role of NPCs varies sharply between games, and within games. In the discussion of this chapter we refer to non-player roles as those that a human could not take, or would not want to take. Thus, all roles in an exclusively multi-player first-person shooter (FPS) such as *Counter-Strike* (Valve Corporation, 2000) are player roles, whereas a typical single-player role-playing

	Player	Non-Player
Win	**Motivation** Games as AI testbeds, AI that challenges players, Simulation-based testing **Examples** Board Game AI (TD-Gammon, Chinook, Deep Blue, AlphaGo, Libratus), Jeopardy! (Watson), StarCraft	**Motivation** Playing roles that humans would not (want to) play, Game balancing **Examples** Rubber banding
Experience	**Motivation** Simulation-based testing, Game demonstrations **Examples** Game Turing Tests (2kBot Prize/Mario),Persona Modelling	**Motivation** Believable and human-like agents **Examples** AI that: acts as an adversary, provides assistance, is emotively expressive, tells a story, ...

Fig. 3.1 Why use AI to play games? The two possible goals (win, experience) AI can aim for and the two roles (player, non-player) AI can take in a gameplaying setting. We provide a summary of motivations and some indicative examples for each of the four AI uses for gameplaying.

game (RPG) such *The Elder Scrolls V: Skyrim* (Bethesda Softworks, 2011) has only one player role, the rest are non-player characters. In general, non-player roles have more limited possibilities than player roles.

In summary, AI could be playing a game to **win** or for the **experience** of play either by taking the role of the **player** or the role of a **non-player** character. This yields four core uses of AI for playing games as illustrated in Fig. 3.1. With these distinctions in mind, we will now look at these four key motivations for building game-playing AI in further detail.

3.1.1 Playing to Win in the Player Role

Perhaps the most common use of AI together with games in academic settings is to play to win, while taking the role of a human player. This is especially common when using games as an AI testbed. Games have been used to test the capabilities and performance of AI algorithms for a very long time, as we discussed in Section 1.2. Many of the milestones in AI and games research have taken the form of some sort of AI program beating the best human player in the world at some games. See, for example, IBM's Deep Blue winning over Garry Kasparov in Chess, Google DeepMind's AlphaGo winning over Lee Sedol [629] and Ke Jie in Go, and IBM's

Watson winning Jeopardy! [201]. All of these were highly publicized events widely seen as confirmations of the increasing capabilities of AI methods. As discussed in Chapter 1, AI researchers are now increasingly turning to video games to find appropriate challenges for their algorithms. The number of active competitions associated with the IEEE CIG and AIIDE conferences is testament to this, as is DeepMind's and Facebook AI Research's choice of *StarCraft II* (Blizzard Entertainment, 2015) as a testbed for their research.

Games are excellent testbeds for artificial intelligence for a number of reasons, as elaborated on in Section 1.3. An important reason is that games are made to test human intelligence. Well-designed games exercise many of our cognitive abilities. Much of the fun we have in playing games comes from learning the games through playing them [351], meaning that well-designed games are also great teachers. This, in turn, means that they offer the kind of gradual skill progression that allows for testing of AI at different capability levels.

There are some reasons besides AI benchmarking for why you might want to use an AI in the place of a human to play games to win. For example, there are some games where you need strong AI to provide a challenge to players. This includes many strategic games of perfect information, such as classic board games, including Chess, Checkers and Go. However, for games with hidden information, it is often easier to provide challenge by simply "cheating", for example, by giving the AI player access to the hidden state of the game or even by modifying the hidden state so as to make it harder to play for the human. For example, in the epic strategy game *Civilization* (MicroProse, 1991), all civilizations can be played by human players. However, playing any Civilization game well under the same conditions as a human is very challenging, and there is, to our knowledge, no AI capable of playing these games as well as a good human player. Therefore, when playing against several computer-controlled civilizations, the game typically cheats by providing these with preferential conditions in various ways.

Another use case for AI that plays to win in a player role is to test games. When designing a new game, or a new game level, you can use a game-playing agent to test whether the game or level is playable, so called **simulation-based testing**. However, in many cases you want the agent to also play the game in a human-like manner to make the testing more relevant; see below on playing for experience.

Historically, the use of AI to play to win in a player role has been so dominant in academic work that some researchers have not even considered other roles for AI in playing games. In game development, on the other hand, this particular motivation for game-playing AI is much more rare; most game-playing AI in existing games is focused on non-player roles and/or playing for experience. This mismatch has historically contributed to the lack of understanding between academia and industry on game AI. In recent years however, there has been a growing understanding of the multitude of motivations for game-playing AI.

3.1.2 Playing to Win in a Non-player Role

Non-player characters are very often designed to *not* offer maximum challenge or otherwise be as effective as possible, but instead to be entertaining or human-like; see below for non-player characters playing for experience. However, there are instances when you want a non-player character to play as well as possible. As mentioned above, strategy games such as *Civilization* (MicroProse, 1991) have an (unanswered) need for high-performing non-cheating opponents, though here we are talking about playing roles that other human players could in principle have taken. Other strategy games, such as *XCOM: Enemy Unknown* (2K Games, 2012), have playing roles that humans would not play, creating a need for NPC AI playing to win.

Other times, creating an NPC playing to win is a necessary precursor to creating an NPC playing for experience. For example, in a racing game, you might want to implement "rubber band AI" where the NPC cars adapt their speed to the human player, so that they are never too far behind or ahead. Doing this is easy, but only if you already have an AI controller that can play the game well, either through actually playing the game well or through cheating in a way that cannot easily be detected. The performance of the controller can then be reduced when necessary so as to match the player's performance.

3.1.3 Playing for Experience in the Player Role

Why would you want an agent that takes the role of a human player, but that does not focus on winning? For example, when you want a **human-like agent**. Perhaps the most important reason for such agents is alluded to above: simulation-based testing. This is important both when designing games and game content manually, and when generating content procedurally; in the latter case, the quality of the game content is often evaluated automatically with the help of an agent playing the game, as discussed in Chapter 4. When trying to see how the game would be played by a human it is therefore important that the agent plays in a human-like manner, meaning that it has performance comparable to a human, has similar reaction speed, makes the same sort of mistakes that a human would do, is curious about and explores the same areas as a human would, etc. If the AI agent plays significantly differently from how a human would play, it might give the wrong information about e.g., whether a game is winnable (it might be winnable but only if you have superhuman reflexes) or whether a game mechanic is used (maybe a human would use it, but not an AI that tries to play optimally).

Another situation where human-like play is necessary is when you want to demonstrate how to play a level to a human player. A common feature of games is some kind of **demo mode**, which shows the game in action. Some games even have a demonstration feature built into the core gameplay mode. For example, *New Super Mario Bros* (Nintendo, 2006) for the Nintendo Wii will show you how to play a particular part of a level if you fail it repeatedly. The game simply takes over the controls and plays for you for about 10 to 20 seconds, and lets you continue afterwards. If all the level content is known beforehand, and there are no other players, such demonstrations can be hardcoded. If some parts of the game are user-designed, or procedurally generated, the game needs to generate these demonstrations itself.

Playing in a "human-like" fashion may seem a rather fuzzy and subjective aim, and it is. There are many ways in which a typical AI agent plays differently from a typical human player. How humans and AIs differ depends on the algorithm used to play the game, the nature of the game itself, and a multitude of other factors. To investigate these differences further, and spur the development of agents that can play in a human-like manner, two different Turing test-like competitions have been held. The 2K BotPrize was held from 2008 to 2013, and challenged competitors to develop agents that could play the FPS *Unreal Tournament 2004* (Epic Games, 2004) in such a way that human participants thought that the bots were human [263, 262, 647]. Similarly, the Turing test track of the Mario AI Competition let people submit playing agents of *Super Mario Bros* (Nintendo, 1985), who were judged by human onlookers as to whether they were human or not [619, 717]. While it takes us too far to go through all of the results of these competitions here, there are some very obvious signs of non-humanness that recur across games for many types of AI agents. These include having extremely fast reactions, switching between actions faster than a human could, not attempting actions which fail (because of having a too good model of the outcome of actions), not doing unnecessary actions (such as jumping when one could just be running) and not hesitating or stopping to think.

Of course, not all players of a game play in the same way. In fact, as discussed further in Chapter 5, if one analyzes a set of play traces of any game one can often find a number of player "archetypes" or "personas", clusters of players who play the game in markedly different ways in terms of e.g., aggression, speed, curiosity and skill. Within work on AI that plays games in human-like styles, there has been work both on learning and mirroring the playstyle of individual players [422, 423, 511, 328, 603] and on learning to play games in the style of one of several personas [267, 269].

3.1.4 Playing for Experience in a Non-player Role

Almost certainly the most common goal for game-playing AI in the game industry is to make non-player characters act, almost always in ways which are not primarily meant to beat the player or otherwise "win" the game (for many NPCs it may not even be defined what winning the game means). NPCs may exist in games for many, sometimes overlapping, purposes: to act as adversaries, to provide assistance and guidance, to form part of a puzzle, to tell a story, to provide a backdrop to the action of the game, to be emotively expressive and so on [724]. The sophistication and behavioral complexity of NPCs likewise vary widely, from the regular left-right movements of the aliens in *Space Invaders* (Midway, 1978) and Koopas in the *Super*

Mario Bros (Nintendo, 1985–2016) series to the nuanced and varied behavior of non-player characters in *Bioshock Infinite* (2K Games, 2013) and the alien in *Alien: Isolation* (Sega, 2014).

Depending on the role of the NPC, very different tasks can be asked of the AI algorithms that control it. (It can certainly be argued that many of the scripts that control NPCs cannot truthfully be described as *artificial intelligence* in any conventional way, but we will stick with the acronym here as it is commonly used for all code that controls non-player characters in the game industry.) In many cases what the game designers look for is **the illusion of intelligence**: for the player to believe that the NPC in some sense is intelligent even though the code controlling it is very simple. Human-likeness in the sense discussed in the previous section might or might not be the objective here, depending on what kind of NPC it is (a robot or a dragon should perhaps not behave in a too human-like manner).

In other cases, the most important feature of an NPC is its **predictability**. In a typical stealth game, a large part of the challenge is for the player to memorize and predict the regularities of guards and other characters that should be avoided. In such cases, it makes sense that the patrols are entirely regular, so that their schedule can be gleaned by the player. Similarly, the *boss monsters* in many games are designed to repeat certain movements in a sequence, and are only vulnerable to the player's attack when in certain phases of the animation cycle. In such cases, too "intelligent" and adaptive behavior would be incompatible with the game design.

It should be noted that even in cases where you would expect to need supple, complex behavior from NPCs, an agent that plays to win might be very problematic. Many high-performing strategies are seen as very **boring** by the player, and prime examples of "unsportsmanlike" behavior. For example, in an experiment with building high-performing AI for a turn-based strategy game, it was found that one of the solutions (based on neuroevolution) was extremely boring to play against, as it simply took a defensive position and attacked any incoming units with long-distance attacks [490]. Similarly, *camping* (staying stationary in a protected position and waiting for enemies to expose themselves to fire) is a behavior that is generally frowned on and often banned in FPS games, but it is often highly effective and easy to learn for an AI (incidentally, real-life military training often emphasizes camping-like tactics—what is effective is often not fun). Another interesting example is the work by Denzinger et al. [165] in which an evolutionary algorithm found that the best way to score a goal in *FIFA 99* (Electronic Arts, 1999) was by forcing a penalty kick. The evolutionary process found a local optimum in the fitness landscape corresponding to a *sweet spot* or *exploit* of the game's mechanics which yielded highly efficient, yet predictable and boring gameplay. Exploiting the game's bugs for winning is a *creative* strategy that is not only followed by AIs but also by human players [381].

3.1.5 Summary of AI Game-Playing Goals and Roles

We argued above that playing to win in the player role has been overemphasized, to the point of neglecting other perspectives, in much of academic research. In the same way, work on AI in the game industry has generally overemphasized playing for experience in a non-player role, to the point of neglecting other perspectives. This has led to an emphasis in the industry on behavior authoring methods such as finite state machines and behavior trees, as AI methods based on search, optimization and learning have been seen as not conducive to playing for experience; a common gripe has been a perceived lack of predictability and authorial control with such methods. However, given a better understanding of what roles AI can be used to play in games, this neglect of methods and perspectives is hopefully coming to an end in both academia and industry.

3.2 Game Design and AI Design Considerations

When choosing an AI method for playing a particular game (in any of the roles discussed in Section 3.1) it is crucial to know the characteristics of the game you are playing and the characteristics of the algorithms you are about to design. These collectively determine what type of algorithms can be effective. In this section we first discuss the challenges we face due to the characteristics of the game per se (Section 3.2.1) and then we discuss aspects of AI algorithmic design (Section 3.2.2) that need to be considered independently of the game we examine.

3.2.1 Characteristics of Games

In this section we discuss a number of characteristics of games and the impact they have on the potential use of AI methods. All characteristics covered are tied to the design of the game but a few (e.g., input representation and forward model) are also dependent on the technical implementation of the game and possibly amenable to change. Much of our discussion is inspired by the book *Characteristics of Games* by Elias et al. [192], which discusses many of these factors from a game design perspective. For illustrative purposes, Fig. 3.2 places a number of core game examples onto the three-dimensional space of **observability**, **stochasticity** and **time granularity**.

3.2.1.1 Number of Players

A good place to start is the number of players a game has. Elias et al. [192] distinguish between:

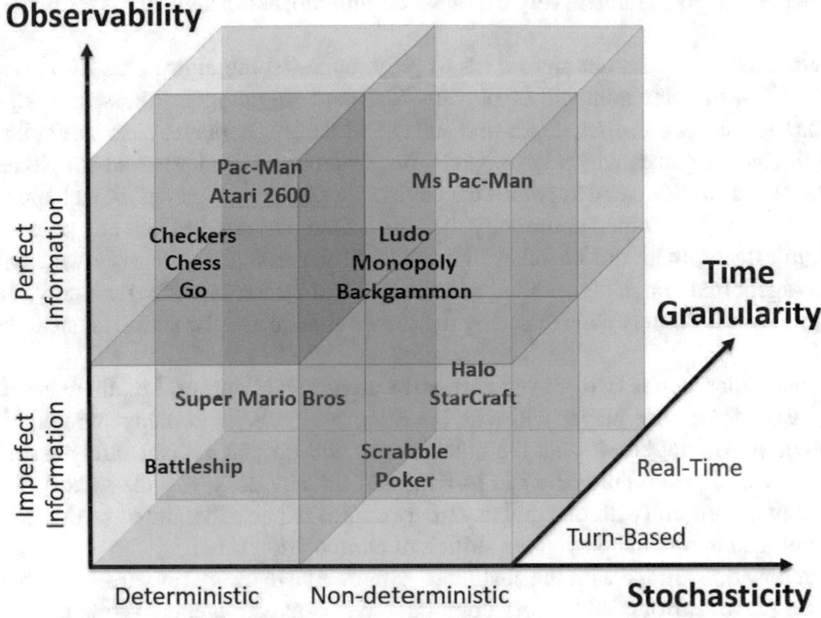

Fig. 3.2 Characteristics of games: game examples across the dimensions of stochasticity, observability and time granularity. Note that the game examples presented are sorted by complexity (action space and branching factor) within each cube. Minimax can *theoretically* solve merely any deterministic, turn-based game of perfect information (red cube in the figure)—in practice, it is still impossible to solve games with substantially large branching factors and action spaces such as Go via Minimax. Any AI method that eventually approximates the Minimax tree (e.g., MCTS) can be used to tackle imperfect information, non-determinism and real-time decision making (see blue cubes in figure). Strictly speaking, *Super Mario Bros* (Nintendo, 1985) involves a small degree of non-determinism only when a player helps creating a particular scene; we can, thus, safely classify the game as deterministic [163].

- **single-player** games, such as puzzles and time-trial racing;
- **one-and-a-half-player** games, such as the campaign mode of an FPS with non-trivial NPCs;
- **two-player** games, such as Chess, Checkers and *Spacewar!* (Russell, 1962); and
- **multi-player** games, such as *League of Legends* (Riot Games, 2009), the *Mario Kart* (Nintendo, 1992–2014) series and the online modes of most FPS games.

The distinction between single-player and one-and-a-half-player games is not a sharp one—there is no clear boundary for how advanced NPCs should be to count as a "half player". In the case of multi-player games, many can be played with only two players, at which point they are effectively two-player games. It is not always the case that other players (or NPCs) are adversarial and try to stop the player—there are many collaborative games, or games where relations between players are complex and have elements of both competition and cooperation. Still, keeping the number of players in mind is very useful when thinking about algorithms for playing games.

When using tree search algorithms to play games, some algorithms fit particularly well with some numbers of players. Standard single-agent tree search algorithms such as breadth-first, depth-first and A* fit the single-player case particularly well (including games which have NPCs, but where those NPCs are so simple and predictable as to be treated as part of the environment). In such games, what happens in the game is determined entirely by the actions the player takes and any potential random effects; there are no other "intentional" players. This fits very well with single-agent tree search algorithms, which are based on the **Markov property**, that the next state is entirely determined by the previous state and the action taken at that point.

A particular case is **two-player zero-sum adversarial games**, i.e., there are exactly two players; one player will win, the other will lose (or perhaps there will be a draw). We do not know what the other player will do, but we can safely assume that she will do everything she can to win, and thereby deny you the victory. The Minimax algorithm (with or without α-β pruning) is perfectly suited to this case, and will lead to optimal play given sufficient computation time.

But how do we cope with the challenge when we have **many players**, or perhaps a single player surrounded by very complicated non-player agents? While it is theoretically possible to expand Minimax to multiple players, this only works if there can be no collusion (or alliances of any kind) between players and the zero-sum nature of the game still remains (which it usually does not). Further, the computational complexity of Minimax quickly gets unmanageable with more than two players, as for every move you take you do not just need to consider the countermove of one player, but of all players. So this approach is rarely workable.

It is more common in the multi-player case to treat the game as a single-player game, but use some kind of model of what the other players do. This could be an assumption that the other players will oppose the player, a learned model based on observed behavior or even a random model. With an appropriate model of what the other players will do, many standard single-player game-playing methods can be used in multi-player settings.

3.2.1.2 Stochasticity

A common way in which many games violate the Markov property is by being **stochastic** (or non-deterministic). In many games, some of what happens is random. As standard digital computer architectures do not allow for "true" randomness, effective randomness is provided by pseudo-random number generators. The word "stochastic" is used to denote processes which cannot be practically predicted, whether they result from true randomness or complex calculations. Games can have varying amounts of stochasticity, from completely deterministic games like Chess to games dominated by stochastic outcomes like Roulette, Ludo, Yahtzee or even Monopoly. It is common for games to have mostly or fully deterministic game mechanics combined with some stochastic element through card-drawing, dice-throwing or some similar mechanism to reduce the possibility of planning. However, stochasticity can occur in essentially any part of a game.

In a game with stochasticity, the outcome of the game is not entirely determined by the actions the players take. In other words, if you play several playthroughs of the same game, taking the same actions at the same points in time, you are not guaranteed the same outcome. This has consequences for AI algorithms. For **tree search** algorithms, it means that we cannot be sure about the state which a sequence of actions will lead to, and therefore about the results of the algorithm. This leads to problems with using many tree search algorithms in their canonical forms, and requires that we add some modifications to address the non-deterministic uncertainty in the forward model. For example, in Monte Carlo tree search modifications such as **determinization** are used, where the different possible outcomes of each action are explored separately [77]. While these algorithm variations can be effective, they generally increase the computational complexity of the base algorithm.

For **reinforcement learning** approaches, including evolutionary reinforcement learning, it means that we have reduced certainty in exactly how good a given strategy/policy is—a good policy may achieve bad outcomes, or a bad policy good outcomes, because of random events in the game. Such outcome uncertainty can be mitigated by evaluating every policy multiple times, though this has significant computational cost. On the other hand, stochasticity can sometimes actually be an advantage when learning policies: a policy which is learned for a stochastic game may be more robust than one learned for a deterministic game, as in the latter case it is possible to learn a very brittle policy that only works for a specific configuration of the game. For example, learning a policy which attacks enemies in specific places at specific times, rather than being able to handle enemies that might arrive from any direction at any time.

While it is very common for digital games to include some form of stochasticity, an interesting case is very early games hardware such as the 1977-vintage Atari 2600, which does not have the facilities for implementing pseudo-random number generators (mainly because it lacks a system clock). If a player takes exactly the same actions at exactly the same times (including the key press that starts the game), exactly the same outcome will be achieved. The Arcade Learning Environment is a widely used game-based AI-benchmark built around an emulator of the Atari 2600 [40]. When training AI agents to play games with no stochasticity, it is entirely possible to learn brittle policies that effectively bypass the complexities of the full game (whether this actually happens, or whether most agents learn more general strategies, is an open question). As we already saw across the various examples of Chapter 2 *Ms Pac-Man* (Namco, 1982) is arguably the most popular non-deterministic arcade game of that era.

3.2.1.3 Observability

Observability is a characteristic that is strongly related to stochasticity. It refers to how much information about the game state is available to the player(s). At one extreme we have classic board games such as Chess, Go and Checkers, where the full board state is always available to the players, and puzzles such as Sudoku and Spelltower. These games have **perfect information**. At the other extreme we can think of classic text adventures such as *Zork* (Personal Software, 1980) or *Colossal Cave Adventure*, where initially very little of the world and its state is revealed to the player and much of the game is about exploring what the world is like. Those games have **hidden information** and therefore only **partial observability**. Many, if not most, computer games have significant hidden information: think of a typical platform game such as *Super Mario Bros* (Nintendo, 1985), or FPS such as the *Halo* series (Microsoft Studios, 2001–2015), where at any point you can only perceive a small part of the game world. Within computer strategy games such as *StarCraft II* (Blizzard Entertainment, 2015) or *Civilization* (MicroProse, 1991) the common term for hidden information is *fog of war*. Even many classic non-digital games have hidden information, including most card games where players keep their hands of cards private (such as Poker) and board games such as *Battleship*.

When developing an AI agent for a game with hidden information, the simplest approach you can follow is to merely ignore the hidden information. At each point in time, just feed the available information to the agent and use that to decide the next action. Doing so actually works quite well in some games, notably action-focused games with linear levels; for example, simple *Super Mario Bros* (Nintendo, 1985) levels can be played well based on only instantaneously available information [706]. However, if you play a strategy game such as *StarCraft* (Blizzard Entertainment, 1998) based on only the available information, you are not even going to see the enemy until it is too late—good play involves active information gathering. Even in *Super Mario Bros* (Nintendo, 1985), complicated levels that feature backtracking require remembering off-screen parts of the level [322]. In a trick-taking card game such as Poker the available information (your own hand) is actually of comparatively less relevance; the core of the game is modeling the hidden information (your adversaries' hands and minds).

Therefore, effective AI for games with partial observability often requires some kind of **modeling of the hidden information**. For some games, notably variants of Poker such as Heads-up limit hold'em, considerable research has been done on game-specific methods for modeling hidden information that includes opponents' play [63]. There are also more generic methods of adding some form of hidden state modeling to existing algorithms, such as Information Set Monte Carlo tree search [146]. Just like when it comes to methods for dealing with stochasticity, these methods typically add considerable computational complexity compared to the base algorithm.

3.2.1.4 Action Space and Branching Factor

When you play the minimalist-masochist mobile game *Flappy Bird* (dotGEARS, 2013), you have a single choice at any point in time: to flap or not to flap. (Flapping is accomplished by touching the screen, and makes the protagonist bird rise in the air.) *Flappy Bird* and similar one-button games, such as *Canabalt* (Beatshapers, 2009), probably have the lowest possible **branching factor**. The branching factor is the number of different actions you can take at any decision point. The branching factor of Flappy Bird is 2: flap or no flap.

For comparison, *Pac-Man* (Namco, 1980) has a branching factor of 4: up, down, left and right. *Super Mario Bros* (Nintendo,1985) has a branching factor of around 32: eight D-pad directions times two buttons (though you may argue that some of these combinations are nonsensical and should not actually be considered). Chess has an average branching factor of 35, whereas Checkers has a somewhat lower branching factor. Go has a whooping 400 for the very first move; as the board is populated, the branching factor decreases, but there are typically a few hundred potential positions to put every stone.

While 400 is a very high branching factor compared to 35 or 2, it dwarfs in comparison to many computer strategy games where multiple units can be moved every turn. Considering each combination of movements of individual units as an action, this means that the branching factor of the game is the product of the branching factor of the individual units. If you have 6 different units that can each take 10 different actions at a given time—a rather conservative estimate compared to typical games of, say, *StarCraft* (Blizzard Entertainment, 1998) or *Civilization* (MicroProse, 1991)—then your branching factor is a million!

But wait, it gets worse. For many games, it is not even possible to enumerate all the actions, as the input space is **continuous**. Think of any modern first-person game played on a computer or console. While it is true that computers do not really capture infinities well, and that "continuous" inputs such as computer mice, touchscreens and thumbsticks (on e.g., XBox and Playstation controllers) actually return a digital number, that number has such fine resolution that it is for all practical purposes continuous. The only way to create a practically enumerable set of actions is to discretize the continuous input space somehow, and strike a compromise between overwhelming branching factors and reducing the input space so much as to not be able to play the game effectively.

The branching factor is a key determinant of the effectiveness of **tree search** algorithms. The complexity of the breadth-first algorithm (for single-player games) and the Minimax algorithm (for adversarial two-player games) for searching to depth d is b^d, where b is the branching factor. In other words, a high branching factor makes it almost impossible to search more than a few steps ahead. This fact has very tangible consequences for which games can be played with tree search methods; for example, Go has an order of magnitude higher branching factor than Chess, and this was arguably the main reason for the very poor performance of all kinds of AI methods on Go for decades (during which the same methods performed well on Chess). Monte Carlo tree search handles high branching factors better because it builds imbalanced trees, but it is by no means immune to the problem. Once the branching factor gets high enough (say, a million, or maybe a billion, depending on the speed of the simulator), it becomes impractical to enumerate even the actions at depth 1 and therefore to use tree search at all.

High branching factors are an issue for **reinforcement learning** algorithms as well, including evolutionary reinforcement learning. Mostly this has to do with the controller/policy representation. If you are using a neural network (or some other function approximator) for representing your policy, you may need to have outputs for each action; alternatively, if you are using the network to assign values to all actions, you need to iterate over them. In both cases, a large number of possible actions carries a cost. Another problem is the exploration-exploitation dilemma: the higher the number of possible actions, the longer it will take to explore them all while learning.

A final comment on branching factors is that for many games, they are not constant. In Chess, you have fewer moves available at the beginning of the game when most of your pieces are blocked, more towards the midgame, and fewer again in the endgame when most pieces may be blocked. In a typical RPG such as those in the *Final Fantasy* series (Square Enix, 1987–2016) the number of available actions increases as the player character accrues items, spells and other possibilities. As mentioned above, the number of available actions in Go decreases as you play.

3.2.1.5 Time Granularity

When discussing branching factors above, we talked about the number of possible actions to take at any "point in time". But how often is that? How often can the player take an action? A fundamental distinction is that between **turn-based** and **real-time** games. Most classic board games are turn-based games. In such games, players take turns, and at each turn a player can take an action, or a specified number of actions. The amount of real time that passes between turns is generally not of any importance inside the game (though tournaments and professional play often incorporate some form of time limit). Real-time games include many popular genres of computer games, such as FPS, racing games and platformers. Even within real-time games, there is considerable variation in how often an in-game action can in practice be taken. At the extreme there is the screen update frequency; the current generation of video games typically strives to have an update frequency of 60 frames per second to ensure a perceived smooth movement, but many games update the screen half as often or even less because of the complexity of rendering complicated scenes. In practice, the number of actions a player character (or any other in-game character) could take per second is usually more limited than that.

To take two examples far apart on the time granularity scale, let us consider two adversarial games: Chess and *StarCraft* (Blizzard Entertainment, 1998). A game of Chess between skilled players on average lasts about 40 turns.[1] In *StarCraft* (Bliz-

[1] http://chess.stackexchange.com/questions/2506/

zard Entertainment, 1998), a highly competitive real-time strategy (RTS) game, professional players often take three to five actions per second (each action is typically executed with a mouse click or a shortcut key). With a typical game lasting 10 to 20 minutes, this means that thousands of actions are taken in a game. But there are not that many more significant events in a game of *StarCraft* (Blizzard Entertainment, 1998) than in a game of Chess—the lead does not change much more often, and grand strategic decisions are not made that much more often. This means that the number of actions between significant game events is much higher in *StarCraft* (Blizzard Entertainment, 1998) than in Chess.

Time granularity affects AI game-playing methods through limiting how far ahead you can look. A given **depth** of search means very different things depending on the time granularity of the game. Ten turns in Chess is enough to execute a whole strategy; ten actions ahead in *StarCraft* (Blizzard Entertainment, 1998) might just be a few seconds, during which the game might not have changed in any significant way. To play *StarCraft* (Blizzard Entertainment, 1998) well using tree search, one would need an exceptional search depth, in the hundreds or thousands of actions, which would clearly be computationally infeasible. One way to address this challenge is to consider **macro-actions** (e.g., as in [525, 524]), which are sets or sequences of smaller, fine-grained, actions.

3.2.2 Characteristics of AI Algorithm Design

In the following, we discuss some important issues in applying AI algorithms to games. These are design choices relating not so much to game design (covered in the previous section), as to AI algorithm design and the constraints under which the algorithm is used. This section expands the discussion about representation and utility covered in Chapter 2 with a focus on game-playing AI.

3.2.2.1 How Is the Game State Represented?

Games differ in what information they present to the player, and how. Text adventures output text, the state of a classic board game can be described by the positions of all board pieces, and graphical video games serve moving graphics together with sound and occasionally outputs such as controller rumble. For digital games, the technical limitations of the hardware on which the game is implemented influences how it is presented; as processor speed and memory capacity increases, the pixel resolution and scene complexity of video games has increased commensurably.

Importantly, the same game can be represented in different ways, and which way it is represented matters greatly to an algorithm playing the game. To take a racing game as an example, the algorithm could receive a first-person view out of the windscreen of the car rendered in 3D, or an overhead view of the track rendering the track and various cars in 2D. It could also simply receive a list of positions and velocities of all cars on the track in the frame of reference of the track (along with a model of the track), or a set of angles and distances to other cars (and track edges) in the frame of reference of the track.

The choices regarding input representation matter a lot when designing a game. If you want to learn a policy for driving a car around a track, and the inputs to the policy are the three continuous variables associated with speed and distance to the left and right edge of the track, learning a decent driving policy is comparatively simple. If your input is instead an unprocessed visual feed—i.e., tens of thousands of pixel values—finding a decent policy is likely to be much harder. Not only is the policy search space in the latter case vastly greater, the proportion of the search space that corresponds to decently-performing policies is likely to be much smaller, as many more nonsensical policies are possible (e.g., turn left if even-numbered pixels are lighter than odd-numbered pixels; this policy does not map to the game state in any sensible way, and if it works it is a fluke). In order to learn to drive well based on visual input—at least in cases where illumination, roadside scenery, etc. vary significantly—you likely need to learn a visual system of some kind. In light of this, most naive policies applied to full visual input would likely not have fared very well. Continuing the car racing example, even in cases where you have very few inputs, how these are represented matters; for example, it is much easier to learn a good driving policy if the inputs are represented in the frame of reference of the car rather than that of the track [707, 714]. A somewhat more comprehensive discussion on ways of representing low-dimensional inputs to neural networks can be found in [567].

In recent years, several groups of researchers have focused on learning policies that use full visual feeds as inputs. For example, Koutnik et al. evolved neural networks to play *The Open Racing Car Simulator* (TORCS) from high-resolution video [353], Kempka et al. used the pixels of the screen as input to a deep Q network that was trained to play a version of *DOOM* (GT Interactive, 1993) [333], and Mnih et al. trained deep networks to play Atari 2600 games using Q-learning [464]. Using the raw pixel inputs is often motivated by giving the AI the same conditions as a human would have, and thus achieving a level playing field between human and AI. Another motivation is that if you want to use your algorithm to play a game "out of the box", without any API or additional engineering to expose the internal state of the game, you will likely have to resort to using the raw visual feed. However, in cases where you have access to the source code of the game or a useful API—as you would almost always have when developing AI for a new game—there is no reason to not utilize the "digested" game state in whatever form makes the task of the AI algorithm easiest. Whether or not to present information that the human player does not have access to, i.e., "cheating", is a separate question.

3.2.2.2 Is There a Forward Model?

A very important factor when designing an AI to play a game is whether there is a simulator of the game, a so-called **forward model**, available. A forward model is a model which, given a state s and an action a, reaches the same state s' as the real game would reach if it it was given a at s. In other words, it is a way of playing the game in simulation, so that consequences of multiple actions can be explored before actually taking some of those actions in the real game. Having a forward model of the game is necessary in order to be able to use any tree search-based approaches to playing a game, as those approaches depend on simulating the outcome of multiple actions.

A very desirable property of a forward model, in addition to that it exists, is that it is **fast**. In order to be able to use a tree search algorithm effectively for control in a real-time game, one would generally need to be able to simulate gameplay at least a thousand times faster than real-time, preferably tens or hundreds of thousands of times faster.

It is very easy to construct a forward model for classic board games such as Chess and Go, as the game state is simply the board state and the rules are very easy to encode. For many video games, constructing a forward model can be done by simply copying (or otherwise reusing) the same code as is used for controlling the game itself, but without waiting for user input or displaying graphics, and without performing all the calculations involved with graphics rendering. For some video games—notably games that were originally implemented for much older hardware, such as classic arcade games that were implemented on 8-bit or 16-bit processors—forward models can be made much faster than real-time, as the core game loop is not that computationally complex. (It might also be possible to do this with some modern games, by running them inside emulators that can replace the graphics routines with dummy code.)

For many games, however, it is impossible or at least very hard to obtain a fast forward model. For most commercial games, the source code is not available, unless you are working at the company that develops the game. Even if the source code is available, current software engineering practices in the game industry make it very hard to extract forward models from game code, as the core control loops are often closely tied up with user interface management, rendering, animation and sometimes network code. A change in software engineering practices to separate the core game loop more cleanly from various input/output functions so that forward models could more easily be built would be one of the most important enablers of advanced AI methods in video games. However, in some cases the computational complexity of the core game loop might still be so high that any forward models built on the core game code would be too slow to be usable. In some of such cases, it might be practical to build and/or learn a simplified or **approximate forward model**, where the state resulting from a series of actions taken in the forward model is not guaranteed to be identical to the state resulting from the same series of actions in the actual game. Whether an approximate forward model is acceptable or not depends on the particular use case and motivation for the AI implementation. Note that a somewhat less than accurate forward model might still be desirable. For example, when there is significant hidden information or stochasticity the AI designer might not want to provide the AI agent with an *oracle* that makes the hidden information observable and tells the agent which random actions will happen. Taking such a design decision might lead to unreasonably good performance and the appearance of cheating.

When a forward model cannot be produced, **tree search** algorithms cannot be applied. It is still possible to manually construct agents, and also to learn agents through **supervised learning** or some form of **reinforcement learning**, such as temporal difference learning or evolutionary reinforcement learning. However, note that while the reinforcement learning approaches in general do not need a complete forward model in the sense that the results of taking any action in any state can be predicted, they still need a way to run the game faster than real-time. If the game cannot be sped up significantly beyond the pace at which it is naturally played, it is going to take the algorithm a very long time to learn to play.

3.2.2.3 Do You Have Time to Train?

A crude but useful distinction in artificial intelligence is between algorithms that try to decide **what to do in a given situation** by examining possible actions and future states—roughly, tree search algorithms of various kinds—and algorithms that **learn a model** (such as a policy) over time—i.e., machine learning. The same distinction exists within AI for playing games. There are algorithms developed that do not need to learn anything about the game, but do need a forward model (tree search); there are algorithms that do not need a forward model, but instead learn a policy as a mapping from state(s) to action (model-free reinforcement learning); and there are algorithms that require both a forward model and training time (model-based reinforcement learning and tree search with adaptive hyperparameters).

What type of algorithm you will want to use depends largely on your motivation for using AI to play games. If you are using the game as a testbed for your AI algorithm, your choice will be dictated by the type of algorithm you are testing. If you are using the AI to enable player experience in a game that you develop—for example, in a non-player role—then you will probably not want the AI to perform any learning while the game is being played, as this risks interfering with the gameplay as designed by the designer. In other cases you are looking for an algorithm that can play some range of games well, and do not have time to retrain the agent for each game.

3.2.2.4 How Many Games Are You Playing?

An aim the AI designer might wish to achieve is that of **general game playing**. Here, we are not looking for a policy for a single game, we are looking for a more generic agent that can play any game that it is presented with—or at least any game from within a particular distribution or genre, and which adheres to a given interface. General game playing is typically motivated by a desire to use games to progress towards artificial *general* intelligence, i.e., developing AI that is not only good at one thing but at many different things [598, 679, 744]. The idea is to avoid overfitting (manually or automatically) a given game and come up with agents that generalize well to many different games; it is a common phenomenon that when developing agents for a particular game, for example, for a game-based AI competition, many special-purpose solutions are devised that do not transfer well to other games [701].

For this reason, it is common to evaluate general game playing agents on unseen games, i.e., games on which they have not been trained and which the designers of the agent were not aware of when developing the agent. There are several frameworks for general game playing, including the General Game Playing Competition [223], the General Video Game AI Competition [528, 527] and the Arcade Learning Environment [40]. These will be discussed later in the chapter.

General video game playing, where AI is developed to play for performance in the player role across many games (ideally all games), can be seen as diametrically opposed to the typical use of AI for playing games in commercial game development, where the AI is playing for experience in a non-player role, and is carefully tuned to a particular game. However, the development of AI methods for general game playing certainly benefits commercial game AI in the end. And even when developing game AI as part of game development, it is good engineering practice to develop methods that are reusable to some extent.

3.3 How Can AI Play Games?

In Chapter 2, we reviewed a number of important AI methods. Most of these methods can be used to play games in one way or another. This section will focus on the core AI methods, and for each family of algorithms it will go through on how they can be used to play games.

3.3.1 Planning-Based Approaches

Algorithms that select actions through planning a set of future actions in a state space are generally applicable to games, and do not in general require any training time. They do require a fast forward model if searching in the game's state space, but not if simply using them for searching in the physical space (path-planning). Tree search algorithms are widely used to play games, either on their own or in supporting roles in game-playing agent architectures.

3.3.1.1 Classic Tree Search

Classic tree search methods, which feature little or no randomness, have been used in game-playing roles since the very beginning of research on AI and games. As mentioned in the introduction of this book the Minimax algorithm and α-β pruning were originally invented in order to play classic board games such as Chess and Checkers [725]. While the basic concepts of adversarial tree search have not really changed since then, there have been numerous tweaks to existing algorithms and some new algorithms. In general, classic tree search methods can easily be applied in games that feature full observability, a low branching factor and a fast forward model. Theoretically they can solve any deterministic game that features full observability for the player (see the red cube in Fig. 3.2); in practice, they still fail in games containing large state spaces.

Best-first search, in particular a myriad variations of the A* algorithm, is very commonly used for **path-planning** in modern video games. When an NPC in a modern 3D FPS or RPG decides how to get from point A to point B, this is typically done using some version of A*. In such instances, search is usually done in (in-game) physical space rather than state space, so no forward model is necessary. As the space is pseudo-continuous, search is usually done on the nodes of a mesh or lattice overlaid on the area to be traversed. Note that best-first search is only used for navigation and not for the full decision-making of the agent; methods such as behavior trees or finite-state machines (which are usually hand-authored) are used to determine *where* to go, whereas A* (or some variation of it) is used to determine *how* to get there. Indeed, in games where player input is by pointing to and clicking at positions to go—think of an overhead brawler like *Diablo* (Blizzard Entertainment, 1996) or an RTS like *StarCraft* (Blizzard Entertainment, 1998)—the execution of the player's order usually involves a path-planning algorithm as well. Recent additions to the family of best-first algorithms include jump point search (JPS), which can improve performance by orders of magnitude compared to standard A* under the right circumstances [662]. Hierarchical pathfinding is its own little research area based on the idea of dividing up an area into subareas and using separate algorithms for deciding how to go between and within the areas. Choosing a path-planning algorithm for a modern video game is usually a matter of choosing the algorithm that works best given the shape of the environments the NPCs (or PCs) are traversing, the particular way a grid or movement graph is overlaid on top of this space, and the demands of the animation algorithm. Generally, one size does not fit all; some textbooks devoted to industry-oriented game AI discuss this in more depth [461].

Beyond path-planning, best-first algorithms such as A* can be used for **controlling** all aspects of NPC behavior. The key to doing this is to search in the state space of the game, not just the physical space. (Obviously, this requires a fast forward model.) To take an example, the winner of the 2009 Mario AI Competition was entirely based on A* search in state space [705]. This competition tasked competitors with developing agents that could play a Java-based clone of the classic platform game *Super Mario Bros* (Nintendo, 1985)—it later evolved into a multi-track competition [322]. While a forward model was not supplied with the original competition software, the winner of the competition, Robin Baumgarten, created one by adapting parts of the core game code. He then built an A* agent which at any point simply tried to get to the right edge of the screen. (An illustration of the agent can be seen in Figs. 3.3 and 2.5.) This worked extremely well: the resulting agent played seemingly optimally, and managed to get to the end of all levels in-

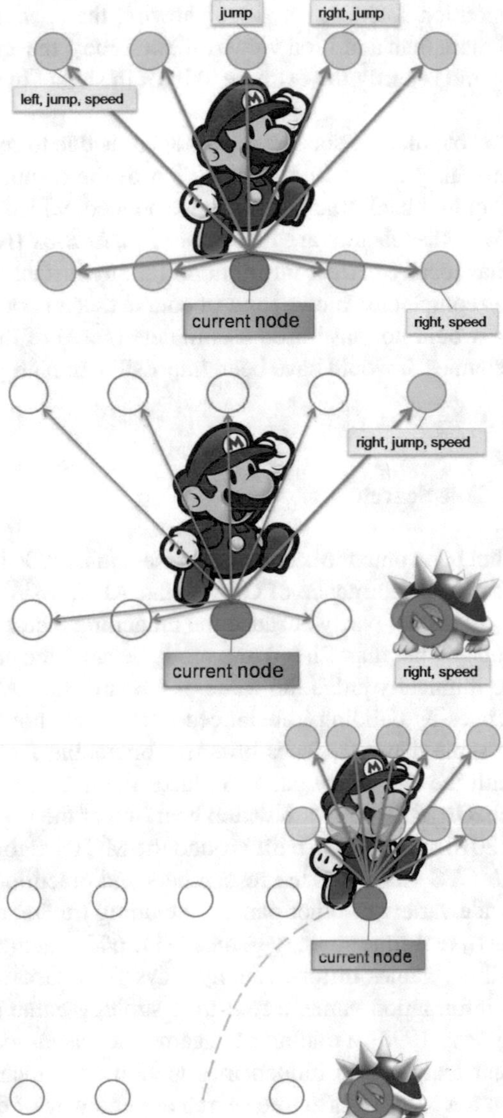

Fig. 3.3 An illustration of the key steps of A* search for playing *Super Mario Bros* (Nintendo, 1985). The agent considers a maximum of nine possible actions at each frame of the game, as a result of combining the jump and speed buttons with moving right or left (top figure). Then the agent picks the action with the highest heuristic value (middle figure). Finally, the Mario agent takes the action (i.e., right, jump, speed in this example), moves to a new state and evaluates the new action space in this new state (bottom figure). More details about the A* agent that won the Mario AI competition in 2009 can be found in [705].

cluded in the competition software. A video showing the agent navigating one of the levels gathered more than a million views on YouTube;[2] the appeal of seeing the agent playing the game is partly the extreme skill of the agent in navigating among multiple enemies.

It is important to note that the success of this agent is due to several factors. One is that the levels are fairly linear; in a later edition of the competition, levels with dead ends which required back-tracking were introduced, which defeated the pure A* agent [322]. Two other factors are that *Super Mario Bros* (Nintendo, 1985) is deterministic and has locally perfect information (at any instant the information in the current screen is completely known) and of course that a good forward model is available: if the A* would not have used a complete model of the game including the movement of enemies, it would have been impossible to plan paths around these enemies.

3.3.1.2 Stochastic Tree Search

The MCTS algorithm burst onto the scene of Go research in 2006 [141, 77], and heralded a quantum leap in performance of Go-playing AI. Classic adversarial search had performed poorly on Go, partly because the branching factor is too high (about an order of magnitude higher than Chess) and partly because the nature of Go makes it very hard to algorithmically judge the value of a board state. MCTS partly overcomes these challenges by building imbalanced trees where not all moves need to be explored to the same depth (reduces effective branching factor) and by doing random rollouts until the end of the game (reduces the need for a state evaluation function). The AlphaGo [629] software which beat two of the best human Go players in the world in 2016 and 2017, is built around the MCTS algorithm.

The success of MCTS on Go has led researchers and practitioners to explore its use for playing a wide variety of other games, including trading card games [746], platform games [294], real-time strategy games [311, 645], racing games [203] and so on. Of course, these games differ in many ways from Go. While Go is a deterministic perfect information game, a real-time strategy game such as *StarCraft* (Blizzard Entertainment, 1998), a trading card game such as *Magic: The Gathering*, or any Poker variant feature both hidden information and stochasticity. Methods such as Information Set Monte Carlo tree search are one way of dealing with these issues, but impose computational costs of their own [146].

Another problem is that in games with fine time granularity, it might take a prohibitively long time for a rollout to reach a terminal state (a loss or a win); in many video games it is possible to take an arbitrary number of actions without winning or losing the game, or even doing something that materially affects the outcome of the game. For example, in *Super Mario Bros* (Nintendo, 1985), most randomly generated action sequences would not see Mario escaping the original screen, but basically pacing back and forth until time runs out, thousands of time steps later.

[2] https://www.youtube.com/watch?v=DlkMs4ZHHr8

One response to this problem is to only roll out a certain number of actions, and if a terminal state is not encountered use a state evaluation function [77]. Other ideas include pruning the action selection so as to make the algorithm search deeper [294]. Given the large number of modifications to all components of the MCTS algorithm, it makes more sense to think of MCTS as a general algorithmic framework rather than as a single algorithm.

Many games could be played either through MCTS, uninformed search (such as breadth-first search) or informed search (such as A*). Deciding which method to use is not always straightforward, but luckily these methods are relatively simple to implement and test. Generally speaking, Minimax can only be used for (two-player) adversarial games whereas other forms of uninformed search are best used for single-player games. Best-first search requires some kind of estimate of a distance to a goal state, but this does not need to be a physical position or the end goal of the game. Varieties of MCTS can be used for both single-player and two-player games, and often outperform uninformed search when branching factors are high.

3.3.1.3 Evolutionary Planning

Interestingly, decision making through planning does not need to be built on tree search. Alternatively, one can use optimization algorithms for planning. The basic idea is that instead of searching for a sequence of actions starting from an initial point, you can optimize the whole action sequence. In other words, you are searching the space of complete action sequences for those that have maximum utility. Evaluating the utility of a given action sequence is done by simply taking all the actions in the sequence in simulation, and observing the value of the state reached after taking all those actions.

The appeal of this idea is that an optimization algorithm might search the plan space in a very different manner compared to a tree search algorithm: all tree search algorithms start from the root of the tree (the origin state) and build a tree from that point. Evolutionary algorithms instead regard the plan as simply a sequence, and can perform mutations or crossover at any point in the string. This could help in guiding the search at different areas of the plan space that a tree search algorithm would explore for the same problem.

While many different optimization algorithms could be used, the few studies on optimization-based planning in games that can be found in the literature use evolutionary algorithms. Perez et al. proposed using evolutionary planning for single-player action games, calling this approach "rolling horizon evolution" [526]. In the particular implementation for the Physical Traveling Salesman Problem (a hybrid between the classic TSP problem and a racing game), an evolutionary algorithm was used to generate a plan every time step. The plan was represented as a sequence of 10-20 actions, and a standard evolutionary algorithm was used to search for plans. After a plan was found, the first step of the plan was executed, just as would be the case with a tree search algorithm. Agents based on evolutionary planning generally perform competitively in the General Video Game AI Competition [528].

Evolutionary planning is particularly promising as a technique for handling very large branching factors, as we have seen that games with multiple independent units (such as strategy games) can have. Justesen et al. [309] applied evolutionary computation to select actions in the turn-based strategy game *Hero Academy* (Robot Entertainment, 2012), calling this approach "online evolution". Given the number of units the player controls and the number of actions available per unit, the branching factor is about one million; therefore, only a single turn ahead was planned. Evolutionary planning was shown to outperform Monte Carlo tree search by a wide margin in that game. Wang et al. [745], and Justesen and Risi [310] later applied variants of this technique to *StarCraft* (Blizzard Entertainment, 1998) tactics. Given the continuous-space nature of the game, the branching factor would be extreme if every possible movement direction for every unit was considered as a separate action. What was evolved was therefore not a sequence of actions, but rather which of several simple scripts (tactics) each unit would use in a given time step (this idea was borrowed from Churchill and Buro, who combined a "portfolio" of scripts with simple tree search [123]). Wang et al. [745] showed that evolutionary planning performed better than several varieties of tree search algorithms in this simple *StarCraft* (Blizzard Entertainment, 1998) scenario.

Evolutionary planning in games is a recent invention, and there are only a limited number of studies on this technique so far. It is not well understood under what conditions this technique performs well, or even really why it performs so well when it does. A major unsolved problem is how to perform evolutionary adversarial planning [586]; whereas planning based on tree search works in the presence of an adversary (for example, see the minimax algorithm), it is not clear how to integrate this into a genotype. Perhaps through competitive coevolution of actions taken by different players? There is, in other words, plenty of scope for further research in this area.

3.3.1.4 Planning with Symbolic Representations

While planning on the level of in-game actions requires a fast forward model, there are other ways of using planning in games. In particular, one can plan in an abstract representation of the game's state space. The field of automated planning has studied planning on the level of symbolic representations for decades [228]. Typically, a language based on first-order logic is used to represent events, states and actions, and tree search methods are applied to find paths from the current state to an end state. This style of planning originated with the STRIPS representation used in Shakey, the world's first digital mobile robot [494]; symbolic planning has since been used extensively in numerous domains.

The horror-themed first-person shooter *F.E.A.R.* (Sierra Entertainment, 2005) became famous within the AI community for its use of planning to coordinate NPC behavior. The game's AI also received nice reviews in the gaming press, partly because the player is able to hear the NPCs communicate with each other about their plan of attack, heightening immersion. In *F.E.A.R.* (Sierra Entertainment, 2005), a STRIPS-like representation is used to plan which NPCs perform which actions (flank, take cover, suppress, fire, etc.) in order to defeat the player character. The representation is on the level of individual rooms, where movement between one room and the next is usually a single action [507]. Using this high-level representation, it is possible to plan much further ahead than would be possible when planning on the scale of individual game actions. Such a representation, however, requires manually defining states and actions.

3.3.2 Reinforcement Learning

As discussed in Chapter 2, a reinforcement learning algorithm is any algorithm that solves a reinforcement learning problem. This includes algorithms from the temporal difference or approximate dynamic programming family (for simplicity, we will refer to such algorithms as *classic* reinforcement learning methods), applications of evolutionary algorithms to reinforcement learning such as neuroevolution and genetic programming, and other methods. In this section, we will discuss both classic methods (including those that involve deep neural networks) and evolutionary methods as they are applied for playing games. Another way of describing the difference between these methods is the difference between *ontogenetic* (which learns during "lifetimes") and *phylogenetic* (which learns between "lifetimes") methods [715].

Reinforcement learning algorithms are applicable to games when there is learning time available. Usually this means plenty of training time: most reinforcement learning methods will need to play a game thousands, or perhaps even millions, of times in order to play it well. Therefore, it is very useful to have a way of playing the game much faster than real-time (or a very large server farm). Some reinforcement learning algorithms, but not all, also require a forward model. Once it has been trained, a reinforcement-learned policy can usually be executed very fast.

It is important to note that the planning-based methods (described in the previous section) for playing games cannot be directly compared with the reinforcement learning methods described in this section. They solve different problems: planning requires a forward model and significant time at each time step; reinforcement learning instead needs learning time and may or may not need a forward model.

3.3.2.1 Classic and Deep Reinforcement Learning

As already mentioned in the introduction of this book, classic reinforcement learning methods were used with games early on, in some cases with considerable success. Arthur Samuel devised an algorithm—which can be said to be the first classic reinforcement learning algorithm—in 1959 to create a self-learning Checkers player. Despite the very limited computational resources of the day, the algorithm learned to play well enough to beat its creator [591]. Another success for classic reinforcement learning in game-playing came a few decades later, when Gerald

Tesauro used the modern formulation of temporal difference learning to teach a simple neural network to play Backgammon, named TD-gammon; it learned to play surprisingly well, after starting with no information and simply playing against itself [689] (TD-gammon is covered in more detail in Chapter 2). This success motivated much interest in reinforcement learning during the 1990s and early 2000s.

However, progress was limited by the lack of good function approximators for the value function (e.g., the Q function). While algorithms such as Q-learning will provably converge to the optimal policy under the right conditions, the right conditions are in fact very restrictive. In particular, they include all state values or {state, action} values that are stored separately, for example, in a table. However, for most interesting games there are far too many possible states for this to be feasible—almost any video game has at least billions of states. This means that the table would be too big to fit in memory, and that most states would never be visited during learning. It is clearly necessary to use a compressed representation of the value function that occupies less memory and also does not require every state to be visited in order to calculate its value. It can, instead, calculate it based on neighboring states that have been visited. In other words, what is needed is a **function approximator**, such as a neural network.

However, using neural networks together with temporal difference learning turns out to be non-trivial. It is very easy to encounter "catastrophic forgetting", where sophisticated strategies are unlearned in favor of degenerate strategies (such as always taking the same action). The reasons for this are complex and go beyond the discussion in this chapter. However, to intuitively understand one of the mechanisms involved, consider what would usually happen for a reinforcement learning agent playing a game. Rewards are very sparse, and the agent will typically see long stretches of no reward, or negative reward. When the same reward is encountered for a long time, the backpropagation algorithm will be trained only with the target value of that reward. In terms of supervised learning, this is akin to training for a long term on a single training example. The likely outcome is that the network learns to only output that target value, regardless of the input. More details on the method of approximating a value function using an ANN can be found in Section 2.8.2.3.

A major success in the use of reinforcement learning of the temporal difference variety together with function approximators came in 2015, when Google DeepMind published a paper where they managed to train deep neural networks to play a number of different games from the classic Atari 2600 games console [464]. Each network was trained to play a single game, with the inputs being the raw pixels of the game's visuals, together with the score, and the output being the controller's directions and fire button. The method used to train the deep networks is deep Q networks, which is essentially standard Q-learning applied to neural networks with many layers (some of the layers used in the architecture were convolutional). Crucially, they managed to overcome the problems associated with using temporal difference techniques together with neural networks by a method called *experience replay*. Here, short sequences of gameplay are stored, and replayed to the network in varying order, in order to break up the long chains of similar states and reward.

This can be seen as akin to batch-based training in supervised learning, using small batches.

3.3.2.2 Evolutionary Reinforcement Learning

The other main family of reinforcement learning methods is evolutionary methods. In particular, using evolutionary algorithms to evolve the weights and/or topology of neural networks (**neuroevolution**) or programs, typically structured as expression trees (**genetic programming**). The fitness evaluation consists in using the neural network or program to play the game, and using the result (e.g., score) as a fitness function.

This basic idea has been around for a long time, but was surprisingly under-explored for a long time. John Koza, a prominent researcher within genetic programming, used an example of evolving programs for playing Pac-Man in his 1992 book [356]. A couple of years later, Pollack and Blair showed that evolutionary computation can be used to train backgammon players using the same setup as Tesauro used in his experiments with TD learning, and with similar results [537]. Outside of games, a community of researchers was forming in the 1990s exploring the idea of using evolutionary computation to learn control strategies for small robots; this field came to be called **evolutionary robotics** [496]. Training robots to solve simple tasks of e.g., navigation, obstacle-avoidance and situational learning has very much in common with training NPCs to play games, in particular two-dimensional arcade-like games [567, 767, 766].

Starting around 2005, a number of advances were made in applying neuroevolution to playing different types of video games. This includes applications to car racing [707, 709, 392, 353], first-person shooters [518], strategy games [79], real-time strategy games [654] and classic arcade games such as Pac-Man [766, 403]. Perhaps the main takeaway from this work is that neuroevolution is extremely versatile, and can be applied to a wide range of games, usually in several different ways for each game. For example, for a simple car racing game it was shown that evolving neural networks that acted as state evaluators, even in combination with a simple one-step lookahead search, substantially outperformed evolving neural networks working as action evaluators (Q functions) [408]. Input representation matters too; as discussed in Section 3.2.2.1, egocentric inputs are generally strongly preferred, and there are additional considerations for individual game types, such as how to represent multiple adversaries [654].

Neuroevolution has seen great success in learning policies in cases where the state can be represented using relatively few dimensions (say, fewer than 50 units in the neural network's input layer), and is often easier to tune and get working than classic reinforcement learning algorithms of the temporal difference variety. However, neuroevolution seems to have problems scaling up to problems with very large input spaces that require large and deep neural networks, such as those using high-dimensional pixel inputs. The likely reason for this is that stochastic search in weight space suffers from the *curse of dimensionality* in a way that gradient descent search (such as backpropagation) does not. Currently, almost all successful examples of learning directly from high-dimensional pixel inputs use deep Q-learning or similar methods, though there are approaches that combine neuroevolution with unsupervised learning, so that controllers are learned that use a compressed representation of the visual feed as input [353].

For more details on the general method of neuroevolution and pointers to the literature the reader is referred to Section 2.8.1, and to the recent survey paper on neuroevolution in games [567].

3.3.3 Supervised Learning

Games can also be played using **supervised learning**. Or rather, policies or controllers for playing games can be learned through supervised learning. The basic idea here is to record traces of human players playing a game and train some function approximator to behave like the human player. The traces are stored as lists of tuples <features, target> where the features represent the game state (or an observation of it that would be available to the agent) and the target is the action the human took in that state. Once the function approximator is adequately trained, the game can be played—in the style of the human(s) it was trained on—by simply taking whatever action the trained function approximator returns when presented with the current game state. Alternatively, instead of learning to predict what action to take, one can also learn to predict the value of states, and use the trained function approximator in conjunction with a search algorithm to play the game. Further details about the potential supervised algorithms that can be used in games are described in Chapter 2.

3.3.4 Chimeric Game Players

While planning, reinforcement learning and supervised learning are fundamentally different approaches to playing games, solving the game-playing problem under different constraints, which does not mean that they cannot be combined. In fact, there are many examples of successful **hybrids** or **chimeras** of approaches from these three broad classes. One example is **dynamic scripting** [650] which can be viewed as a form of a **learning classifier system** [363] in that it involves a rule-based (here called script-based) representation coupled with reinforcement learning. Dynamic scripting adjusts the importance of scripts via reinforcement learning at runtime and is based on the current game state and immediate rewards obtained. Dynamic scripting has seen several applications in games including fighting games [417] and real-time strategy games [409, 154]. The approach has been used mainly for AI that adapts to the skills of the player, thereby aiming at the experience of the player and not necessarily at wining the game.

Another good example is AlphaGo. This extremely high-performing Go-playing agent actually combines planning through search, reinforcement learning and supervised learning [629]. At the core of the agent is a Monte Carlo tree search algorithm which searches in the state space (planning). Rollouts, however, are combined with evaluations from a neural network which estimates the value of states, and node selection is informed by a position estimation network. Both the state network and position network are initially trained on databases of games between grandmasters (supervised learning), and later on further trained by self-play (reinforcement learning).

3.4 Which Games Can AI Play?

Different games pose different challenges for AI playing, in the same way they pose different challenges for human playing. Not only are there differences in what kind of access the AI player has to the games, but also between different game types: a policy for playing Chess is unlikely to be proficient at playing the games in the *Grand Theft Auto* (Rockstar Games, 1997–2013) series. This section is organized according to game genres, and for each game genre it discusses what the particular cognitive, perceptual, behavioral and kinesthetic challenges games of that genre generally pose, and then gives an overview of how AI methods have been used to play that particular game genre. It also includes several extended examples, giving some detail about particular implementations. Once again it is important to note that the list is not inclusive of all possible game genres AI can play as a player or non-player character; the selection is made on the basis of popularity of game genres and the available published work on AI for playing games in each genre.

3.4.1 Board Games

As discussed in the introduction of this book, the earliest work on AI for playing games was done in classic board games, and for a long time that was the only way in which AI was applied to playing games. In particular, Chess was so commonly used for AI research that it was called the "drosophila of artificial intelligence" [194], alluding to the use of the common fruit fly as a model organism in genetics research. The reasons for this seem to have been that board games were simple to implement, indeed possible at all to implement on the limited computer hardware available in the early days of AI research, and that these games were seen to require something akin to "pure thought". What a game such as Chess or Go does require is **adversarial planning**. Classic board games typically place no demand at all on perception, reactions, motor skills or estimation of continuous movements, meaning that their skill demands are particularly narrow, especially compared to most video games.

Most board games have very simple discrete state representations and deterministic forward models—the full state of the game can often be represented in less than 100 bytes, and calculating the next state of the game is as simple as applying a small set of rules—and reasonably small branching factors. This makes it very easy to apply **tree search**, and almost all successful board game-playing agents use some kind of tree search algorithm. As discussed in the sections on tree search in Chapter 2, the Minimax algorithm was originally invented in the context of playing Chess. Decades of research concentrated on playing Chess (with a lesser amount of research on Checkers and Go), with specific conferences dedicated to this kind of research, led to a number of algorithmic advances that improved the performance of the Minimax algorithm on some particular board game. Many of these have limited applicability outside of the particular game they were developed on, and it would take us too far to go into these algorithmic variations here. For an overview of advances in Chess playing, the reader is referred to [98].

In Checkers the reigning human champion was beaten by the Chinook software in 1994 [594] and the game was *solved* in 2007, meaning that the optimal set of moves for both players was found (it is a draw if you play optimally) [593]; in Chess, Garry Kasparov was famously beaten by Deep Blue in 1997 [98]. It took until 2016 for Google DeepMind to beat a human Go champion with their AlphaGo software [629], mainly because of the algorithmic advances necessary. Whereas Chess and Checkers can be played effectively with some variation of the Minimax algorithm combined with relatively shallow state evaluations, the larger branching factor of Go necessitated and spurred the development of MCTS [77].

While MCTS can be utilized to play board games without a state evaluation function, supplementing that algorithm with state and action evaluation functions can massively enhance the performance, as seen in the case of AlphaGo, which uses deep neural networks for state and action evaluation. On the other hand, when using some version of Minimax it is necessary to use state evaluation functions as all interesting board games (more complex than Tic-Tac-Toe) have too large state spaces to be searched until the end of the game in acceptable time. These evaluation functions can be manually constructed, but in general it is a very good idea to use some form of learning algorithm to learn their parameters (even though the structure of the function is specified by the algorithm designer). As discussed above, Samuel was the first to use a form of reinforcement learning to learn a state evaluation function in a board game (or any kind of game) [591], and Tesauro later used TD learning to very good effect in Backgammon [689]. Evolutionary computation can also be used to learn evaluation functions, for example, Pollack showed that co-evolution could perform well on Backgammon using a very similar setup to Tesauro [537]. Noteworthy examples of strong board game players based on evolved evaluation functions are *Blondie24* [207] and *Blondie25* [208], a Checkers- and a Chess-playing program respectively. The evaluation functions were based on five-layered deep convolutional networks, and *Blondie25* in particular performed well against very strong Chess players.

While classic board games such as Go and Chess have existed for hundreds or even thousands of years, the past few decades have seen a rejuvenation of board game design. Many of the more recently designed board games mix up the formula of classic board games with design thinking from other game genres. A good example is *Ticket to Ride* (Days of Wonder, 2004), which is a board game that includes elements of card games, such as variable numbers of players, hidden information and stochasticity (in the draw of the cards). For these reasons, it is hard to construct well-performing AI players based on standard tree-search methods; the best known agents include substantial domain knowledge yet perform poorly compared to human players [160]. Creating generic well-performing agents for this type of game is an interesting research challenge.

Given the simplicity of using tree search for board game playing, it is not surprising that every approach we have discussed so far builds on one tree search algorithm or another. However, it is possible to play board games without forward models—usually with results that are "interesting" rather than good. For example, Stanley and Miikkulainen developed a "roving eye" approach to playing Go, where an evolved neural network self-directedly scans the Go board and decides where to place the next piece [656]. Relatedly, it is reportedly possible for the position evaluation network of AlphaGo to play a high-quality game of Go on its own, though it naturally plays stronger if combined with search.

3.4.2 Card Games

Card games are games centered on one or several decks of cards; these might or might not be the standard 52-card French deck which is commonly used in classic card games. Most card games involve players possessing different cards that change ownership between players, or between players and the deck, or other positions on the table. Another important element of most card games is that some cards are visible to the player who possesses them but not to other players. Therefore, almost all card games feature a large degree of **hidden information**. In fact, card games are perhaps the type of games where hidden information most dominates gameplay.

For example, take the classic card game *Poker*, which is currently very popular in its *Texas hold 'em* variety. The rules are relatively simple: the player which at the end of a few rounds holds the best cards (the "best hand") wins. Between the rounds, the player can exchange a number of cards for fresh cards drawn from the deck. If there were perfect information, i.e., all players could see each others' hands, this would be an uninteresting game that could be played according to a lookup table. What makes Texas hold 'em—and similar Poker variants—challenging and interesting is that each player does not know what cards the other players have. The cognitive challenges of playing these games involve acting in the absence of information, which implies inferring the true game state from incomplete evidence, and potentially affecting other players' perception of the true game state. In other words, a game of Poker is largely about guessing and bluffing.

A key advance in playing Poker and similar games is the Counterfactual Regret Minimization (CFR) algorithm [797]. In CFR, algorithms learn by self-play in a similar fashion to how temporal difference learning and other reinforcement learning algorithms have been used in perfect information games like Backgammon and Checkers. The basic principle is that after every action, when some hidden state has been revealed, it computes the alternative reward of all other actions that could have been taken, given the newly revealed information. The difference between the reward attained from the action that was actually taken and the best action that could have been taken is called the *regret*. The policy is then adjusted so as to minimize the regret. This is done iteratively, slowly converging on a policy that is optimal in the sense that it loses as little as possible over a large number of games. However, for games as complex as Texas hold 'em, simplifications have to be done in order to use the CFR algorithm in practice.

DeepStack is a recent agent and algorithm that has reached world-class performances in Texas hold 'em [467]. Like CFR, DeepStack uses self-play and recursive reasoning to learn a policy. However, it does not compute an explicit strategy before play. Instead, it uses tree search in combination with a state value approximation to select actions at each turn. In this sense, it is more like the heuristic search of AlphaGo (but in a setting with plenty of imperfect information) than like the reinforcement-learned policy of TD-gammon.

Another, much more recent, card game which is drawing increasing interest from the research community is *Hearthstone* (Blizzard Entertainment, 2014); see Fig. 3.4. This is a collectible card game in the tradition of *Magic: The Gathering*, but with somewhat simpler rules and only played on computers. A game of Hearthstone takes place between two players, with each player having a deck of 30 cards. Each card represents either a creature or a spell. Each player has a handful of cards (< 7) in hand (invisible for the other player), and at each turn draws a new card and has the option of playing one or more cards. Creature cards convert to creatures that are placed on the player's side of the table (visible for both players), and creatures can be used to attack the opponent's creatures or player character. Spells have a multiplicity of different effects. The hundreds of different cards in the game, the possibility of choosing to take multiple actions each turn, the long time taken to play a game (20 to 30 turns is common), the presence of stochasticity and of course the hidden information (mainly what cards are in the opponent's hand) conspire to make *Hearthstone* (Blizzard Entertainment, 2014) a hard game to play for both humans and machines.

Perhaps the simplest approach to playing *Hearthstone* (Blizzard Entertainment, 2014) is to simply ignore the hidden information and play each turn in a greedy fashion, i.e., search the space of possible actions within a single turn and choose the one that optimizes some criterion such as health point advantage at the end of that turn, given the available information only. Agents that implement such greedy policies are included with some open source Hearthstone simulators, such as *Metastone*.[3] Standard tree search algorithms such as Minimax or MCTS are generally ineffective here (as in Poker) because of the very high degree of hidden information. One approach to constructing high-performing agents is instead to hand-code

[3] http://www.demilich.net/

Fig. 3.4 A screenshot from *Hearthstone* (Blizzard Entertainment, 2014) displaying a number of different creature or spell cards available in the game. Image obtained from Wikipedia (fair use).

domain knowledge, for example, by building an ontology of cards and searching in an abstract symbolic space [659].

Unlike in Poker, where the player has no control over what cards it is dealt, in *Hearthstone* (Blizzard Entertainment, 2014) the player can also construct a deck with which to play the game. This adds another level of challenge to playing the game: in addition to choosing what action(s) to take at each turn, the successful player must also construct what allows her to implement her strategy. The composition of the deck effectively constrains what strategy can be chosen, and then implemented tactically through action selection. While these two levels interplay—a strong player takes the composition of the deck and the strategy it affords into account when choosing a move, and vice versa—it is also true that the problems of deck building and action selection can to some extent be treated separately, and implemented in different agents. One approach to deck building is to use evolutionary computation. The deck is seen as the genome, and the fitness function involves simple heuristic agents using the deck for playing [218]. A similar approach has also been used in the multi-player card game *Dominion* [416].

3.4.3 Classic Arcade Games

Classic arcade games, of the type found in late 1970s and early 1980s arcade cabinets, home video game consoles and home computers, have been commonly used as AI benchmarks within the last decade. Representative platforms for this game type

(a) *Track & Field* (Konami, 1983) is a game about athletics. The game screenshot depicts the start of the 100 m dash. Image obtained from Wikipedia (fair use).

(b) In *Tapper* (Bally Midway, 1983) the player controls a bartender who serves drinks to customers. Image obtained from Wikipedia (fair use).

Fig. 3.5 *Track & Field* (Konami, 1983), *Tapper* (Bally Midway, 1983), and most classic arcade games require rapid reactions and precision.

are the Atari 2600, Nintendo NES, Commodore 64 and ZX Spectrum. Most classic arcade games are characterized by movement in a two-dimensional space (sometimes represented isometrically to provide the illusion of three-dimensional movement), heavy use of graphical logics (where game rules are triggered by intersection of sprites or images), continuous-time progression, and either continuous-space or discrete-space movement.

The cognitive challenges of playing such games vary by game. Most games require fast reactions and precise timing, and a few games, in particular early sports games such as *Track & Field* (Konami, 1983) and *Decathlon* (Activision, 1983), rely almost exclusively on speed and reactions (Fig. 3.5(a)). Very many games require prioritization of several co-occurring events, which requires some ability to predict the behavior or trajectory of other entities in the game. This challenge is explicit in e.g., *Tapper* (Bally Midway, 1983)—see Fig. 3.5(b)—but also in different ways part of platform games such as *Super Mario Bros* (Nintendo, 1985), shooting galleries such as *Duck Hunt* (Nintendo, 1984) or *Missile Command* (Atari Inc., 1980) and scrolling shooters such as *Defender* (Williams Electronics–Taito, 1981) or *R-type* (Irem, 1987). Another common requirement is navigating mazes or other complex environments, as exemplified most clearly by games such as *Pac-Man* (Namco, 1980), *Ms Pac-Man* (Namco, 1982), *Frogger* (Sega, 1981) and *Boulder Dash* (First Star Software, 1984), but also common in many platform games. Some games, such as *Montezuma's Revenge* (Parker Brothers, 1984), require long-term planning involving the memorization of temporarily unobservable game states. Some games feature incomplete information and stochasticity, others are completely deterministic and fully observable.

3.4.3.1 Pac-Man and Ms Pac-Man

Various versions and clones of the classic *Pac-Man* (Namco, 1981) game have been frequently used in both research and teaching of artificial intelligence, due to the depth of challenge coupled with conceptual simplicity and ease of implementation. In all versions of the game, the player character moves through a maze while avoiding pursuing ghosts. A level is won when all pills distributed throughout the level are collected. Special power pills temporarily give the player character the power to consume ghosts rather than being consumed by them. As seen in Chapter 2, the differences between the original *Pac-Man* (Namco, 1981) and its successor *Ms Pac-Man* (Namco, 1982) may seem minor but are actually fundamental; the most important is that one of the ghosts in *Ms Pac-Man* (Namco, 1982) has non-deterministic behavior, making it impossible to learn a fixed sequence of actions as a solution to the game. The appeal of this game to the research community is evidenced by a recent survey covering over 20 years of active AI research using these two games as testbeds [573].

Several frameworks exist for Pac-Man-based experimentation, some tied to competitions. The Pac-Man screen capture competition is based around the Microsoft Revenge of Arcade version of the original game, and does not provide a forward model nor facilities for speeding up the game [404]. The Ms Pac-Man vs Ghost Team competition framework is written in Java and includes both a forward model and ability to speed up the game significantly; it also includes an interface for controlling the ghost team rather than Ms Pac-Man, the player character [574]. The Atari 2600 version of *Ms Pac-Man* (Namco, 1982) is available as part of the ALE framework.[4] There is also a Python-based Pac-Man framework used for teaching AI at UC Berkeley.[5]

As expected, the performance of AI players varies depending on the availability of a forward model, which allows the simulation of ghost behavior. The screen capture-based competition, which does not offer a forward model, is dominated by heuristic approaches (some of them involve pathfinding in the maze without taking ghost movement into account), which perform at the level of beginner human players [404]. It has been observed that even searching one step ahead, and using a state evaluator based on an evolved neural network, can be an effective method for playing the game [403]. Of course, searching deeper than a single ply yields additional benefits; however, the stochasticity introduced in *Ms Pac-Man* (Namco, 1982) poses challenges even in the presence of a forward model. MCTS has been shown to work well in this case [590, 524]. Model-free approaches to reinforcement learning have also been used for playing the game with some success [57]. In general, the best competitors in the Ms Pac-Man vs Ghost Team Competition play at the level of intermediate-skill human players [574]. At the moment of writing this book *Ms Pac-Man* (Namco, 1982) is reported to be practically solved (reaching the maximum possible score of 999,990 points) by the Microsoft Maluuba team. The

[4] www.arcadelearningenvironment.org

[5] http://ai.berkeley.edu/project_overview.html

team used an RL technique called hybrid reward architecture [738] which decomposes the reward function of the environment into different RL problems (a set of reward functions) that a corresponding number of agents need to solve. Each agent selects its actions by considering the aggregated Q values for each action across all agents.

Pac-Man (Namco, 1980) can also be played for experience rather than performance. In a series of experiments, neural networks that controlled the ghosts in a clone of the game were evolved to make the game more entertaining for human players [766]. The experiment was conceptually based on Malone's definition of fun in games as challenge, curiosity and fantasy dimensions [419] and sought to find ghost behavior that maximized these traits. In particular, the fitness was composed of three factors: 1) the appropriate level of challenge (i.e., when the game is neither too hard nor too easy), 2) the diversity of ghost behavior, and 3) the ghosts' spatial diversity (i.e., when ghosts behavior is explorative rather than static). The fitness function used to evolve interesting ghost behaviors was cross-validated via user studies [770].

3.4.3.2 Super Mario Bros

Versions and clones of Nintendo's landmark platformer *Super Mario Bros* (Nintendo, 1985) have been extremely popular for AI research, including research on game playing, content generation and player modeling (research using this game is described in several other parts of this book). A large reason for this is the *Mario AI Competition*, which was started in 2009 and included several different tracks focused on playing for performance, playing in a human-like manner and generating levels [322, 717]. The software framework for that competition[6] was based on *Infinite Mario Bros* (Notch, 2008), a Java-based clone of *Super Mario Bros* (Nintendo, 1985) featuring simple level generation [706, 705]. Different versions of the competition software, generally referred to as the Mario AI Framework or Mario AI Benchmark, have since been used in many dozens of research projects. In the following, we will for simplicity refer to methods for playing various versions of *Super Mario Bros* (Nintendo, 1985), *Infinite Mario Bros* or the Mario AI Framework/Benchmark simply as playing "Mario".

The first version of the Mario AI could be simulated thousands of times faster than real-time, but did not include a forward model. Therefore the first attempts to learn a Mario-playing agent was through learning a function from a state observation directly to Mario actions [706]. In that project, neural networks were evolved to guide Mario through simple procedurally generated levels. The inputs were the presence or absence of environment features or enemies in a coarse grid centered on Mario, and the outputs were interpreted as the button presses on the Nintendo controller (up, down, left, right). See Fig. 3.6 for an illustration of the state representation. A standard feedforward MLP architecture was used for the neural network,

[6] http://julian.togelius.com/mariocompetition2009/

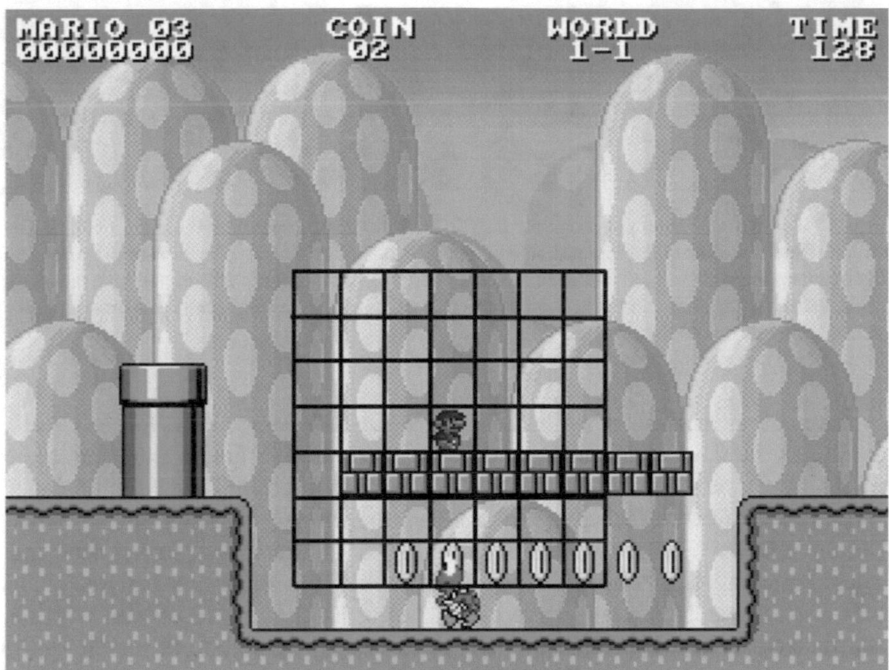

Fig. 3.6 The inputs to the Mario-playing neural network are structured as a Moore neighborhood centered on Mario. Each input is 1 if the corresponding tile is occupied by a tile (such as ground) that Mario cannot pass through, and 0 otherwise. In another version of the experiment, a second set of inputs was added where an input was 1 if there was an enemy at the corresponding tile. Image adapted from [706].

and the fitness function was simply how far the controller was able to progress on each level. Using this setup and a standard evolution strategy, neural networks were evolved that could win some levels but not all, and generally played at the strength of a human beginner.

However, as is so often the case, having a forward model makes a big difference. The first Mario AI Competition, in 2009, was won by Robin Baumgarten, who constructed a forward model for the game by reusing some of the open-source game engine code [705]. Using this model, he constructed an agent based on A* search in state space. At each time frame, the agent searches for the shortest path towards the right edge of the screen, and executes the first action in the resulting plan. As the search utilizes the forward model and therefore takes place in state space rather than just physical space, it can incorporate the predicted movements of the (deterministic) enemies in its planning. This agent was able to finish all the levels used in the 2009 Mario AI Competition, and produces behavior that appears optimal in terms of time to complete levels (it does not focus on collecting coins or killing enemies). See Section 2.3.2 for an explanation of the algorithm and a figure illustrating its use in Mario.

Fig. 3.7 An example level generated for the 2010 Mario AI Competition. Note the overhanging structure in the middle of the screenshot, creating a dead end for Mario; if he chooses to go beneath the overhanging platform, he will need to backtrack to the start of the platform and take the upper route instead after discovering the wall at the end of the structure. Agents based on simple A* search are unable to do this.

Given the success of the A*-based agent in the 2009 competition, the next year's edition of the competition updated the level generator so that it generated more challenging levels. Importantly, the new level generator created levels that included "dead ends", structures where Mario can take the wrong path and if so must backtrack to take the other path [322]. See Fig. 3.7 for an example of such a dead end. These structures effectively "trap" agents that rely on simple best-first search, as they end up searching a very large number of paths close to the current position, and time out before finding a path that backtracks all the way to beginning of the structure. The winner of the 2010 Mario AI Competition was instead the REALM agent [56]. This agent uses an evolved rule-based system to decide sub-goals within the current segment of the level, and then navigates to these sub-goals using A*. REALM successfully handles the dead ends that were part of the levels in the 2010 competition, and is as far as we know the highest-performing Mario-playing agent there is.

Other search algorithms beyond A* have been tried for playing Mario, including Monte Carlo tree search [294]. It was found that the standard formulation of MCTS did not perform very well, because the algorithm did not search deep enough and because the way the average reward of a branch is calculated results in risk-averse behavior. However, with certain modifications to remedy these problems, an MCTS variant was found that could play Mario as well as a pure A* algorithm. In a follow-up experiment, noise was added to the Mario AI Benchmark and it was found that MCTS handled this added noise much better than A*, probably because MCTS relies on statistical averaging of the reward whereas A* assumes a deterministic world.

All of the above work has been focused on playing for performance. Work on playing Mario for experience has mostly focused on imitating human playing styles, or otherwise creating agents that play Mario similarly to a human. To further this research, a Turing test track of the Mario AI Competition was created [619]. In this track, competitors submitted agents, and their performance on various levels was recorded. Videos of the agents playing different levels where shown to human spectators along with videos of other humans playing the same levels, and the spectators were asked to indicate which of the videos were of a human player. Agents were scored based on how often they managed to fool humans, similarly to the setup of the original Turing test. The results indicated simple heuristic solutions that included hand-coded routines for such things as sometimes standing still (giving the impression of "thinking about the next action") or occasionally misjudging a jump can be very effective in giving the appearance of human playing style. Another way of providing human-like behavior is to explicitly **mimic** humans by learning from play traces. Ortega et al. describe a method for creating Mario-playing agents in the style of particular human players [511]: evolve neural networks where the fitness function is based on whether the agent would perform the same action as the human, when faced with the same situation. This was shown to generate more human-like behavior than evolving the same neural network architecture with a more straightforward fitness function. In a similar effort to create human-like Mario AI players Munoz et al. [469] used both play traces and information about the player's eyes position on the screen (obtained via gaze tracking) as inputs of an ANN, which was trained to approximate which keyboard action is to be performed at each game step. Their results yield a high prediction accuracy of player actions and show promise towards the development of more human-like Mario controllers based on information beyond gameplay data.

3.4.3.3 The ALE Framework

The Arcade Learning Environment (ALE) is an environment for general game-playing research based on an emulation of the classic video game console *Atari 2600* [40]. (While the environment can technically accommodate other emulators as well, the Atari 2600 emulator is the one that has been used in practice, to the point that the ALE framework is sometimes simply referred to as "Atari".) The Atari 2600 is a console from 1976 with 128 bytes of RAM, maximum 32 kilobytes of ROM per game and no screen buffer, posing severe limitations on the type of games that could be implemented on the system [466]. ALE provides an interface for agents to control games via the standard joystick input, but does not provide any processed version of the internal state; instead, it provides the 160×210 pixel screen output to the agent, which will need to parse this visual information somehow. There is a forward model, but it is relatively slow and generally not used.

Some of the early work using ALE used neuroevolution; in particular a study compared several neuroevolution algorithms on 61 Atari games [251]. They found that they could use the popular neuroevolution algorithm NEAT to evolve decent-quality players for individual games, provided that these algorithms were given positions of in-game objects as recognized by a computer vision algorithm. The HyperNEAT algorithm, an indirect encoding neuroevolution algorithm that can create arbitrarily large networks, was able to learn agents that could play based on the raw pixel inputs, even surpassing human performance in three of the tested games. In that paper, the neuroevolution approaches generally performed much better than the classic reinforcement learning methods tried.

Later, ALE was used in Google DeepMind's research on deep Q learning, which was reported in a *Nature* paper in 2015 [464]. As detailed in Chapter 2, the study showed that by training a deep neural network (five layers, where the first two are convolutional) with Q-learning augmented with experience replay, human-level playing performance could be reached in 29 out of 49 tested Atari games. That research spurred a flurry of experiments in trying to improve the core deep reinforcement formula presented in that paper.

It is worth noting that in almost all of the ALE work focuses on learning neural networks (or occasionally other agent representations) for individual games. That is, the network architecture and input representation is the same across all games, but the parameters (network weights) are learned for a single game and can only play that game. This seems to be at odds with the idea of general game playing, i.e., that you could learn agents that play not a single game, but any game you give them. It could also be noted that ALE itself is better suited for research into playing individual games than for research on general game playing, as there is only a limited number of Atari 2600 games, and it is highly non-trivial to create new games for this platform. This makes it possible to tune architectures and even agents to individual games.

3.4.3.4 General Video Game AI

The General Video Game AI (GVGAI) competition is a game-based AI competition that has been running since 2014 [528]. It was designed partly as a response to a trend seen in many of the existing game-based AI competitions, e.g., those organized at the CIG and AIIDE conferences, that submissions were getting increasingly game-specific by incorporating more and more domain knowledge. A central idea of GVGAI is therefore that submissions to the competition are tested on *unseen* games, i.e., games that have not been released to the competitors before and which are therefore impossible to tailor the submissions to. At the time of writing, the GVGAI repository includes around 100 games, with ten more games being added for every competition event.

To simplify the development of games in GVGAI, a language called the Video Game Description Language (VGDL) was developed [181, 597]. This language allows for concise specification of games in a Python-like syntax; a typical game description is 20-30 lines, with levels specified in separate files. Given the underlying assumptions of 2D movement and graphical logics, most of the games in GVGAI corpus are remakes of (or inspired by) classic arcade games such as *Frogger* (Sega, 1981) (see Fig. 3.8(b)), *Boulder Dash* (First Star Software, 1984) or *Space Invaders* (Taito, 1978) (see Fig. 3.8(a)), but some are versions of modern indie games such as *A Good Snow Man Is Hard to Build* (Hazelden and Davis, 2015).

The original track of the GVGAI competition, for which the most results are available, is the single-player planning track. Here, agents are given a fast forward model of the game, and 40 milliseconds to use it to plan for the next action. Given these conditions, it stands to reason that planning algorithms of various kinds would rule the day. Most of the top performers in this track have been based on variations on MCTS or MCTS-like algorithms, such as the Open Loop Expectimax

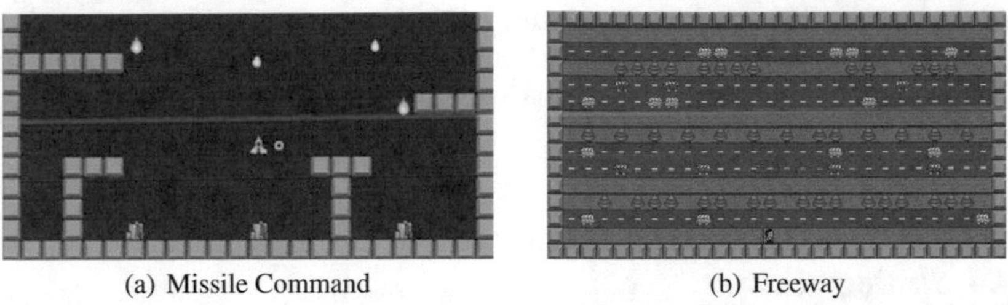

(a) Missile Command (b) Freeway

Fig. 3.8 Two example games that have been used in the GVGAI competition.

Tree Search algorithm [528] or MCTS with options search [161]. One surprisingly high-performing agent uses Iterative Width search, where the core idea is to build a propositional database of all the facts and then use this to prune a breadth-first search algorithm to only explore those branches which provide a specified amount of novelty, as measured by the size of the smallest set of facts seen for the first time [389]. Agents based on evolutionary planning also perform well, but not as well as those based on stochastic tree search or tree search with novelty pruning.

While some agents are better than others overall, there is clear non-transitivity in the rankings, in the sense that the best algorithms for one particular game may not be the best for another game—in fact, there seem to be patterns where families of algorithms perform better on families of games [213, 59]. Given these patterns, a natural idea for achieving higher playing performance is to use **hyper-heuristics** or **algorithm selection** to select at runtime which algorithm to use for which game, an approach which has seen some success so far [453].

Two other GVGAI tracks are related to gameplay, namely the two-player planning track and the learning track. The two-player planning track resembles the single-player planning track but features a number of two-player games, some of which are cooperative and some competitive. At the time of writing, the best agents in this track are slightly modified versions of single-player track agents that make naive assumptions about the behavior of the other player [216]; it is expected that agents with more sophisticated player models will eventually do better. The learning track, by contrast, features single-player games and provides players with time to learn a policy but does *not* provide them with a forward model. It is expected that algorithms such as deep reinforcement learning and neuroevolution will do well here, but there are as yet no results; it is conceivable that algorithms that learn a forward model and then perform tree search will dominate.

3.4.3.5 Other Environments

In addition to ALE and GVGAI, there are several other environments that can be used for AI experimentation with arcade-style games. The *Retro Learning Environment* is a learning environment similar in concept to ALE, but instead based on an emulation of the Super Nintendo console [45]. A more general, and less focused, system is *OpenAI Universe*,[7] which acts as a unified interface to a large number of different games, ranging from simple arcade games to complex modern adventure games.

3.4.4 Strategy Games

Strategy games, particularly **computer strategy games**, are games where the player controls multiple characters or units, and the objective of the game is to prevail in some sort of conquest or conflict. Usually, but not always, the narrative and graphics reflect a military conflict, where units may be e.g., knights, tanks or battleships. The perhaps most important distinction within strategy games is between **turn-based** and **real-time** strategy games, where the former leave plenty of time for the player to decide which actions to take each time, and the latter impose a time pressure. Well-known turn-based strategy games include epic strategy games such as the *Civilization* (MicroProse, 1991) and the *XCOM* (MicroProse, 1994) series, as well as shorter games such as *Hero Academy* (Robot Entertainment, 2012). Prominent real-time strategy games include *StarCraft* I (Blizzard Entertainment, 1998) and II (Blizzard Entertainment, 2010), the *Age of Empires* (Microsoft Studios, 1997–2016) series and the *Command and Conquer* (Electronic Arts, 1995–2013) series. Another distinction is between single-player games that focus on exploration, such as the *Civilization* games, and multi-player competitive games such as *StarCraft* (Blizzard Entertainment, 1998–2015). Most, but not all, strategy games feature hidden information.

The cognitive challenge in strategy games is to lay and execute complex plans involving multiple units. This challenge is in general significantly harder than the planning challenge in classical board games such as Chess mainly because multiple units must be moved at every turn; the number of units a player controls can easily exceed the limits of short-term memory. The planning horizon can be extremely long, where for example in *Civilization V* (2K Games, 2010) decisions you make regarding the building of individual cities will affect gameplay for several hundred turns. The order in which units are moved can significantly affect the outcome of a move, in particular because a single action might reveal new information during a move, leading to a prioritization challenge. In addition, there is the challenge of predicting the moves of one or several adversaries, who frequently have multiple units. For real-time strategy games, there are additional perceptual and motoric challenges related to the speed of the game. This cognitive complexity is mirrored in the computational complexity for agents playing these games—as discussed in Section 3.2.1.4, the branching factor for a strategy game can easily reach millions or more.

[7] https://universe.openai.com/

The massive search spaces and large branching factors of strategy games pose serious problems for most search algorithms, as just searching a single turn forward might already be infeasible. One way of handling this is to decompose the problem, so unit acts on its own; this creates one search problem for each unit, with the branching factor equivalent to the branching factor of the individual unit. This has the advantage of being tractable and the disadvantage of preventing coordinated action among units. Nevertheless, heuristic approaches where units are treated separately are used in the built-in AI in many strategy games. Not coincidentally the built-in AI of many strategy games is generally considered inadequate.

In research on playing strategy games, some solutions to this involve cleverly sub-sampling the space of turns so that standard search algorithms can be used. An example of such an approach is an MCTS variant based on decomposition through Naive Sampling [503]. Another approach is non-linear Monte Carlo, which was applied to *Civilization II* (MicroProse, 1996) with very promising results [65]. The basic idea here is to sample the space of turns (where each turn consists of actions for all units) randomly, and get an estimate for the value of each turn by performing a rollout (take random actions) until a specified point. Based on these estimates, a neural network was trained to predict the value of turns; regression can then be used to search for the turn with the highest predicted value.

But planning does not need to be based on tree search. Justesen et al. applied online evolutionary planning to *Hero Academy* (Robot Entertainment, 2012), a two-player competitive strategy game with perfect information and a relatively low number of moves per turn (five in the standard setting) [309]. Each chromosome consisted of the actions to take during a single turn, with the fitness function being the material difference at the end of the turn. It was found that this approach vastly outperformed MCTS (and other tree search algorithms) despite the shallow search depth of a single turn, likely because the branching factor made it impossible for the tree search algorithms to sufficiently explore even the space of this single turn.

Evolution has also been used to create agents that play strategy games. For example, NEAT-based macro-management controllers were trained for the strategy game *Globulation 2* (2009). In that study, however, the NEAT controller does not aim to win but rather play for the experience; in particular, it is evolved to take macro-actions (e.g., build planning, battle planning) in order to provide a balanced game for all players [499]. AI approaches based on artificial evolution that also have been used as playtesting mechanisms for a number of other strategy games [588, 297].

3.4.4.1 StarCraft

The original *StarCraft*, released in 1998 by Blizzard Entertainment, is still widely played competitively, a testament to its strong game design and in particular its unusual depth of challenge (see Fig. 3.10). It is usually played with the *Brood War* expansion, and referred to as *SC:BW*. The existence of *Brood War API (BWAPI)*,[8] an

[8] https://github.com/bwapi/bwapi

interface for playing the game with artificial agents, has enabled a thriving community of researchers and hobbyists working an AI for *StarCraft* (Blizzard Entertainment, 1998). Several competitions are held annually based on *SC:BW* and BWAPI,[9] including competitions at the IEEE CIG and AIIDE conferences [87]. *TorchCraft* is an environment built on top of SC:BW and BWAPI to facilitate machine learning, especially deep learning, research using *StarCraft* [681]. In parallel, a similar API was recently released for interfacing AI agents with the follow-up game *Starcraft II* (Blizzard Entertainment, 2010), which is mechanically and conceptually very similar but with numerous technical differences. Given the existing API and competitions, almost all existing research has been done using *SC:BW*.

As *SC:BW* is such a complex game and the challenge of playing it well is so immense, most research focuses on only part of the problem, most commonly through playing it at some level of abstraction. It is common to divide the different levels of decision-making in *SC:BW* (and similar real-time strategy games) into three levels, depending on the time scale: Strategy, Tactics and Micro-Management (see Fig. 3.9). So far, no agents have been developed that can play a complete game at the level of even an intermediate human player; however, there has been significant progress on several sub-problems of the very formidable problem of playing this game.

For a fuller overview of research on AI for playing *SC:BW*, the reader is referred to a recent survey [504]. Below, we will exemplify some of the research done in this space, with no pretense of complete coverage.

On the **micro** level AI plays out over the timescale of usually less than a minute, where time between taking actions is typically on the order of a second. When focusing on this most low-level form of *SC:BW* battle, one need not consider base building, research, fog of war, exploration and many other aspects of the full *SC:BW* game. Usually, two factions face off, each with a set of a few or a few dozen units. The goal is to destroy the opponent's units. This game mode can be played in the actual game, which does not allow for significant speed-up and does not provide a forward model, or in the simulator SparCraft [123], which does provide a forward model. (There is a also a Java version of SparCraft, called *JarCraft* [311].).

In the model-free scenario, agents must be based on either hand-crafted policies or policies learned through reinforcement learning or supervised learning, without the luxury of a forward model. Hand-crafted policies can be implemented e.g., based on potential fields, where different units are attracted to or repelled by other units in order to create effective combat patterns [242], or on fuzzy logic [541]. When it comes to machine learning methods for model-free scenarios, standard [622] and deep reinforcement learning [729] have been used with some effect to learn policies. Typically, the problem is decomposed so that a single Q-function is learned that is then applied to each unit separately [729].

Using the SparCraft simulator, we can do more because of the availability of a forward model. Churchill and Buro developed a simple approach to dealing with the excessive branching factor called Portfolio Greedy Search [123]. The core idea is

[9] http://www.starcraftai.com/

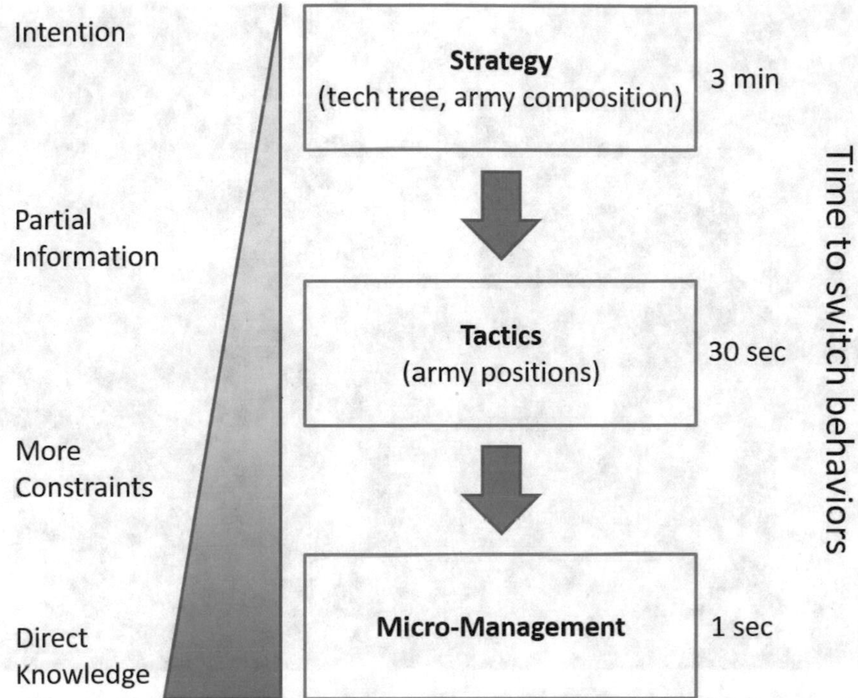

Fig. 3.9 Three different levels of decision making in *StarCraft* (Blizzard Entertainment, 1998). The width of the triangle represents amount of information whereas its colored gradient illustrates the degree of partial observability. For instance, at the highest strategic level both the level of observability and available information to the player are relatively low. On the other end of the triangle, at the lowest level of micro-management the player must consider the type, position, and other dynamic properties of each of the units she controls; that information is mostly observable. For reference, a full game of *StarCraft* (Blizzard Entertainment, 1998) often takes around 20 minutes, though there is considerable variation. The image is reproduced with permission from Gabriel Synnaeve.

that instead of selecting between actions for each unit, a small number (a portfolio) of simple heuristics called scripts are used, and a greedy search algorithm is used to assign one of these scripts to each unit. This approach prunes the branching drastically, but limits the space of discoverable policies to those that can be described as combinations of the scripts. Subsequently, it was shown that the portfolio of scripts idea can be combined with MCTS with good results [311]. Even better results can be obtained by doing portfolio selection through evolutionary planning [745]; it is likely that these ideas generalize to many strategy games and other games with high branching factors due to their controlling of many units simultaneously.

Moving to the other end of the micro-management-tactics-strategy continuum, large-scale **strategy** adaptation remains a very hard problem. Existing *SC:BW* bots are rarely able to implement multiple strategies, let alone adapt their strategy based on how the game progresses. In order to do this, it is necessary to create a model of

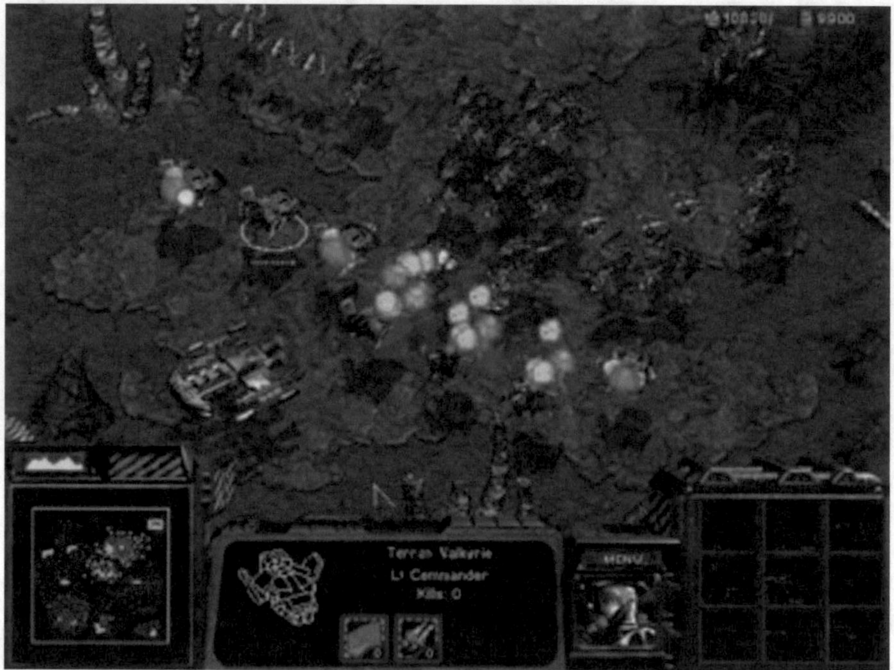

Fig. 3.10 A screenshot from *StarCraft: Brood War* (Blizzard Entertainment, 1998) displaying a number of different units available in the game. At the time of writing, playing well this real-time strategy game is considered one of the next grand challenges for AI research. Image obtained from Wikipedia (fair use).

what the opponent is trying to do based on limited evidence. Here, pioneering work by Weber and Mateas focused on mining logs of *SC:BW* matches to predict what strategy will be taken by a player from early-game actions [750].

A few more ambitious attempts have been made to create complete agents that can handle strategy, tactics and micro-management in a principled fashion. For example, Synnaeve and Bessière built an agent based on Bayesian programming that is able to perform reasonably well [680].

3.4.5 Racing Games

Racing games are games where the player is tasked with controlling some kind of vehicle or character so as to reach a goal in the shortest possible time, or as to traverse as far as possible along a track in a given time. Usually the game employs a first-person perspective, or a vantage point from just behind the player-controlled vehicle. The vast majority of racing games take a continuous input signal as a steering input, similar to a steering wheel. Some games, such as those in the *Forza Mo-*

Fig. 3.11 A screenshot from TORCS which has been predominately used in the Simulated Car Racing Championship. Image obtained from https://sourceforge.net/projects/torcs/ (fair use).

torsport (Microsoft Studios, 2005–2016) or *Real Racing* (Firemint and EA Games, 2009–2013) series, allow for complex input including gear stick, clutch and handbrake, whereas more arcade-focused games such as those in the *Need for Speed* (Electronic Arts, 1994–2015) series typically have a simpler set of inputs and thus lower branching factor. Racing games such as those in the *WipeOut* (Sony Computer Entertainment Europe, 1995–2012) and *Mario Kart* (Nintendo, 1992–2017) series introduce additional elements, such as weapons that can be used to temporarily incapacitate competitors' vehicles.

While the cognitive challenges in playing racing games may appear simple, most racing games actually require multiple simultaneous tasks to be executed and have significant skill depth. At the most basic level, the agent needs to control for the position of the vehicle and adjust the acceleration or braking, using fine-tuned continuous input, so as to traverse the track as fast as possible. Doing this optimally requires at least short-term planning, one or two turns (of the track) forward. If there are resources to be managed in the game, such as fuel, damage or speed boosts, this requires longer-term planning. When other vehicles are present on the track, there is an adversarial planning aspect added, in trying to manage or block overtaking; this planning is often done in the presence of hidden information (position and resources of other vehicles on different parts of the track) and under considerable time pressure, and benefits from models of the adversarial drivers.

One relatively early commercial game application that stands out is the AI for *Forza Motorsport* (Microsoft Studios, 2005), which was marketed under the name *Drivatar* [259]. The Drivatar agents are built on a form of supervised lazy learning. To train the agents, humans drive a number of racing tracks, which are composed of a number of segments; all tracks in the game need to be composed of segments drawn from the same "alphabet". During driving, the agent selects the driving commands that most closely approximate the racing line taken by the players on the relevant segment. This approach was successful in realizing personalized driving agents, i.e., agents that could drive new tracks in the style of human players they had been trained on, but posed restrictions on the design of the tracks.

There are various approaches to training agents to drive without supervision, through reinforcement learning. A sequence of papers has shown how neuroevolution can be used to train agents that drive a single track in the absence of other cars as well as a good human driver [707], how incremental evolution can be used to train agents with sufficiently general driving skills to drive unseen tracks [709], and how competitive co-evolution can be used to adversarially train agents to drive more or less aggressively in the presence of other cars [708]. In all these experiments, the weights of a relatively small fixed-topology network were trained with an evolution strategy. The inputs to the network were the speed of the car and a handful of rangefinder sensors that returned the distance to the edges of the track, or other cars. The low dimensionality of the resulting network enabled high-performing networks to be found relatively easily.

The *Simulated Car Racing Championship*, which has run annually since 2007, is partly based on this work, and uses a similar sensor model. The first year, the competition was based on simple 2D racing game, and the winner of the competition was a controller based on fuzzy logic [710]. In 2008, the competition software was rebuilt around TORCS, a 3D racing game with a reasonably sophisticated physics model [393] (see Fig. 3.11). In the following years, a large number of competitors submitted agents based on various different architectures to the competition, including evolutionary computation, temporal difference learning, supervised learning and simple hand-coded rule-based systems [393]. A general trend has been observed over the course of the competition that the winning agents incorporate more and more domain knowledge in the form of hand-coded mechanisms, with learning algorithms generally only used for tuning parameters of these mechanisms. The best agents, such as *COBOSTAR* [90] or *Mr. Racer* [543] generally drive as well as or better than a good human driver when driving alone on a track, but still struggle with overtaking and other forms of adversarial driving.

As discussed above, the Simulated Car Racing Championship provides information in a form that is relatively easy to map to driving commands, making the learning of at least basic driving strategy (but not fine-tuning) relatively easy. However, some authors have attempted learning to drive from raw pixel data. Early work on this topic includes that by Floreano et al., who evolved a neural network with a movable "retina" to drive in a simple simulated environment. The output of the neural network included both driving commands and commands for how to move the retina, and only the relatively few pixels in the retina were used as inputs to the network [206]. Later, Koutnik et al. managed to evolve controllers that used higher-dimensional input by evolving the networks in compressed weight space; essentially, the parameters of a JPEG encoding of the network connections was evolved, allowing evolutionary search to work effectively in the space of large neural networks [353]. Supervised learning of deep networks has also been applied to visual driving, yielding high-performing TORCS drivers that learn from examples [795].

The examples above do not make use of a forward model of any kind. However, car dynamics are relatively simple to model, and it is easy to create a fast approximate model for racing games. Given such a model, standard tree search algorithms can easily be applied to car control. For example, Fischer et al. showed that MCTS coupled with a simple forward model can produce decent performance in TORCS [203].

3.4.6 Shooters and Other First-Person Games

First-person shooters constitute an important genre of video games ever since the success of *DOOM* (GT Interactive, 1993) and *Wolfenstein 3D* (Apogee Software and FormGen, 1992) in the early 1990s. While a basic tenet of an FPS would seem to be that the world is observed through a first-person point of view, there are games that are generally recognized as FPSes, such as the *Gears of War* (Microsoft Studios, 2006–2016) series, which have the camera positioned slightly behind and/or above the player. Similarly, the word "shooter" signifies that the games revolve around shooting projectiles with some kind of weapon. On that basis, a game such as *Portal* (Electronic Arts, 2007) can be seen as an FPS though it is debatable whether the player implement is actually a weapon.

Shooters are often seen as fast-paced games where speed of perception and reaction is crucial, and this is true to an extent, although the speed of gameplay varies between different shooters. Obviously, quick reactions are in general not a problem for a computer program, meaning that an AI player has a certain advantage over a human by default. But there are other cognitive challenges as well, including orientation and movement in a complex three-dimensional environment, predicting actions and locations of multiple adversaries, and in some game modes also team-based collaboration. If visual inputs are used, there is the added challenge of extracting relevant information from pixels.

There has been some early work on optimizing parameters for existing agents in order to improve their efficiency [127], but extensive work on AI for FPS games was spurred by two competitions: first, the **2K BotPrize** and more recently **VizDoom**.

3.4.6.1 Unreal Tournament 2004 and the 2K BotPrize

Unreal Tournament 2004 (UT2k4) (Epic Games, 2004) is an FPS which was released in 2004, with what was at the time state-of-the-art graphics and gameplay. While the game itself has not been open sourced, a team based at the Charles University in Prague created *Pogamut*, a Java-based API that allows for simple control of the game [222]. Pogamut supplies the agent with an object-based information interface, which the agent can query about the locations of objects and characters, and

(a) The judges' room

(b) The players' room

Fig. 3.12 The first 2K BotPrize competition held in Perth, Australia, on 17 December 2008, as part of the 2008 IEEE Symposium on Computational Intelligence and Games.

also provides convenience functions for executing actions such as firing a projectile towards a specific point.

Some work using UT2k4 tries to achieve high-performing agents for one or several in-game tasks, using techniques such as neuroevolution. For example, van Hoorn et al. subdivided the task of playing UT2k4 into three sub-tasks: shooting, exploring and path-following [734]. Using an earlier approach [698], which combines neuroevolution with the subsumption architecture of Rodney Brooks [70], they then evolved neural networks for each of these tasks in succession. The resulting agent was able to play some game scenarios relatively effectively.

However, the main use of the UT2k4 benchmark has been in the **2K BotPrize** (see Fig. 3.12). This competition, which ran from 2008 to 2014, stands out among game-based AI competitions for not focusing on playing for performance, but rather for playing for experience. Specifically, it was a form of Turing test, where submitted agents were judged not by how well they survived firefights with other agents, but by whether they could fool human judges (who in later configurations of the competition also participated in the game) that they were humans [262, 263, 264].

The winners of the final 2K BotPrize in 2014 were two teams whose bots managed to convince more than half of the human judges that they (the bots) were human. The first winning team, UT^2, from the University of Texas at Austin, is primarily based on neuroevolution through multiobjective evolution [603]. It consists of a number of separate controllers, where most of these are based on neural networks; at each frame, it cycles through all of these controllers, and uses a set of priorities to decide the outputs of which controller will command various aspects of the agent. In addition to neural networks, some controllers are built on different principles, in particular the Human Retrace Controller, which uses traces of human players to help the agent navigate out of stuck positions. The second winner, *MirrorBot* by Mihai Polceanu, is built around the idea of observing other players in the game and mirroring their behavior [535].

3.4.6.2 Raw Screen Inputs and the Visual Doom AI Challenge

The **VizDoom** framework [333] is build around a version of the classic *DOOM* (GT Interactive, 1993) FPS game that allows researchers to develop AI bots that play the game using only the screen buffer. VizDoom was developed as an AI testbed by a team of researchers at the Institute of Computing Science, Poznan University of Technology (see Fig. 3.13). The framework includes several tasks of varying complexity, from health pack collection and maze navigation to all-out deathmatch. An annual competition based on VizDoom is held at the IEEE CIG conference since 2016, and the framework is also included in *OpenAI Gym*,[10] a collection of games which can be used for AI research.

Most of the published work on VizDoom has been based on deep reinforcement learning with convolutional neural networks, given that method's proven strength in learning to act based on raw pixel inputs. For example, *Arnold*, a well-performing agent in the first VizDoom competition, is based on deep reinforcement learning of two different networks, one for exploration and one for fighting [115].

But it is also possible to use evolutionary computation to train neural network controllers. As the very large input size requires huge networks, which do not work well with evolutionary optimization in general, it is necessary to compress the information somehow. This can be done by using an autoencoder trained on the visual stream as the game is played; the activations of the bottleneck layer of the autoencoder can then be used as inputs to a neural network that decides about the actions, and the weights of the neural network can be evolved [12]. Previous attempts at evolving controllers acting on visual input in the related game *Quake* (GT Interactive, 1996) have met with only limited success [519].

3.4.7 Serious Games

The genre of **serious games**, or **games with a purpose** beyond entertainment, has become a focus domain of recent studies in game AI. One could argue that most existing games are serious by nature as they incorporate some form of learning for the player during play. Games such as *Minecraft* (Mojang, 2011), for example, were not designed with a particular learning objective in mind; nevertheless they have been used broadly in classrooms for science education. Further, one could argue that serious games do not have a particular genre of their own; games may have a purpose regardless of the genre they were designed on. Strictly speaking the design of serious games involves a particular set of learning objectives. Learning objectives may be educational objectives such as those considered in STEM education—a popular example of such a game is the *Dragonbox* (WeWantToKnow, 2011) series that teaches primary school students equation solving skills, and basic addition and subtraction skills. (A serious academic effort on game-based STEM

---

[10] https://gym.openai.com/

Fig. 3.13 A screenshot from the VizDoom framework which—at the moment of writing—is used in the Visual Doom AI Challenge. The framework is giving access to the depth of the level (enabling 3D vision). Image obtained from http://vizdoom.cs.put.edu.pl/ with permission.

education is the narrative-centered *Crystal Island* game series for effective science learning [577, 584].) The learning objective can, instead, be the training of social skills such as conflict resolution and social inclusion through games; *Village Voices* [336] (see Fig. 3.14(a)), *My Dream Theater* [100] (see Fig. 3.14(b)), and *Prom Week* [447] are examples of such soft skill training games. Alternatively, the aim could be that war veterans suffering from post-traumatic stress disorder are trained to cope with their cognitive-behavioral manifestations when faced with in-game stressors in games such as *StartleMart* [272, 270] and *Virtual Iraq* [227]. The learning objective could also be that of soliciting collective intelligence for scientists. A number of scientific games have recently led to the discovery of new knowledge via crowd-playing (or else human computation); arguably one of the most popular scientific discovery games is *Foldit* [138] through which players collectively discovered a novel algorithm for protein folding.

The cognitive and emotional skills required to play a serious game largely depend on the game and the underlying learning objectives. A game about math would normally require computation and problem solving skills. A game about stress inoculation and exposure therapy would instead require cognitive-behavioral coping mechanisms, metacognition and self-control of cognitive appraisal. The breadth of

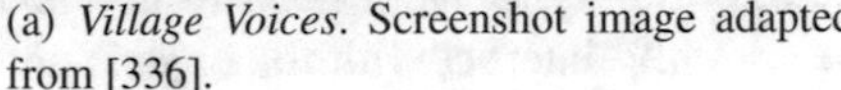

(a) *Village Voices*. Screenshot image adapted from [336].

(b) *My Dream Theater*. Screenshot image adapted from [99].

Fig. 3.14 The *Village Voices* and the *My Dream Theater* games for conflict resolution. *Village Voices* realizes experiential learning of conflict in social, multi-player settings, whereas *My Dream Theater* offers a single-player conflict management experience. In the first game the AI takes the role of modeling conflict and generating appropriate quests for the players, whereas—more relevant to the aims of this chapter—in *My Dream Theater* AI takes the role of controlling expressive agent (NPC) behaviors.

cognitive and emotional skills required from players is as wide as the number of different learning objectives a serious game can integrate into its design.

Many serious games have NPCs and AI can help in making those NPCs believable, human-like, social and expressive. AI in serious games is generally useful for modeling NPC behavior and playing the game as an NPC but not for winning; rather for the experience of play. Whether the game is for education, health or simulation purposes NPC agents need to act believably and emotively in order to empower learning or boost the engagement level of the game. Years of active research have been dedicated on this task within the fields of affective computing and virtual agents. The usual approach followed is the construction of top-down (ad-hoc designed) agent architectures that represent various cognitive, social, emotive and behavioral abilities. The focus has traditionally being on both the modeling of the agents behavior but also on its appropriate expression under particular contexts. A popular way of constructing a computational model of agent behavior is to base it on a theoretical cognitive model such as the OCC model [512, 183, 16, 189, 237], which attempts to effect human-like decision making, appraisal and coping mechanisms dependent on a set of perceived stimuli. For the interested reader, Marsella et al. [428] cover the most popular computational models of emotion for agents in a thorough manner.

Expressive and believable conversational agents such as *Greta* [534] and *Rea* [105] or virtual humans [674] that embody affective manifestations can be considered in the design of serious games. The use of such character models has been dominant in the domains of intelligent tutoring systems [131], embodied conversational agents [104, 16], and affective agents [238] for educational and health purposes. Notable examples of such agent architecture systems include the work of Lester's group in the *Crystal Island* game series [578, 577, 584], the expressive agents of *Prom Week* [447] and *Façade* [441], the agents of the *World of Minds* [190] game, and the FAtiMA [168] enabled agents of *My Dream Theater* [100].

3.4.8 Interactive Fiction

While there are several variations, games within the **interactive fiction** genre normally contain a fantasy world consisting of smaller areas such as rooms; however, a simulated environment is not a necessity. Importantly, players need to use text commands to play the game. The player can normally interact with the objects and available game characters, collect objects and store them in her inventory, and solve various puzzles. Games of this genre are also named *text-based adventure* games or often associated with text-based *role-playing games*. Popular examples include the games of the *Zork* series (Infocom, 1979–1982) and *Façade* [441].

In this game genre, AI can play the role of understanding text as coming from players in a natural language format. In other words, the game AI can feature **natural language processing** (NLP) for playing the game as the player's companion or as an opponent. Further NLP can be used as an input for the generation of a dialog, a text or a story in an interactive fashion with the player. It is normally the case that text-based input is used to drive a story (interactive narrative) which is often communicated via embodied conversational agents and is represented through the lens of a virtual camera. Similarly to traditional cinematography, both the camera position and the communicated narrative contribute to the experience of the viewer. Opposed to traditional cinematography, however (but similarly to interactive drama), the story in games can be influenced by the player herself. It goes without saying that research in text-based games is naturally interwoven with research in believable conversational agents (as covered in the previous section), computational and **interactive narrative** [693, 441, 562, 792] and **virtual cinematography** [252, 193, 300, 84, 15, 578]. The discussion on the interplay between interactive narrative and virtual cinematography is expanded in Chapter 4 and in particular in the section dedicated to narrative generation.

Work on text-based AI in games starts from the early language-based interaction with Eliza [751] and the Z-Machine used by text adventure games such as *Zork I* (Infocom, 1980), to *Façade* [438, 441] and to recent word2vec [459] approaches (e.g., its TensorFlow implementation [2]) for playing Q & A games [340] and text-based adventure games [352]. It is important to note that beyond the use of AI to understand natural language we can use AI to play text-based games. A notable recent example of an agent that manages to handle both tasks is the one developed by Kostka et al. [352], named *Golovin*. The *Golovin* agent uses related corpora such as fantasy books to create language models (via word2vec [459]) that are appropriate to this game domain. To play the game the agent uses five types of command generators: battle mode, gathering items, inventory commands, general actions and movement. *Golovin* is validated on 50 interactive fiction games demonstrating comparable performance to the current state of the art. Another example is the agent of Narasimhan et al. [475] that plays *multi-user dungeon* games, a form of multiplayer or collaborative interactive fiction. Their agent converts text representations to state representations using Long Short-Term Memory (LSTM) networks. These representations feed a deep Q network which learns approximate evaluations for each action in a given game state [475]. Their approach significantly outperforms other baselines in terms of number of completed quests in small-, and even medium-sized, games. For the interested reader a dedicated annual competition on text-based adventure game AI was initiated in conjunction with the IEEE CIG conference in 2016.[11] Participants of the competition submit agents that play games for the Z-Machine.

Examples of useful development tools for text-based games include the *Inform* series of design systems and programming languages, which are inspired by the Z-machine and have led to the development of several text-based games and interactive fiction based on natural language. Notably, *Inform 7*[12] was used for the design of the *Mystery House Possessed* (Emily Short, 2005) game.

3.4.9 Other Games

The list of games AI can play is not limited to the genres covered above. While the genres we covered in more detail are, in our opinion, the most representative there are AI studies focusing on other game genres, a number of which we outline below.

A game type with an increasing interest is **casual games** due to their growing popularity and accessibility via mobile devices in recent years. Casual games are often simple and are designed as short episodes of play (levels) to allow flexibility with respect to gameplay time. This feature gives the player the ability to conclude an episode in a short period of time without needing to save the game. As a result, the player can engage on a single level for a few seconds, or play a number of levels throughout the day, or instead repeatedly play new levels up to hours of gameplay. The game skills required to play casual games depend on the genre of the casual game which can vary from puzzle such as *Bejeweled* (PopCap Games, 2001), *Angry Birds* (Chillingo, 2009), and *Cut the Rope* (Chillingo, 2010), to adventure such as *Dream Chronicles* (KatGames, 2007), to strategy such as *Diner Dash* (PlayFirst, 2004), to arcade such as *Plants vs. Zombies* (PopCap Games, 2009) and *Feeding Frenzy* (PopCap Games, 2004), to card and board games such as *Slingo Quest* (funkitron, Inc., 2006).

Notable academic efforts on casual games include the work of Isaksen et al. [288, 289] where an AI agent is built to test the difficulty of *Flappy Bird* (dotGEARS, 2013) levels. The baseline AI player follows a simple pathfinding algorithm that performs well in completing the *Flappy Bird* levels. To imitate human

[11] http://atkrye.github.io/IEEE-CIG-Text-Adventurer-Competition/

[12] http://inform7.com/

play, however, the AI player features elements of human motor skills such as precision, reaction time, and actions per second. Another example on that line of work is the use of AI agents to test the generation of game levels in a variant of *Cut the Rope* (Chillingo, 2010). The AI agents featured in the *Ropossum* authoring tool both perform automatic playtesting and also optimize the playability of a level using first-order logic [614]. The level generation elements of *Ropossum* are further discussed in the next chapter. Another casual game that has recently attracted the interest of game AI research is *Angry Birds* (Chillingo, 2009). The game has an established AI competition [560], named the *Angry Birds AI Competition*,[13] that runs since 2012 mainly in conjunction with the International Joint Conference on Artificial Intelligence. AI approaches in *Angry Birds* (Chillingo, 2009) have so far focused mainly on planning and reasoning techniques. Examples include a qualitative spatial reasoning approach which evaluates level structural properties and game rules, and infers which of these are satisfied for each building block of the level [796]. The usefulness of each level building block (i.e., how good it is to hit it) is then computed based on these requirements. Other approaches model discrete knowledge about the current game state of *Angry Birds* (Chillingo, 2009) and then attempt to satisfy the constraints of the modeled world [92] based on extensions of answer set programming [69].

Beyond casual games the genre of **fighting games** has received a considerable amount of interest from both academic and industrial players. Fighting games require cognitive skills mostly related to kinesthetic control and spatial navigation but also related to reaction times and decision making, both of which need to be fast [349]. Popular approaches for fighting games include classic reinforcement learning—in particular, the SARSA algorithm for on-policy learning of Q values which are represented by linear and ANN function approximators—as applied by the Microsoft Research team in the *Tao Feng: Fist of the Lotus* (Microsoft Game Studios, 2003) game [235]. Reinforcement learning has also been applied with varying degrees of success for adaptive difficulty adjustment in fighting games [158, 561, 27]; i.e., AI that plays for the experience of the player. Evolutionary reinforcement learning variants have also been investigated for the task [437]. A notable effort in the fighting games AI research scene has been on the Java-based fighting game *FightingICE* provided by the *Fighting Game AI Competition*[14] [402] (see Fig. 3.15). The competition is organized by the Ritsumeikan University in Japan and has run since 2013 with the aim to derive the best possible fighting bot; i.e., AI that plays to win. Approaches using *FightingICE* vary from dynamic scripting [417], to k-nearest neighbor [760], to Monte Carlo tree search [790], to neuroevolution [357], among others. So far MCTS-based approaches appear to be advantageous on winning in fighting games.

The last game we will cover in this section is *Minecraft* (Mojang, 2011) for its unique properties as a testbed for game AI research. *Minecraft* (Mojang, 2011) is a **sandbox** game played in a 3D procedurally generated world that players can navi-

[13] https://aibirds.org/

[14] http://www.ice.ci.ritsumei.ac.jp/~ftgaic/

Fig. 3.15 A screenshot from the Java-based *FightingICE* framework for fighting games.

gate through. The game features game mechanics that enable players to build constructions out of cubes (see Fig. 3.16) but it does not have a specific goal for the player to accomplish. Beyond exploration and building, players can also gather resources, craft objects and combat opponents. The game has sold more than 121 million copies across all platforms [51], making it the second best-selling video game of all time, only behind *Tetris* (Alexey Pajitnov and Vladimir Pokhilko, 1984).[15] The 3D open-world nature of *Minecraft* (Mojang, 2011) and the lack of specific goals provides players ultimate freedom to play and explore the world in dissimilar ways. The benefits of playing *Minecraft* (Mojang, 2011) appear to be many and some of them have already been reported by the educational research community [479]. For example, the blocks available in the game can be arranged to produce any object a player might think of, thereby fostering the player's creativity [479] and diagrammatic lateral thinking [774]. Further, the blocks' functionalities that can be combined and extended may result in new knowledge for the players that is acquired gradually. Moreover, the simple stylized voxel-based graphics of the game allow the player to concentrate on the gameplaying and exploration tasks within a simple, yet aesthetically pleasing, environment. Overall, the multitude of reasons that make *Minecraft* (Mojang, 2011) so appealing to millions of players are also the reasons that make the game a great testbed for AI and game AI research. In particular, the game offers an AI player an open world ready to be explored with numerous possibilities for open-ended play. Further, in-game tasks for an AI agent vary from

[15] https://en.wikipedia.org/wiki/List_of_best-selling_video_games

Fig. 3.16 A screenshot from *Minecraft* (Mojang, 2011) showcasing a city hall structure made by the MCFRArchitect Build Team. Image obtained from Wikipedia (fair use).

exploration and seeking treasure to making objects and building structures, alone or as a team of agents.

A notable and recent effort on the use of *Minecraft* (Mojang, 2011) for AI is *Project Malmo* [305] which is supported by Microsoft Research. Project Malmo is a Java-based AI experimentation platform built as a game mod of the original game and designed to support research within the areas of robotics, computer vision, machine learning, planning, and multi-agent systems and general game AI [305]. Note that the open-source platform is accessible via GitHub.[16] Early experiments with AI in *Project Malmo* include the application of deep neural networks for navigating the 3D mazes [465] and for fighting the opponents of the game [726]. Beyond *Project Malmo*, it is also worth noting that various other mods of the game have been used directly for teaching robotics [11]—including algorithms for maze navigation and planning—and teaching general AI methods [38].

3.5 Further Reading

The methods used for playing games have an extensive body of literature that is covered in detail both in Chapter 2 and here. The different game genres that we covered in this chapter contain a corresponding literature the interested reader could use as a starting point for further exploration.

[16] https://github.com/Microsoft/malmo#getting-started

3.6 Exercises

Prior to moving to the next chapter about generating content herein we provide a general exercise that allows you to apply any of the algorithms covered in Chapter 2 in a single domain. The purpose of the exercise is for you to get familiar with those methods before testing them in more complicated and complex domains and tasks.

As discussed above, the *Ms Pac-man vs Ghost Team* competition is a contest held at several AI conferences around the world in which AI controllers for Ms Pac-Man and the ghost team compete for the highest ranking. For this exercise, you will have to **develop a number of Ms Pac-Man AI players** to compete against the ghost team controllers included in the software package. This is a simulator entirely written in Java with a well-documented interface. While the selection of two to three different agents within a semester period has been shown to be a good educational practice we will leave the final number of Ms Pac-Man agents to be developed on you or your class instructor.

The website of the book contains code for the game and a number of different sample code classes in Java to get you started. All AI methods covered in Chapter 2 are applicable to the task of controlling Ms Pac-Man. You might find out, however, that some of them are more relevant and efficient than others. So which methods would work best? How do they compare in terms of performance? Which state representation should you use? Which utility function is the most appropriate? Do you think you can make the right implementation decisions so that your Ms Pac-Man plays at a professional level? How about at a world champion, or even at a superhuman level?

3.6.1 Why Ms Pac-Man?

Some of our readers might object to this choice of game and think that there should be more interesting games AI can play with. While *Ms Pac-Man* (Namco, 1982) is very old it is arguably a classic game that is still fun and challenging to play, as well as a simple testbed to start experimenting with the various AI methods and approaches introduced in this chapter and the previous one. *Ms Pac-Man* (Namco, 1982) is simple to understand and play but is not easy to master. This controversial element of play simplicity combined with problem complexity makes *Ms Pac-Man* (Namco, 1982) the ideal testbed for trying out different AI methodologies for controlling the main character, Ms Pac-Man. Another exciting feature of this game is its non-deterministic nature. Randomness not only augments the fun factor of the game but it also increases the challenge for any AI approach considered. As discussed, another argument for the selection of *Ms Pac-Man* (Namco, 1982) is that the game and its variants have been very well studied in the literature, tested through several game AI competitions, but also covered for years on AI (or game AI) courses in several universities across the globe. The reader is also referred to a recent survey paper by Rohlfshagen et al. [573], covering over 20 years of research in Pac-Man and a YouTube video[17] about the importance of Pac-Man in game AI research.

3.7 Summary

In this chapter we discussed the different roles AI can take and the different characteristics games and AI methods have, the various methods that are available for playing games and the different games it can play. In particular, AI can either play to win or play in order to create a particular experience for a human player or observer. The former goal involves maximizing a utility that maps to the game performance whereas the latter goal involves objectives beyond merely winning such as engagement, believability, balance and interestingness. AI as an actor can take the role of either a player character or a non-player character that exists in a game. The **characteristics** of games an AI method needs to consider when playing include the number of players, the level of stochasticity of the game, the amount of observability available, the action space and the branching factor, and the time granularity. Further, when we design an algorithm to play a game we also need to consider algorithmic aspects such as the state representation, the existence of a forward model, the training time available, and the number of games AI can play.

The above roles and characteristics were detailed in the first part of the chapter as they are important and relevant regardless of the AI method applied. When it comes to the **methods** covered in this chapter, we focused on tree-search, reinforcement learning, supervised learning, and hybrid approaches for playing games. In a sense, we tailored the methods outlined in Chapter 2 to the task of gameplaying. The chapter concluded with a detailed review of the studies and the relevant methodologies on a *game genre* basis. Specifically, we saw how AI can play board, card, arcade, strategy, racing, shooter, and serious games, interactive fiction, and a number of other game genres such as casual and fighting games.

[17] https://www.youtube.com/watch?t=49&v=w5kFmdkrIuY

Chapter 4
Generating Content

Procedural content generation (PCG) [616] is an area of game AI that has seen an explosive growth of interest. While games that incorporate some procedurally generated content have existed since the early 1980s—in particular, the dungeon crawler *Rogue* (Toy and Wichmann, 1980) and the space trading simulator *Elite* (Acornsoft, 1984) are early trailblazers—research interest in academia has really picked up within the second half of the last decade.

Simply put, PCG refers to methods for generating game content either autonomously or with only limited human input. Game **content** is that which is contained in a game: levels, maps, game rules, textures, stories, items, quests, music, weapons, vehicles, characters, etc. Typically, NPC behavior and the game engine itself are not thought of as content. Probably the most common current usage of PCG is for generating levels and terrains, but in the future we might see widespread generation of all kinds of content, possibly even complete games.

Seeing PCG from an AI perspective, content generation problems are AI problems where the solutions are content artifacts (e.g., levels) that fulfill certain constraints (e.g., being playable, having two exits) and/or maximize some metrics (e.g., length, difference in outcomes between different strategies). And as we will see, many AI methods discussed in Chapter 2 can be used for PCG purposes, including evolutionary algorithms and neural networks. But there are also a number of methods that are commonly used for PCG that are not typically thought of as AI; some of these will be presented in this chapter.

This chapter is devoted to methods for generating game content, as well as paradigms for how to incorporate them into games. We start with discussing why you want to use PCG at all—just like for playing games, there are some very different motivations for generating content. Then in Section 4.2 we present a general taxonomy for PCG methods and possible roles in games. Next, Section 4.3 summarizes the most important methods for the generation of content. Shifting attention to which roles PCG can take in games Section 4.4 discusses ways of involving designers and players in the generative process. Section 4.5 takes the perspective of content types, and presents examples of generating some common and uncommon types of game content. Finally, Section 4.6 discusses how to evaluate PCG methods.

G. N. Yannakakis and J. Togelius, *Artificial Intelligence and Games*, https://doi.org/10.1007/978-3-319-63519-4_4

4.1 Why Generate Content?

Perhaps the most obvious reason to generate content is that it could remove the need to have a human designer or artist creating that content. Humans are expensive and slow, and it seems we need more and more of them all the time. Ever since computer games were invented, the number of person-months that go into the development of a successful commercial game has increased more or less constantly.[1] It is now common for a game to be developed by hundreds of people over a period of several years. This leads to a situation where fewer games are profitable, and fewer developers can afford to develop a game, leading in turn to less risk-taking and less diversity in the games marketplace. Many of the costly employees necessary in this process are designers and artists rather than programmers. A game development company that could replace some of the artists and designers with algorithms would have a competitive advantage, as games could be produced **faster** and **cheaper** while preserving quality. (This argument was made forcefully by legendary game designer Will Wright in his talk "The Future of Content" at the 2005 Game Developers Conference, a talk which helped reinvigorate interest in procedural content generation.) Figure 4.1 illustrates a cost breakdown of an average AAA game and showcases the dominance of artwork and marketing in that process. Art, programming, and debugging constitute around 50% of the cost of an AAA game. Essentially, PCG can assist in the processes of art and content production, thus directly contributing to the reduction of around 40% of a game's cost.

Of course, threatening to put them out of their jobs is not a great way to sell PCG to designers and artists. It is also true that at the current stage of technology, we are far from being able to replace all that a designer or artist can do. We could therefore turn the argument around: content generation, especially embedded in intelligent design tools, can augment the **creativity** of individual human creators. Humans, even those of the "creative" vein, tend to imitate each other and themselves. Algorithmic approaches might come up with radically different content than a human would create, through offering an unexpected but valid solution to a given content generation problem. Some evidence for this is already available in the literature [774]. This could make it possible for small teams without the resources of large companies, and even for hobbyists, to create content-rich games by freeing them from worrying about details and drudge work while retaining overall directorship of the games. PCG can then provide a way for *democratizing* game design by offering reliable and accessible ways for everyone to make better games in less time.

Both of these arguments assume that what we want to make is something like the games we have today. But PCG methods could also enable completely **new types of games**. To begin with, if we have software that can generate game content at the speed it is being "consumed" (played), there is in principle no reason why games need to end. For everyone who has ever been disappointed by their favorite game

[1] At least, this is true for "AAA" games, which are boxed games sold at full price worldwide. The recent rise of mobile games seems to have made single-person development feasible again, though average development costs are rising on that front too.

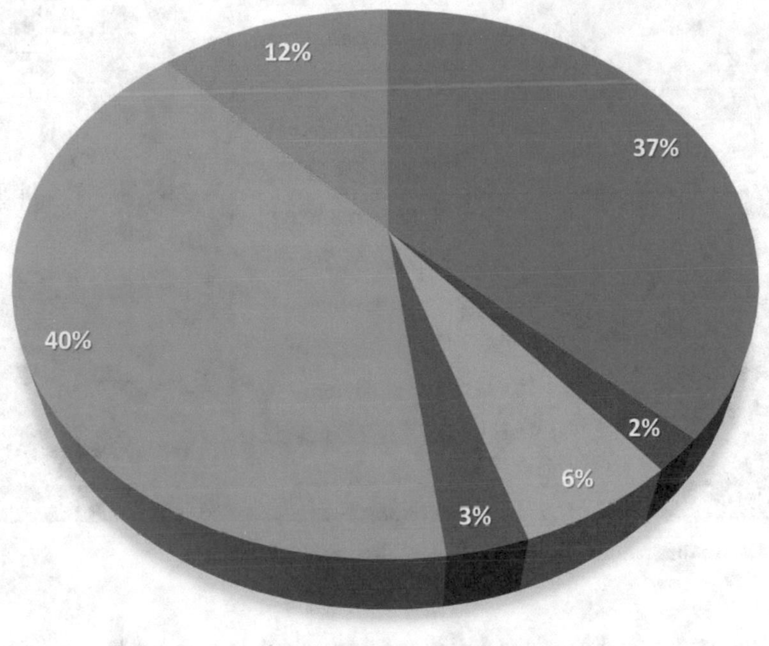

Fig. 4.1 Average cost breakdown estimate of AAA game development. Data source: http://monstervine.com/2013/06/chasing-an-industry-the-economics-of-videogames-turned-hollywood/.

not having any more levels to clear, characters to meet, areas to explore, etc., this is an exciting prospect.

Even more excitingly, the newly generated content can be tailored to the tastes and needs of the player playing the game. By combining PCG with player modeling, for example through measuring and using neural networks to model the response of players to individual game elements, we can create **player-adaptive games** that seek to maximize the enjoyment of players (see the roles of PCG in Section 4.4). The same techniques could be used to maximize the learning effects of a serious game, or perhaps the addictiveness of a "casual" game.

Finally, a completely different but no less important reason for developing PCG methods is to **understand design and creativity**. Computer scientists are fond of saying that you do not really understand a process until you have implemented it in code (and the program runs). Creating software that can competently generate game content could help us understand the process by which we "manually" generate the content, and clarify the affordances and constraints of the design problem we are addressing. This is an iterative process, whereby better PCG methods can lead

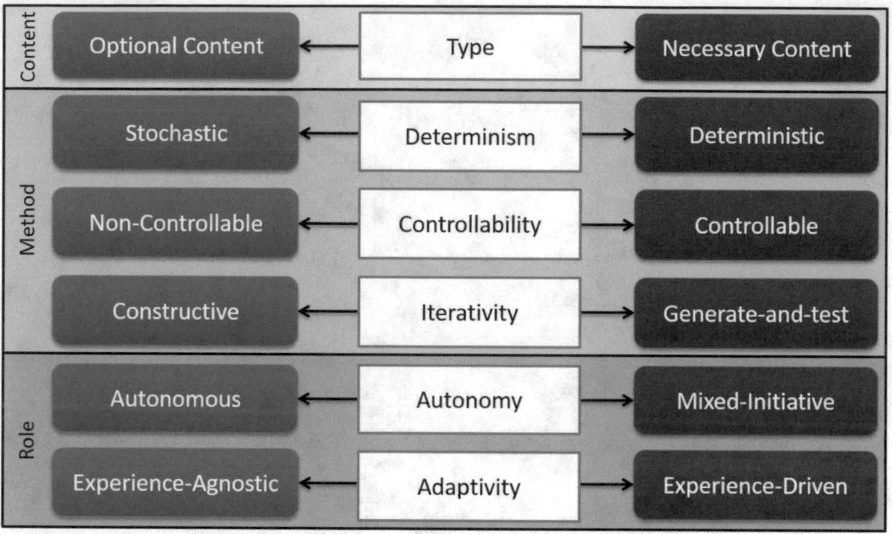

Fig. 4.2 An illustration of the PCG taxonomy discussed in Section 4.2.

to better understanding of the design process, which in turn can lead to better PCG algorithms enabling a co-creativity process between designers and algorithms [774].

4.2 Taxonomy

With the variety of content generation problems, methods and approaches that are available, it helps to have a structure that can highlight the differences and similarities between them. In the following, we introduce a revised version of the taxonomy of PCG that was originally presented by Togelius et al. [720]. It consists of a number of dimensions, where an individual method or solution should usually be thought of as lying somewhere on a continuum between the ends of that dimension. Beyond the taxonomy for the **content** types (Section 4.2.1) and the properties of the PCG **methods**, which we cover in Section 4.3, we provide an outline of the **roles** a PCG algorithm can take, which we cover extensively in Section 4.4.

4.2.1 Taxonomy for Content

We identify the type of the generated outcome as the sole dimension that relates to content in this taxonomy (see also Fig. 4.2).

4.2.1.1 Type: Necessary Versus Optional

PCG can be used to generate **necessary** game content that is required for the completion of a level, or it can be used to generate **optional** content that can be discarded or exchanged for other content. The main distinguishing feature between necessary and optional content is that necessary content should always be correct while this condition does not hold for optional content. Necessary needs to be consumed or passed as the player makes their way through the game, whereas optional content can be avoided or bypassed. An example of optional content is the generation of different types of weapons in first-person shooter games [240] or the auxiliary reward items in *Super Mario Bros* (Nintendo, 1985). Necessary content can be the main structure of the levels in *Super Mario Bros* (Nintendo, 1985), or the collection of certain items required to pass to the next level.

4.2.2 Taxonomy for Methods

PCG algorithms can be classified according to a number of properties such as their level of controllability, determinism and so on. This section outlines the three dimensions across which PCG methods can be classified. Figure 4.2 offers an illustration of the taxonomy that we discuss in this section.

4.2.2.1 Determinism: Stochastic Versus Deterministic

Our first distinction with regards to PCG methods concerns the amount of randomness in content generation. The right amount of variation in outcomes between different runs of an algorithm with identical parameters is a design decision. **Stochasticity**[2] allows an algorithm (such as an evolutionary algorithm) to offer great variation, necessary for many PCG tasks, at the cost of controllability. While content diversity and expressivity are desired properties of generators, the effect of randomness on the final outcomes can only be observed and controlled after the fact. Completely **deterministic** PCG algorithms, on the other hand, can be seen as a form of data compression. A good example of deterministic PCG is the first-person shooter *.kkrieger* (.theprodukkt 2004), which manages to compress all of its textures, objects, music and levels together with its game engine in just 96 KB of storage space.

[2] Strictly speaking there is a distinction between a stochastic and a **non-deterministic** process in that the former has a defined random distribution whereas the latter does not. For the purposes of this book, however, we use the two terms interchangeably.

4.2.2.2 Controllability: Controllable Versus Non-controllable

The generation of content by PCG can be controlled in different ways. The use of a random seed is one way to gain control over the generation space; another way is to use a set of parameters that control the content generation along a number of dimensions. Random seeds were used when generating the world in *Minecraft* (Mojang, 2011), which means the same world can be regenerated if the same seed is used [755]. A vector of content features was used in [617] to generate levels for *Infinite Mario Bros* (Persson, 2008) that satisfy a set of feature specifications.

4.2.2.3 Iterativity: Constructive Versus Generate-and-Test

A final distinction may be made between algorithms that can be called **constructive** and those that can be described as **generate-and-test**. A constructive algorithm generates the content once, and is done with it; however, it needs to make sure that the content is correct or at least "good enough" as it is being constructed. An example of this approach is using fractals or cellular automata to generate terrains or grammars to generate levels (also refer to the corresponding PCG method sections below). A generate-and-test algorithm, instead, incorporates both a generate and a test mechanism. After a candidate content instance is generated, it is tested according to some criteria (e.g., is there a path between the entrance and exit of the dungeon, or does the tree have proportions within a certain range?). If the test fails, all or some of the candidate content is discarded and regenerated, and this process continues until the content is good enough. A popular PCG framework that builds upon the generate-and-test paradigm is the search-based [720] approach discussed in Section 4.3.

4.2.3 Taxonomy of Roles

In this section we identify and briefly outline the four possible roles a PCG algorithm can take in the game design process classified across the dimensions of autonomy and player-based adaptivity. The various PCG roles are illustrated in Fig. 4.2 and are extensively discussed in Section 4.4.

4.2.3.1 Autonomy: Autonomous Versus Mixed-Initiative

The generative process that does not consider any input from the human designer is defined as **autonomous** PCG whereas **mixed-initiative** PCG refers to the process that involves the human designer in the creative task. Both roles are discussed in further detail in Section 4.4.

4.2.3.2 Adaptivity: Experience-Agnostic Versus Experience-Driven

Experience-agnostic content generation refers to the paradigm of PCG where content is generated without taking player behavior or player experience into account, as opposed to **experience-driven** [783], adaptive, personalized or player-centered content generation where player interaction with the game is analyzed and content is created based on a player's previous behavior. Most commercial games tackle PCG in a generic, experience-agnostic way, while experience-driven PCG has been receiving increasing attention in academia. Recent extensive reviews of PCG for player-adaptive games can be found in [783, 784].

4.3 How Could We Generate Content?

There are many different algorithmic approaches to generating content for games. While many of these methods are commonly thought of as AI methods, others are drawn from graphics, theoretical computer science or even biology. The various methods differ also in what types of content they are suitable to generate. In this section, we discuss a number of PCG methods that we consider important and include search-based, solver-based, and grammar-based methods but also cellular automata, noise and fractals.

4.3.1 Search-Based Methods

The **search-based** approach to PCG [720] has been intensively investigated in academic PCG research in recent years. In search-based procedural content generation, an evolutionary algorithm or some other stochastic search or optimization algorithm is used to search for content with the desired qualities. The basic metaphor is that of design as a search process: a good enough solution to the design problem exists within some space of solutions, and if we keep iterating and tweaking one or many possible solutions, keeping those changes which make the solution(s) better and discarding those that are harmful, we will eventually arrive at the desired solution. This metaphor has been used to describe the design process in many different disciplines: for example, Will Wright—designer of *SimCity* (Electronic Arts, 1989) and *The Sims* (Electronic Arts, 2000)—described the game design process as search in his talk at the 2005 Game Developers Conference. Others have previously described the design process in general, and in other specialized domains such as architecture, the design process can be conceptualized as search and implemented as a computer program [757, 55].

The core components of the search-based approach to solving a content generation problem are the following:

- A **search algorithm**. This is the "engine" of a search-based method. Often relatively simple evolutionary algorithms work well enough, however sometimes there are substantial benefits to using more sophisticated algorithms that take e.g., constraints into account, or that are specialized for a particular content representation.
- A **content representation**. This is the representation of the artifacts you want to generate, e.g., levels, quests or winged kittens. The content representation could be anything from an array of real numbers to a graph to a string. The content representation defines (and thus also limits) what content can be generated, and determines whether effective search is possible; see also the discussion about representation in Chapter 2.
- One or more **evaluation functions**. An evaluation function is a function from an artifact (an individual piece of content) to a number indicating the quality of the artifact. The output of an evaluation function could indicate e.g., the playability of a level, the intricacy of a quest or the aesthetic appeal of a winged kitten. Crafting an evaluation function that reliably measures the aspect of game quality that it is meant to measure is often among the hardest tasks in developing a search-based PCG method. Refer also to the discussion about utility in Chapter 2.

Let us look at some of the choices for **content representations**. To take a very well-known example, a level in *Super Mario Bros* (Nintendo, 1985) might be represented in any of the following ways:

1. Directly, as a level map, where each variable in the genotype corresponds to one "block" in the phenotype (e.g., bricks, question mark blocks, etc.).
2. More indirectly, as a list of the positions and properties of the different game entities such as enemies, platforms, gaps and hills (an example of this can be found in [611]).
3. Even more indirectly, as a repository of different reusable patterns (such as collections of coins or hills), and a list of how they are distributed (with various transforms such as rotation and scaling) across the level map (an example of this can be found in [649]).
4. Very indirectly, as a list of desirable properties such as the number of gaps, enemies, or coins, the width of gaps, etc. (an example of this can be found in [617]).
5. Most indirectly, as a random number seed.

While it clearly makes no sense to evolve random number seeds (it is a representation with no locality whatsoever) the other levels of abstraction can all make sense under certain circumstances. The fundamental tradeoff is between more direct, more fine-grained representations with potentially higher locality (higher correlation of fitness between neighboring points in search space) and less direct, more coarse-grained representations with probably lower locality but smaller search spaces. Smaller search spaces are in general easier to search. However, larger search spaces would (all other things being equal) allow for more different types of content to be expressed, or in other words increase the *expressive range* of the generator.

The third important choice when designing a search-based PCG solution is the **evaluation function** known as the *fitness function*. The evaluation function assesses all candidate solutions, and assigns a score (a fitness value or evaluation value) to each candidate. This is essential for the search process; if we do not have a good evaluation function, the evolutionary process will not work as intended and will not find good content. It could be argued that designing an entirely accurate content quality evaluation is "AI-complete" in the sense that to really understand what a human finds fun, you actually have to be a human or understand humans in depth. However, like for other problems in various areas of AI, we can get pretty far with well-designed domain-specific heuristics. In general, the evaluation function should be designed to model some desirable quality of the artifact, e.g., its playability, regularity, entertainment value, etc. The design of an evaluation function depends to a great extent on the designer and what she thinks are the important aspects that should be optimized and how to formulate that.

In search-based PCG, we can distinguish between three classes of evaluation functions: direct, simulation-based, and interactive.

- **Direct** evaluation functions map the generated content (or features extracted from it) directly to a content quality value and, in that sense, they base their fitness calculations directly on the phenotype representation of the content. No simulation of the gameplay is performed during the mapping. Some direct evaluation functions are hand-coded, and some are learned from data. Direct evaluation functions are fast to compute and often relatively easy to implement, but it is sometimes hard to devise a direct evaluation function for some aspects of game content. Example features include the placement of bases and resources in real-time strategy games [712], the size of the ruleset in strategy games [415] or the current mood of the game scene based on visual attention theory [185]. The mapping between features and fitness might be contingent on a model of the playing style, preferences or affective state of players. An example of this form of fitness evaluation can be seen in the study done by Shaker et al. [617, 610] for personalizing player experience using models of players to give a measure of content quality. In that study, the authors trained neural networks to predict the player experience (such as challenge, frustration and enjoyment) of players given a playing style and the characteristics of a level as inputs. These trained neural networks can then be used as fitness functions by searching for levels that, for example, maximize predicted enjoyment while minimizing frustration. Or the exact opposite, if you wish.
- **Simulation-based** evaluation functions use AI agents that play through the generated content in order to estimate its quality. Statistics are calculated about the agents' behavior and playing style and these statistics are then used to score game content. The type of the evaluation task determines the area of proficiency of the AI agent. If content is evaluated on the basis of playability, e.g., the existence of a path from the start to the end in a maze or a level in a 2D platform game, then AI agents should be designed that excel in reaching the end of the game. On the other hand, if content is optimized to maximize a particular player experience, then an AI agent that imitates human behavior should generally be adopted. An

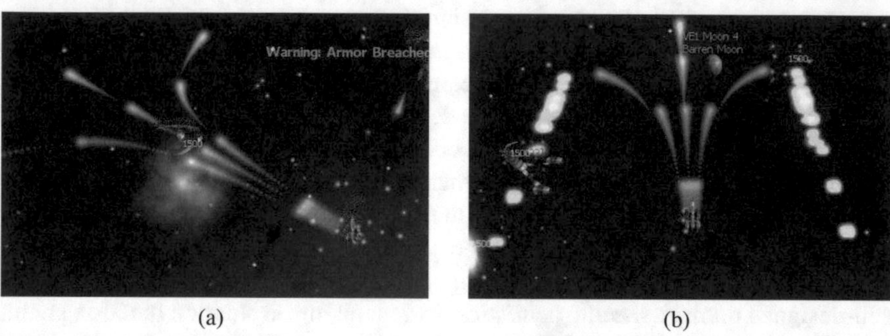

(a) (b)

Fig. 4.3 Examples of evolved weapons in *Galactic Arms Races*. Images obtained from www.aigameresearch.org.

example study that implements a human-like agent for assessing content quality is presented in [704] where neural-network-based controllers are trained to drive like human players in a car racing game and then used to evaluate the generated tracks. Each track generated is given a fitness value according to statistics calculated while the AI controller is playing. Another example of a simulation-based evaluation function is measuring the average fighting time of bots in a first-person shooter game [103]. In that study, levels were simply selected to maximize the amount of time bots spent on the level before killing each other.

- **Interactive** evaluation functions evaluate content based on interaction with a human, so they require a human "in the loop". Examples of this method can be found in the work by Hastings et al. [250], who implemented this approach by evaluating the quality of the personalized weapons evolved implicitly based on how often and how long the player chooses to use these weapons. Figure 4.3 presents two examples of evolved weapons for different players. Cardamone et al. [102] also used this form of evaluation to score racing tracks according to the users' reported preferences. Ølsted et al. used the same approach to design levels for first-person shooter games [501]. The first case is an example of an *implicit* collection of data, i.e., that the players did not answer direct questions about their preferences, while players' preferences were collected *explicitly* in the second. The problem with explicit data collection is that, if not well integrated, it requires the gameplay session to be interrupted. This method, however, provides a reliable and accurate estimator of player experience, as opposed to implicit data collection, which is usually noisy and based on assumptions. Hybrid approaches are sometimes employed to mitigate the drawbacks of these two methods by collecting information across multiple modalities such as combining player behavior with eye gaze and/or skin conductance. Example applications of this approach can be found in biofeedback-based camera viewpoint generation [434], level generation [610] and visuals generation in physical interactive games [771].

Search-based methods have extremely broad applicability, as evolutionary computation can be used to construct almost any kind of game content. However, this generality comes with a number of drawbacks. One is that it is generally a rather slow method, requiring the evaluation of a large number of candidate content items. The time required to evolve a good solution can also not be precisely predicted, as there is no runtime guarantee for evolutionary algorithms. This might make search-based PCG solutions unsuitable for time-critical content generation, such as when you only have a few seconds to serve up a new level in a game. It should also be noted that the successful application of search-based PCG methods relies on judicious design choices when it comes to the particular search algorithm, representation and evaluation function.

As we will see below, there are several algorithms that are generally better suited than evolutionary algorithms for generating game content of specific types. However, none of them have the **versatility** of the search-based approach, with abilities to incorporate all kinds of objectives and constraints. As we will see, many other algorithms for content generation can also be combined with search-based solutions, so that evolutionary algorithms can be used to search the parameter space of other algorithms.

4.3.2 Solver-Based Methods

While the search-based approach to content generation means using one or several objective functions, perhaps in conjunction with constraints, to specify for a randomized search algorithm such as an evolutionary algorithm what to look for, there is another approach to PCG which is also based on the idea of content generation as search in a space of artifacts. **Solver-based methods** for PCG use constraint solvers, such as those used in logic programming, to search for content artifacts that satisfy a number of constraints.

Constraint solvers allow you to specify a number of constraints in a specific language; some solvers require you to specify the constraints mathematically, others use a logic language for specification. Behind the scenes they can be implemented in many different ways. While there are some solvers that are based on evolutionary computation, it is more common to use specialized methods, such as reducing the problem to a SAT (satisfiability) problem and using a SAT solver to find solutions. Many such solvers progress not through evaluating whole solutions, but searching in spaces of partial solutions. This has the effect of iteratively pruning the search space: eliminating parts of the search space repeatedly until only viable solutions are left. This marks a sharp difference with the search-based paradigms, and also suggests some differences in the use cases for these classes of algorithms. For example, while evolutionary algorithms are *anytime algorithms*, i.e., they can be stopped at any point and some kind of solution will be available (though a better solution would probably have been available if you had the algorithm run for longer), this is not always the case with constraint satisfaction methods—though the time taken until a viable solution is found can be very low, depending on how many constraints need to be satisfied. Unlike with evolutionary algorithms, it is possible to prove bounds on the worst-case complexity of SAT solvers and the algorithms that depend on them. However, while the worst-case complexity is often shockingly high, such algorithms can be very fast when applied (judiciously) in practice.

An example of solver-based methods is the level generator for Smith and Whitehead's *Tanagra* mixed-initiative platform level design tool [642]. At the core of the tool is a constraint-based generator of platform game levels. The constraint solver uses a number of constraints on what constitutes a solvable platform level (e.g., maximum jump length, and distance from jumps to enemies) as well as constraints based on aesthetic considerations, such as the "rhythm" of the level, to generate new platform game levels or level segments. This generation happens very fast and produces good results, but is limited to fairly linear levels. Another example is the work by El-Nasr et al. on procedural generation of lighting [188, 185]. The system developed in those studies configures and continuously modulates the lighting in a scene with the aim to enhance player experience by using constraint non-linear optimization to select the best lighting configuration.

One approach to constraint solving which can be particularly useful for PCG because of its generality is Answer Set Programming (ASP) [638]. ASP builds on AnsProlog, a constraint programming language which is similar to Prolog [69]. Complex sets of constraints can be specified in AnsProlog, and an ASP solver can then be used to find all models (all configurations of variables) that satisfy the constraints. For PCG purposes, the model (the set of parameters) can be used to describe a world, a story or similar, and the constraints can specify playability or various aesthetic considerations. An example of an ASP application for level and puzzle generation is the *Refraction* game (see Fig. 4.4). For a good introduction to and overview of the use of ASP in PCG you may refer to [638, 485].

Generally solver-based methods can be suitable when the whole problem can be encoded in the language of the constraint solver (such as AnsProlog). It is generally complicated to include simulation-based tests or any other call to the game engine inside a constraint satisfaction program. An alternative if simulation-based tests need to be performed is evolutionary algorithms, which can also be used to solve constraint satisfaction problems. This allows a combination of fitness values and constraints to drive evolution [376, 382, 240].

4.3.3 Grammar-Based Methods

Grammars are fundamental structures in computer science that also have many applications in procedural content generation. In particular, they are very frequently used for producing plants, such as trees, which are commonly used in many different types of games. However, grammars have also been used to generate missions and dungeons [173, 174], rocks [159], underwater environments [3] and caves [424]. In these cases, grammars are used as a constructive method, creating content with-

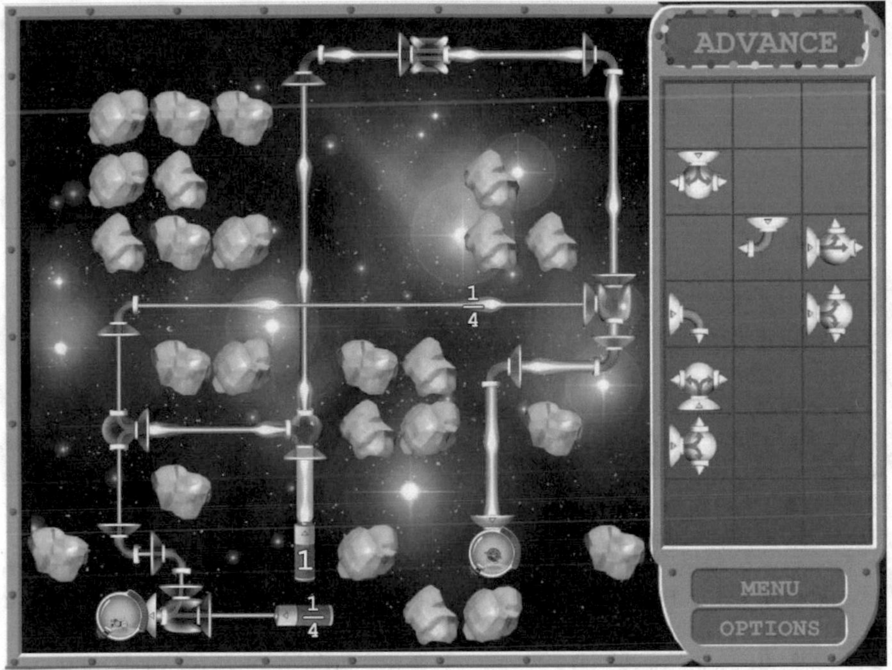

Fig. 4.4 A screenshot from the *Refraction* educational game. A solver-based PCG method (ASP) is used to generate the levels and puzzles of the game. Further details about the application of ASP to *Refraction* can be found in [635, 89].

out any evaluation functions or re-generation. However, grammar methods can also be used together with search-based methods, so that the grammar expansion is used a genotype-to-phenotype mapping.

A (formal) **grammar** is a set of **production rules** for rewriting strings, i.e., turning one string into another. Each rule is of the form (symbol(s)) $\rightarrow$ (other symbol(s)). Here are some example production rules:

1. $A \rightarrow AB$
2. $B \rightarrow b$

Expanding a grammar is as simple as going through a string, and each time a symbol or sequence of symbols that occurs in the *left-hand side* of a rule is found, those symbols are replaced by the *right-hand side* of that rule. For example, if the initial string is A, in the first rewriting step the A would be replaced by AB by rule 1, and the resulting string will be AB. In the second rewriting step, the A would again be transformed to AB and the B would be transformed to b using rule 2, resulting in the string ABb. The third step yields the string $ABbb$ and so on. A convention in grammars is that upper-case characters are nonterminal symbols, which are on the left-hand side of rules and therefore rewritten further, whereas lower-case characters are terminal symbols which are not rewritten further.

Starting with the axiom A (in L-systems the seed strings are called axioms) the first few expansions look as follows:

A
AB
ABA
ABAAB
ABAABABA
ABAABABAABAAB
ABAABABAABAABABAABABA
ABAABABAABAABABAABABAABAABABAABAAB

This particular grammar is an example of an **L-system**. L-systems are a class of grammars whose defining feature is parallel rewriting, and was introduced by the biologist Aristid Lindenmayer in 1968 explicitly to model the growth of organic systems such as plants and algae [387]. With time, they have turned out to be very useful for generating plants in games as well as in theoretical biology.

One way of using the power of L-systems to generate 2D (and 3D) artifacts is to interpret the generated strings as instructions for a turtle in **turtle graphics**. Think of the turtle as moving across a plane holding a pencil, and simply drawing a line that traces its path. We can give commands to the turtle to move forwards, or to turn left or right. For example, we can define the L-system alphabet {F, +, -, [,]} and then use the following key to interpret the generated strings:

- F: move forward a certain distance (e.g., 10 pixels).
- $+$: turn left 30 degrees.
- $-$: turn right 30 degrees.
- [: push the current position and orientation onto the stack.
-]: pop the position and orientation off the stack.

Bracketed L-systems can be used to generate surprisingly plant-like structures. Consider the L-system defined by the single rule $F \to F[-F]F[+F][F]$. Figure 4.5 shows the graphical interpretation of the L-system after 1, 2, 3 and 4 rewrites starting from the single symbol F. Minor variations of the rule in this system generate different but still plant-like structures, and the general principle can easily be extended to three dimensions by introducing symbols that represent rotation along the axis of drawing. For this reason, many standard packages for generating plants in game worlds are based on L-systems or similar grammars. For a multitude of beautiful examples of plants generated by L-systems refer to the book *The Algorithmic Beauty of Plants* by Prusinkiewicz and Lindenmayer [542].

There are many extensions of the basic L-system formalism, including non-deterministic L-systems, which can help with increasing diversity of the generated content, and context-sensitive L-systems, which can produce more complicated patterns. Formally specifying L-systems can be a daunting task, in particular as the mapping between the axiom and rules on the one hand and the results after expansion on the other are so complex. However, search-based methods can be used to

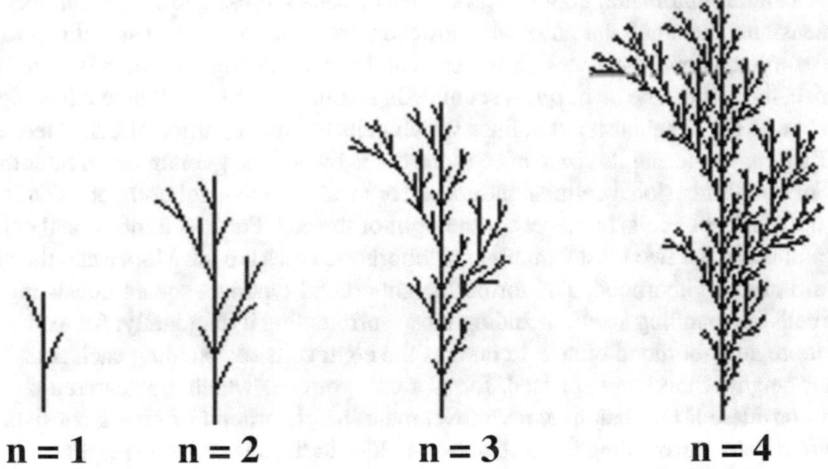

Fig. 4.5 Four rewrites of the bracketed L-system $F \to F[-F]F[+F][F]$.

find good axioms or rules, using for example desired height or complexity of the plant in the evaluation function [498].

4.3.4 Cellular Automata

A cellular automaton (plural: **cellular automata**) is a discrete model of computation which is widely studied in computer science, physics, complexity science and even some branches of biology, and can be used to computationally model biological and physical phenomena such as growth, development, patterns, forms, or even emergence. While cellular automata (CA) have been the subject of extensive study, the basic concept is actually very simple and can be explained in a sentence or two: cellular automata are a set of cells placed on a grid that change through a number of discrete time steps according to a set of rules; these rules rely on the current state of each cell and the state of its neighboring cells. The rules can be applied iteratively for as many time steps as we desire. The conceptual idea behind cellular automata was introduced by Stanislaw Ulam and John von Neumann [742, 487] back in the 1940s; it took about 30 more years, however, for us to see an application of cellular automata that showed their potential beyond basic research. That application was the two-dimensional cellular automaton designed in Conway's *Game of Life* [134]. The *Game of Life* is a zero-player game; its outcome is not influenced by the player's input throughout the game and it is solely dependent on its initial state (which is determined by the player).

A cellular automaton contains a number of cells represented in any number of dimensions; most cellular automata, however, are either one-dimensional (vectors) or two-dimensional (matrices). Each cell can have a finite number of **states**; for instance, the cell can be *on* or *off*. A set of cells surrounding each cell define its **neighborhood**. The neighborhood defines which cells around a particular cell affect the cell's future state and its size can be represented by any integer number greater than 1. For one-dimensional cellular automata, for instance, the neighborhood is defined by the number of cells to the left or the right of the cell. For two-dimensional cellular automata, the two most common neighborhood types are the **Moore** and the **von Neumann** neighborhood. The former neighborhood type is a square consisting of the cells surrounding a cell, including those surrounding it diagonally; for example, a Moore neighborhood of size 1 contains the eight cells surrounding each cell. The latter neighborhood type, instead, forms a cross of cells which are centered on the cell considered. For example, a von Neumann neighborhood of size 1 consists of the four cells surrounding the cell, above, below, to the left and to the right.

At the beginning of an experiment (at time $t = 0$) we initialize the cells by assigning a state for each one of them. At each time step t we create a new generation of cells according to a rule or a mathematical function which specifies the new state of each cell given the current state of the cell and the states of the cells in its neighborhood at time $t-1$. Normally, the rule for updating the state of the cells remains the same across all cells and time steps (i.e., it is static) and is applied to the whole grid.

Cellular automata have been used extensively in games for modeling environmental systems like heat and fire, rain and fluid flow, pressure and explosions [209, 676] and in combination with influence maps for agent decision making [678, 677]. Another use for cellular automata has been for thermal and hydraulic erosion in procedural terrain generation [500]. Of particular interest for the purposes of this section is the work of Johnson et al. [304] on the generation of infinite cave-like dungeons using cellular automata. The motivation in that study was to create an infinite cave-crawling game, with environments stretching out endlessly and seamlessly in every direction. An additional design constraint was that the caves are supposed to look organic or eroded, rather than having straight edges and angles. No storage medium is large enough to store a truly endless cave, so the content must be generated at runtime, as players choose to explore new areas. The game does not scroll but instead presents the environment one screen at a time, which offers a time window of a few hundred milliseconds in which to create a new room every time the player exits a room.

The method introduced by Johnson et al. [304] used the following four parameters to control the map generation process:

- a percentage of rock cells (inaccessible areas) at the beginning of the process;
- the number of CA generations (iterations);
- a neighborhood threshold value that defines a rock;
- the Moore neighborhood size.

(a) A random map

(b) A map generated with cellular automata

Fig. 4.6 Cave generation: Comparison between a CA and a randomly generated map. The CA parameters used are as follows: the CA runs for four generations; the size of the Moore neighborhood considered is 1; the threshold value for the CA rule is 5 ($T = 5$); and the percentage of rock cells at the beginning of the process is 50% (for both maps). Rock and wall cells are represented by white and red color respectively. Colored areas represent different tunnels (floor clusters). Images adapted from [304].

In the dungeon generation implementation presented in [304], each room is a 50×50 grid, where each cell can be in one of two states: *empty* or *rock*. Initially, the grid is empty. The generation of a single room is as follows.

1. The grid is "sprinkled" with rocks: for each cell, there is probability (e.g., 0.5) that it is turned into rock. This results in a relatively uniform distribution of rock cells.
2. A number of CA generations (iterations) are applied to this initial grid.
3. For each generation the following simple rule is applied to all cells: a cell turns into rock in the next iteration if at least T (e.g., 5) of its neighbors are rock, otherwise it will turn into free space.
4. For aesthetic reasons the rock cells that border on empty space are designated as "wall" cells, which are functionally rock cells but look different.

The aforementioned simple procedure generates a surprisingly lifelike cave-room. Figure 4.6 shows a comparison between a random map (sprinkled with rocks) and the results of a few iterations of the cellular automaton. But while this process generates a single room, a game would normally require a number of connected rooms. A generated room might not have any openings in the confining rocks, and there is no guarantee that any exits align with entrances to the adjacent rooms. Therefore, whenever a room is generated, its immediate neighbors are also generated. If there is no connection between the largest empty spaces in the two rooms, a tunnel is drilled between those areas at the point where they are least separated. A few more iterations of the CA algorithm are then run on all nine neighboring rooms

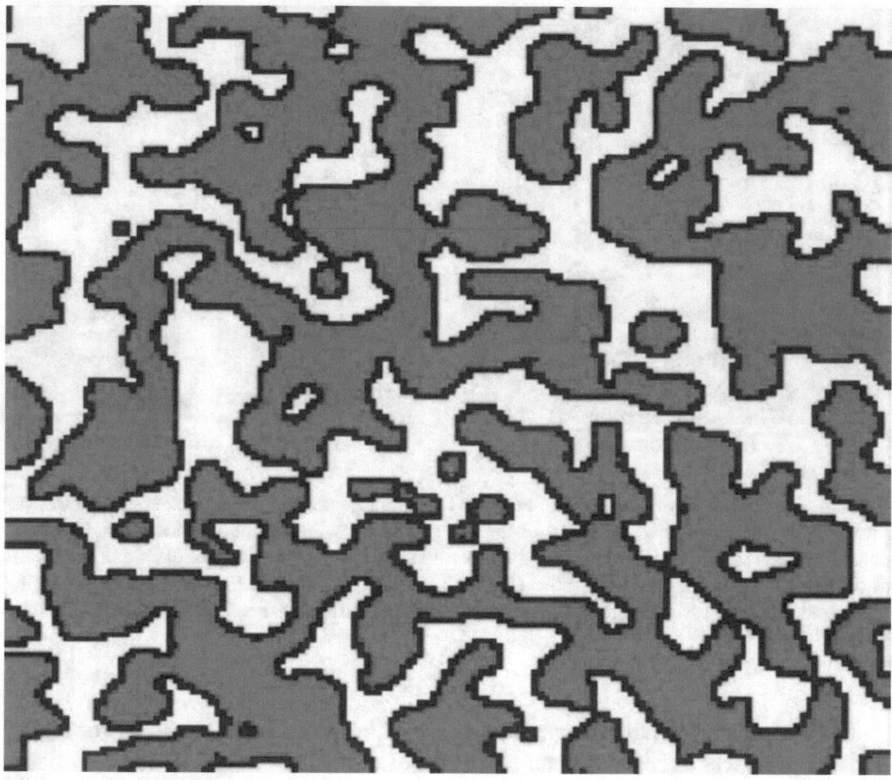

Fig. 4.7 Cave generation: a 3×3 base grid map generated with CA. Rock and wall cells are represented by white and red color respectively; gray areas represent floor. Moore neighborhood size is 2, T is 13, number of CA iterations is 4, and the percentage of rock cells at the initialization phase is 50%. Image adapted from [304].

together, to smooth out any sharp edges. Figure 4.7 shows the result of this process, in the form of nine rooms that seamlessly connect. This generation process is extremely fast, and can generate all nine rooms in less than a millisecond on a modern computer. A similar approach to that of Johnson et al. is featured in the *Galak-Z* (17-bit, 2016) game for dungeon generation [9]. In that game cellular automata generate the individual rooms of levels and the rooms are tied together via a variation of a Hilbert curve, which is a continuous fractal space-filling curve [261]. *Galak-Z* (17-bit, 2016) shows an alternative way of combining CA with other methods for achieving the desired map generation result.

In summary, CA are very fast constructive methods that can be used effectively to generate certain kinds of content such as terrains and levels (as e.g., in [304]), but they can also be potentially used to generate other types of content. The greatest benefits a CA algorithm can offer to a game content generator is that it depends on a small number of parameters and that it is intuitive and relatively simple to grasp and implement. However, the algorithm's constructive nature is the main cause for its disadvantages. For both designers and programmers, it is not trivial to fully understand the impact that a single parameter may have on the generation process, since each parameter affects multiple features of the generated output. While the few parameters of the core algorithm allow for a certain degree of controllability, the algorithm cannot guarantee properties such as playability or solvability of levels. Further, it is not possible to design content that has specific requirements, e.g., a map with a certain connectivity, since gameplay features are disjoint from the control parameters of the CA. Thus, any link between the CA generation method and gameplay features would have to be created through a process of trial and error. In other words, one would need to resort to preprocessing or a generate-and-test approach.

4.3.5 Noise and Fractals

One class of algorithms that are very frequently used to generate heightmaps and textures are **noise** algorithms, many of which are **fractal** algorithms, meaning that they exhibit scale-invariant properties. Noise algorithms are usually fast and easy to use but lack in controllability.

Both textures and many aspects of terrains can fruitfully be represented as two-dimensional matrices of real numbers. The width and height of the matrix map to the x and y dimensions of a rectangular surface. In the case of a texture, this is called an **intensity map**, and the values of cells correspond directly to the brightness of the associated pixels. In the case of terrains, the value of each cell corresponds to the height of the terrain (over some baseline) at that point. This is called a **heightmap**. If the resolution with which the terrain is rendered is greater than the resolution of the heightmap, intermediate points on the ground can simply be interpolated between points that do have specified height values. Thus, using this common representation, any technique used to generate noise could also be used to generate terrains, and vice versa—though they might not be equally suitable.

It should be noted that in the case of terrains, other representations are possible and occasionally suitable or even necessary. For example, one could represent the terrain in three dimensions, by dividing the space up into *voxels* (cubes) and computing the three-dimensional voxel grid. An example is the popular open-world game *Minecraft* (Mojang, 2011), which uses unusually large voxels. Voxel grids allow structures that cannot be represented with heightmaps, such as caves and overhanging cliffs, but they require a much larger amount of storage.

Fractals [180, 500] such as **midpoint displacement algorithms** [39] are in common use for real-time map generation. Midpoint displacement is a simple algorithm for generating two-dimensional landscapes (seen from the side) by repeatedly subdividing a line. The procedure is as follows: start with a horizontal line. Find the midpoint of that line, and move the line up or down by a random amount, thus breaking the line in two. Now do the same thing for both of the resulting lines, and so on for as many steps as you need in order to reach sufficient resolution. Ev-

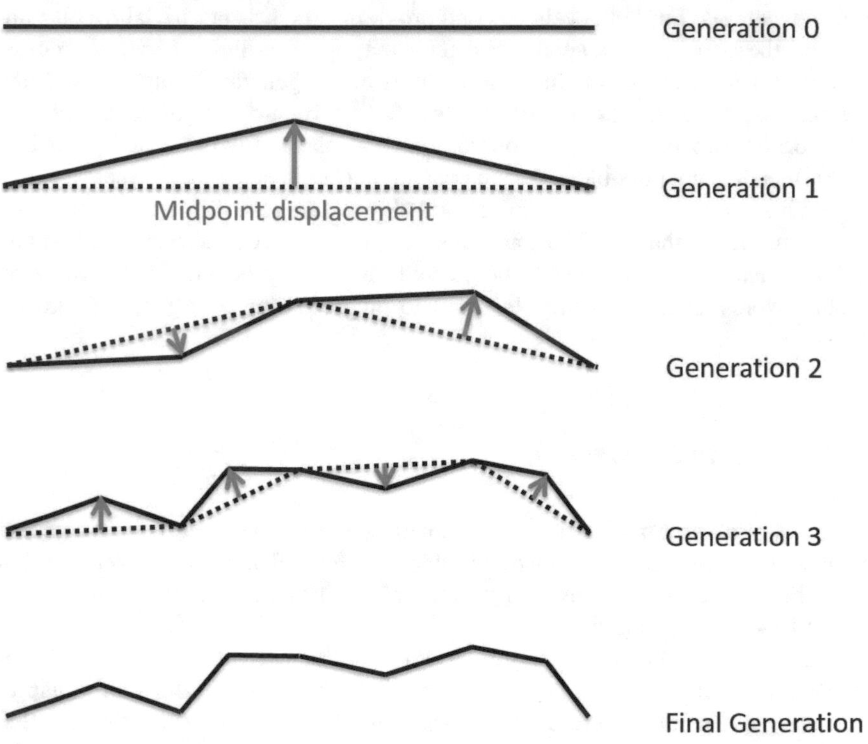

Fig. 4.8 The Midpoint Displacement algorithm visualized.

ery time you call the algorithm recursively, lower the range of the random number generator somewhat (see Fig. 4.8 for an example).

A useful and simple way of extending the midpoint displacement idea to two dimensions (and thus creating two-dimensional heightmaps which can be interpreted as three-dimensional landscapes) is the **Diamond-Square algorithm** (also known as "the cloud fractal" or "the plasma fractal" because of its frequent use for creating such effects) [210]. This algorithm uses a square 2D matrix with width and height $2^n + 1$. To run the algorithm you normally initialize the matrix by setting the values of all cells to 0, except the four corner values which are set to random values in some chosen range (e.g., $[-1, 1]$). Then you perform the following steps:

1. *Diamond step*: Find the midpoint of the four corners, i.e., the most central cell in the matrix. Set the value of that cell to the average value of the corners. Add a random value to the middle cell.

2. *Square step*: Find the four cells in between the corners. Set each of those to the average value of the two corners surrounding it. Add a random value to each of these cells.

Call this method recursively for each of the four subsquares of the matrix, until you reach the resolution limit of the matrix ($3 * 3$ sub-squares). Every time you call the method, reduce the range of the random values somewhat. The process is illustrated in Fig. 4.9.

There are many more advanced methods for generating fractal noise, with different properties. One of the most important is **Perlin noise**, which has some benefits over Diamond Square [529]. These algorithms are covered thoroughly in books that focus on texturing and modeling from a graphics perspective [180].

4.3.6 Machine Learning

An emerging direction in PCG research is to train generators on **existing content**, to be able to produce more content of the same type and style. This is inspired by the recent results in deep neural networks, where network architectures such as **generative adversarial networks** [232] and **variational autoencoders** [342] have attained good results in learning to produce images of e.g., bedrooms, cats or faces, and also by earlier results where both simpler learning mechanisms such as Markov chains and more complex architectures such as recurrent neural networks have learned to produce text and music after training on some corpus.

While these kinds of generative methods based on machine learning work well for some types of content—most notably music and images—many types of game content pose additional challenges. In particular, a key difference between game content generation and procedural generation in many other domains is that most game content has strict structural constraints to ensure playability. These constraints differ from the structural constraints of text or music because of the need to play games in order to experience them. A level that structurally prevents players from finishing it is not a good level, even if it is visually attractive; a strategy game map with a strategy-breaking shortcut will not be played even if it has interesting features; a game-breaking card in a collectible card game is merely a curiosity; and so on. Thus, the domain of game content generation poses different challenges from that of other generative domains. The same methods that can produce "mostly correct" images of bedrooms and horses, that might still have a few impossible angles or vestigial legs, are less suitable for generating mazes which must have an exit. This is one of the reasons why machine learning-based approaches have so far only attained limited success in PCG for games. The other main reason is that for many types of game content, there simply isn't enough existing content to train on. This

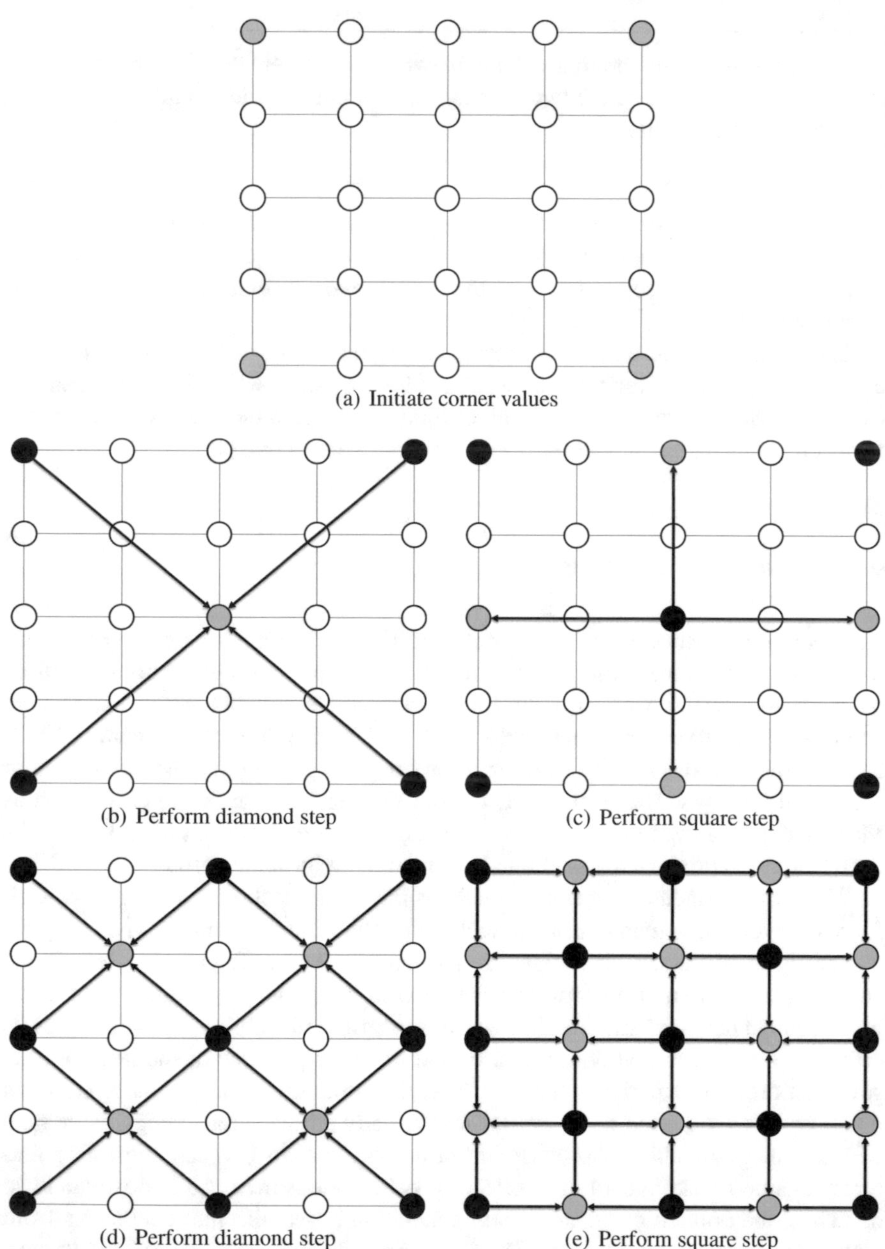

(a) Initiate corner values
(b) Perform diamond step
(c) Perform square step
(d) Perform diamond step
(e) Perform square step

Fig. 4.9 The Diamond-Square algorithm visualized in five steps. Adapted from a figure by Christopher Ewin, licensed under CC BY-SA 4.0.

is, however, an active research direction where much progress might be achieved in the next few years.

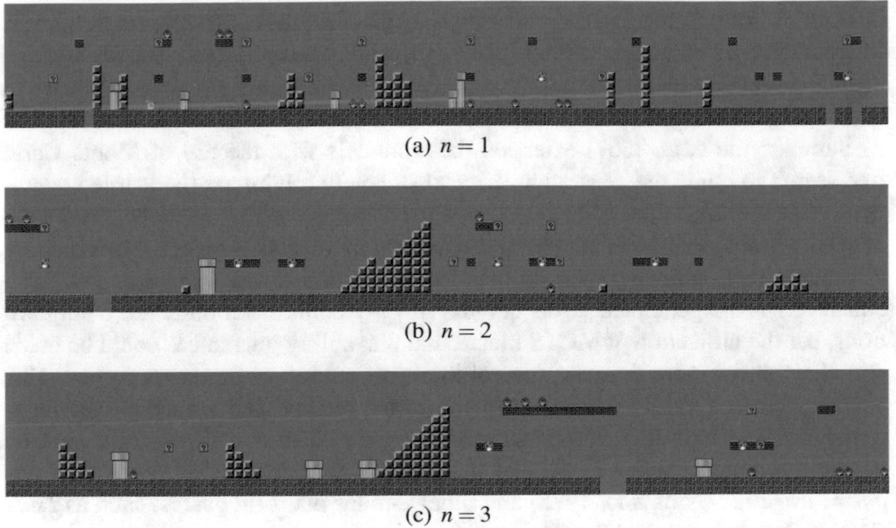

(a) $n = 1$

(b) $n = 2$

(c) $n = 3$

Fig. 4.10 Mario levels reconstructed by n-grams with n set to 1, 2, and 3, respectively.

The core difference between PCG via machine learning and approaches such as search-based PCG is that the content is created *directly* (e.g., via sampling) from models which have been trained on game content. While some search-based PCG approaches use evaluation functions that have been trained on game content—for instance, the work of Shaker et al. [621] or Liapis et al. [373]—the actual content generation is still based on search. Below, we present some examples of PCG via machine learning; these particular PCG studies built on the use of ***n*-grams**, **Markov models** and **artificial neural networks**. For more examples of early work along these lines, see the recent survey paper [668].

4.3.6.1 *n*-grams and Markov Models

For content that can be expressed as one- or two-dimensional discrete structures, such as many game levels, methods based on Markov models can be used. One particularly straightforward Markov model is the n-gram model, which is commonly used for text prediction. The n-gram method is very simple—essentially, you build conditional probability tables from strings and sample from these tables when constructing new strings—and also very fast.

Dahlskog et al. trained n-gram models on the levels of the original *Super Mario Bros* (Nintendo, 1985) game, and used these models to generate new levels [156]. As n-gram models are fundamentally one-dimensional, these levels needed to be converted to strings in order for n-grams to be applicable. This was done through dividing the levels into vertical "slices," where most slices recur many times throughout the level [155]. This representational trick is dependent on there being a large amount of redundancy in the level design, something that is true in many games. Models were trained using various levels of n, and it was observed that while $n = 0$ creates essentially random structures and $n = 1$ creates barely playable levels, $n = 2$ and $n = 3$ create rather well-shaped levels. See Fig. 4.10 for examples of this.

Summerville et al. [667] extended these models with the use of Monte Carlo tree search to guide the generation. Instead of solely relying on the learned conditional probabilities, they used the learned probabilities during rollouts (generation of whole levels) that were then scored based on an objective function specified by a designer (e.g., allowing them to bias the generation towards more or less difficult levels). The generated levels could still only come from observed configurations, but the utilization of MCTS meant that playability guarantees could be made and allowed for more designer control than just editing of the input corpus. This can be seen as a hybrid between a search-based method and a machine learning-based method. In parallel, Snodgrass and Ontañón trained two-dimensional Markov Chains—a more complex relative of the n-gram—to generate levels for both *Super Mario Bros* (Nintendo, 1985) and other similar platform games, such as *Lode Runner* (Brøderbund, 1983) [644].

4.3.6.2 Neural Networks

Given the many uses of **neural networks** in machine learning, and the many different neural network architectures, it is little wonder that neural networks are also highly useful for machine learning-based PCG. Following on from the *Super Mario Bros* (Nintendo, 1985) examples in the previous section, Hoover et al. [277] generated levels for that same game by extending a representation called functional scaffolding for musical composition (FSMC) that was originally developed to compose music. The original FSMC representation posits 1) music can be represented as a function of time and 2) musical voices in a given piece are functionally related [276]. Through a method for evolving neural networks called NeuroEvolution of Augmenting Topologies [655], additional musical voices are evolved to be played simultaneously with an original human-composed voice. To extend this musical metaphor and represent *Super Mario Bros* (Nintendo, 1985) levels as functions of time, each level is broken down into a sequence of tile-width columns. The height of each column extends the height of the screen. While FSMC represents a unit of time by the length of an eighth-note, a unit of time in this approach is the width of each column. At each unit of time, the system queries the ANN to decide a height to place a tile. FSMC then inputs a musical note's pitch and duration to the ANNs. This approach translates pitch to the height at which a tile is placed and duration to the number of times a tile-type is repeated at that height. For a given tile-type or musical voice, this information is then fed to a neural network that is trained on two-thirds of the existing human-authored levels to predict the value of a tile-type at each column. The idea is that the neural network will learn hidden relationships between the tile-types in the human-authored levels that can then help humans construct entire levels from as little starting information as the layout of a single tile-type.

Of course, machine learning can also be used to generate other types of game content that are not levels. A fascinating example of this is *Mystical Tutor*, a design assistant for *Magic: The Gathering* cards [666]. In contrast to some of the other generators that aim to produce complete, playable levels, Mystical Tutor acknowledges that its output is likely to be flawed in some ways and instead aims to provide inspirational raw material for card designers.

4.4 Roles of PCG in Games

The generation of content algorithmically may take different roles within the domain of games. We can identify two axes across which PCG roles can be placed: players and designers. We envision PCG systems that consider designers while they generate content or they operate interdependently of designers; the same applies for players. Figure 4.11 visualizes the key roles of PCG in games across the dimensions of designer initiative and player experience.

Regardless of the generation method used, game genre or content type PCG can act either *autonomously* or as a *collaborator* in the design process. We refer to the former role as **autonomous** generation (Section 4.4.2) and the latter role as **mixed-initiative** (Section 4.4.1) generation. Further, we cover the **experience-driven** PCG role by which PCG algorithms consider the player experience in whatever they try to generate (Section 4.4.3). As a result, the generated content is associated to the player and her experience. Finally, if PCG does not consider the player as part of the generation process it becomes **experience-agnostic** (Section 4.4.4).

PCG techniques can be used to generate content in **runtime**, as the player is playing the game, allowing the generation of endless variations, making the game infinitely replayable and opening the possibility of generating player-adapted content, or **offline** during the development of the game or before the start of a game session. The use of PCG for offline content generation is particularly useful when generating complex content such as environments and maps; several examples of that was discussed at the beginning of the chapter. An example of the use of runtime content generation can be found in the game *Left 4 Dead* (Valve, 2008), a first-person shooter game that provides dynamic experience for each player by analyzing player behavior on the fly and altering the game state accordingly using PCG techniques [14, 60]. A trend related to runtime content generation is the creation and sharing of player-generated content. Some games such as *LittleBigPlanet* (Sony Computer Entertainment, 2008) and *Spore* (Electronic Arts, 2008) provide a content editor (level editor in the case of *LittleBigPlanet* and the *Spore* Creature Creator) that allows the players to edit and upload complete creatures or levels to a central online server where they can be downloaded and used by other players. With respect to the four different roles of PCG in games, runtime generation is possible in the autonomous and experience-agnostic roles, it is always the case in the experience-driven role whereas it is impossible in the mixed-initiative role. On the other hand, the offline generation of content can occur both autonomously and in

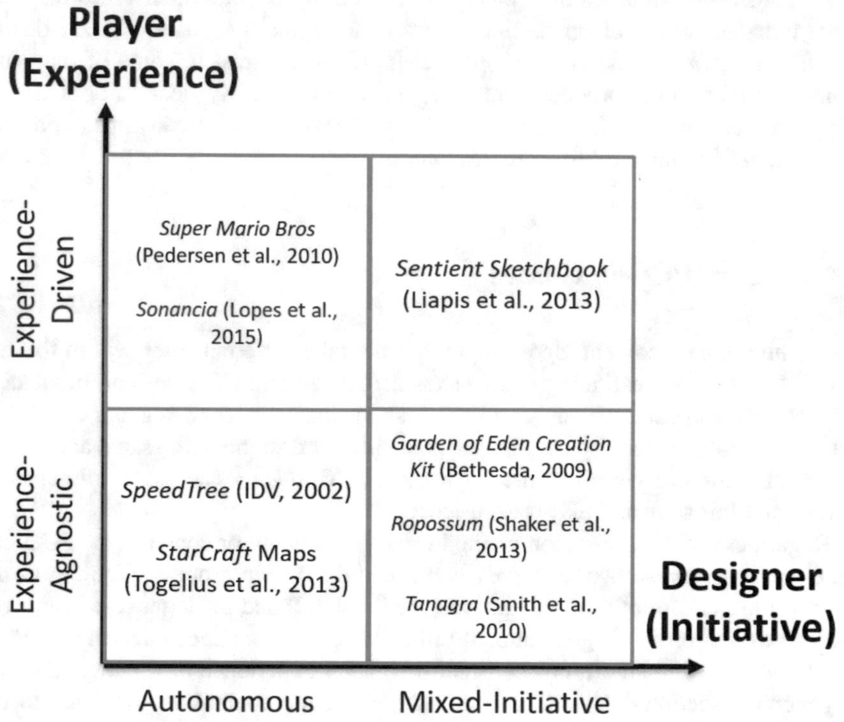

Fig. 4.11 The four key PCG roles in games across the dimensions of designer initiative and player experience. For each combination of roles the figure lists a number of indicative examples of tools or studies covered in this chapter.

an experience-agnostic manner. Further it is exclusively the only way to generate content in a mixed-initiative fashion whereas it is not relevant for experience-driven generation as this PCG role occurs in runtime by definition.

The following four subsections outline the characteristics of each of the four PCG roles with a particular emphasis on the mixed-initiative and the experience-driven roles that have not yet covered in length in this chapter.

4.4.1 Mixed-Initiative

AI-assisted game design refers to the use of AI-powered tools to support the game design and development process. This is perhaps the AI research area which is most promising for the development of better games [764]. In particular, AI can assist in the creation of game content varying from levels and maps to game mechanics and narratives.

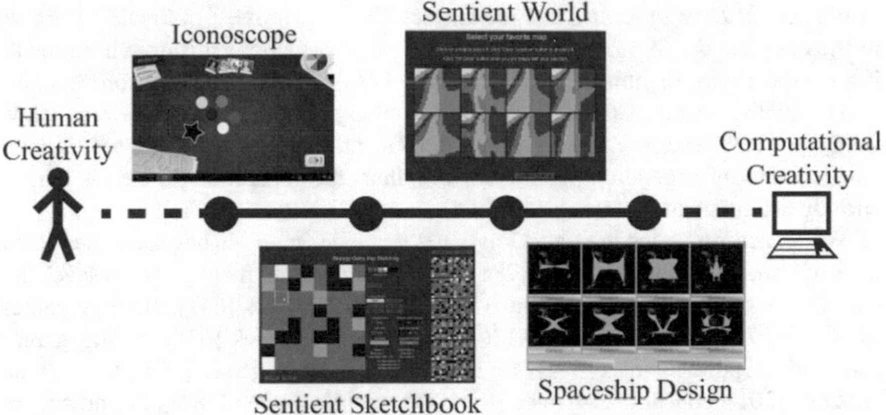

Fig. 4.12 The mixed-initiative spectrum between human and computer initiative (or creativity) across a number of mixed-initiative design tools discussed in this section. *Iconoscope* is a mixed-initiative drawing game [372], *Sentient Sketchbook* is a level editor for strategy games [379], Sentient World is mixed-initiative map editor [380] and Spaceship Design is a mixed-initiative (mostly computer initiative) tool powered by interactive evolution [377]. Adapted from [375].

We identify AI-assisted game design as the task of creating artifacts via the interaction of a human initiative and a computational initiative. The computational initiative is a PCG process and thus, we discuss this co-design approach under the heading of procedural content generation. Although the term **mixed-initiative** lacks a concrete definition [497], in this book we define it as the process that considers both the human and the computer *proactively* making content contributions to the game design task although the two initiatives do not need to contribute to the same degree [774]. Mixed-initiative PCG thus differs from other forms of co-creation, such as the collaboration of multiple human creators or the interaction between a human and non-proactive computer support tools (e.g., spell-checkers or image editors) or non-computer support tools (e.g., artboards or idea cards). The initiative of the computer can be seen as a continuum between the *no initiative* state, leaving full control of the design to the designer and having the computer program simply carry out the commands of the human designer, to the *full initiative* state which yields an autonomously creative system. Any state between the two is also possible as we will see in the examples below and as depicted in Fig. 4.12.

4.4.1.1 Game Domains

While the process of AI-assisted game design is applicable to any creative facets within game design [381] it is level design that has benefited the most from it. Within commercial-standard game development, we can find AI-based tools that allow varying degrees of computer initiative. On one end of the scale, level editors such as the *Garden of Eden Creation Kit* (Bethesda, 2009) or game engines such as the *Unreal Development Kit* (Epic Games 2009) leave most of creative process to the designer but they, nevertheless, boost game development through automating interpolations, pathfinding and rendering [774]. On the other end of the computer initiative scale, PCG tools specialized on e.g., vegetation—*SpeedTree* (IDV, 2002)—or FPS levels—*Oblige* (Apted, 2007)—only require the designer to set a small amount of generation parameters and, thus, the generation process is almost entirely autonomous.

Within academia, the area of AI-assisted game design tools has seen significant research interest in recent years [785] with contributions mainly to the level design task across several game genres including platformers [641], strategy games [380, 379, 774, 378] (see Fig. 4.13(a)), open world games [634], racing games [102], casual puzzle games [614] (see Fig. 4.13(b)), horror games [394], first-person shooters [501], educational games [89, 372], mobile games [482], and adventure games [323]. The range of mixed-initiative game design tools expands to tools that are designed to assist with the generation of complete game rulesets such as the MetaGame [522], the RuLearn [699] and the Ludocore [639] tools to tools that are purposed to generate narratives [480, 673] and stories within games [358].

4.4.1.2 Methods

Any PCG approach could potentially be utilized for mixed-initiative game design. The dominant methods that have so far being used, however, rely on evolutionary computation following the search-based PCG paradigm. Even though evolution, at fist sight, does not appear to be the most attractive approach for real-time processing and generation of content, it offers great advantages associated, in particular, with the stochasticity of artificial evolution, diversity maintenance and potential for balancing multiple design objectives. Evolution can be constrained to the generation of playable, usable, or, in general, content of particular *qualities* within desired design specifications. At the same time, it can incorporate metrics such as novelty [382] or surprise [240], for maximizing the *diversity* of generated content and thus enabling a change in the creative path of the designer [774]. Evolution can be computationally costly, however, and thus, interactive evolution is a viable and popular alternative for mixed-initiative evolutionary-based generation (e.g., see [102, 380, 377, 501]).

Beyond artificial evolution, another class of algorithms that is relevant for mixed-initiative content generation is constraint solvers and constraint optimization. Methods such as answer set programming [383, 69] have been used in several AI-assisted level design tools including *Tanagra* [641] for platformers and *Refraction* [89] for educational puzzle games. Artificial neural networks can also perform certain tasks in a mixed-initiative manner, such as performing “autocomplete” or level repair [296] through the use of deep learning approachers such as stacked autoencoders. The goal here is to provide a tool that “fills in” parts of a level that the human designer does not want or have time to create, and correcting other parts to achieve further consistency.

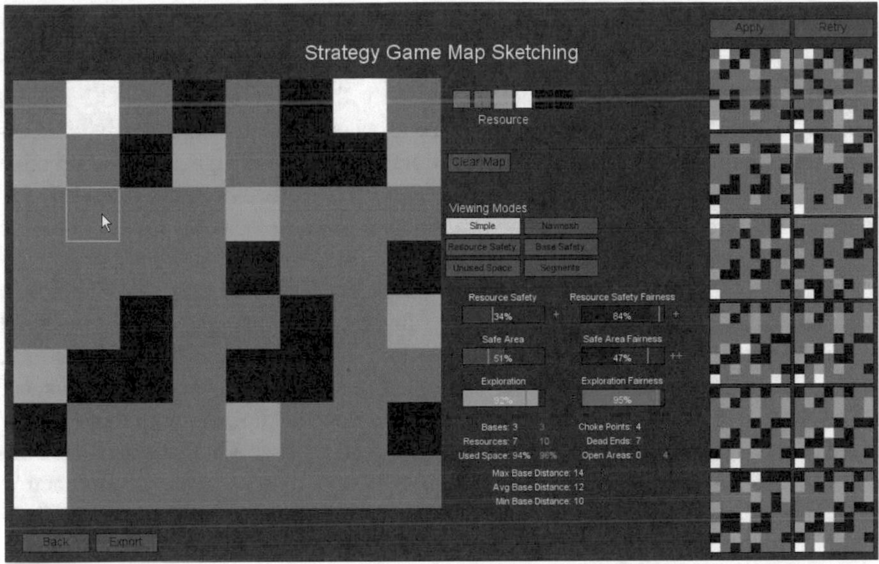

(a) *Sentient Sketchbook*

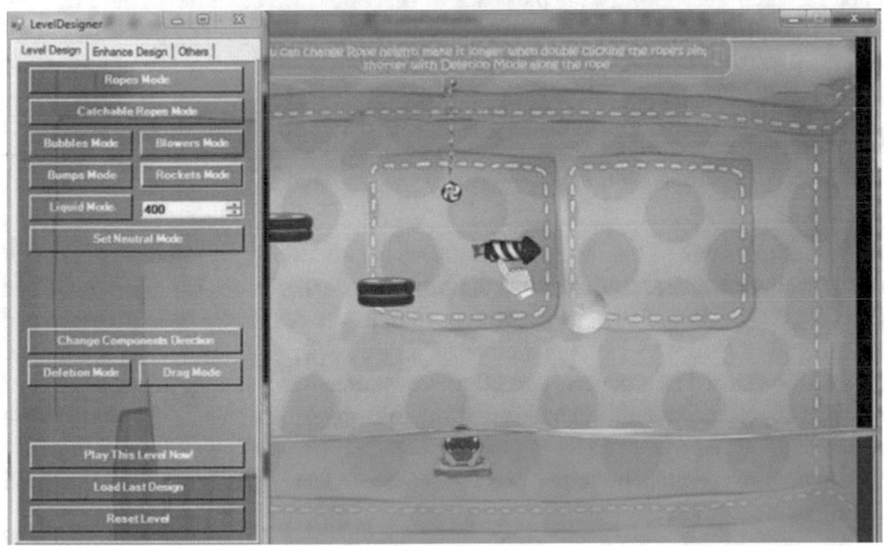

(b) *Ropossum*

Fig. 4.13 Examples of mixed-initiative level design tools. *Sentient Sketchbook* (a) offers map sketch suggestions to the designer via artificial evolution (see rightmost part of the image); the suggestions are evolved to either maximize particular objectives of the map (e.g., balance) or are evolved to maximize the novelty score of the map. In *Ropossum* (b) the designer may select to design elements of *Cut the Rope* (Chillingo, 2010) game levels; the generation of the remaining elements are left to evolutionary algorithms to design.

4.4.2 Autonomous

The role of autonomous generation is arguably the most dominant PCG role in games. The earlier parts of this chapter are already dedicated to extensive discussions and studies of PCG systems that do not consider the designer in their creative process. As a result we will not cover this PCG role in further detail here. What is important to be discussed, however, is the fuzzy borderline between mixed-initiative and autonomous PCG systems. It might be helpful, for instance, to consider autonomous PCG as the process by which the role of the designer starts and ends with an offline setup of the algorithm. For instance, the designer is only involved in the parameterization of the algorithm as in the case of *SpeedTree* (IDV, 2002). One might wish, however, to further push the borderline between autonomous and mixed-initiative generation and claim that generation is genuinely autonomous only if the creative process reconsiders and adapts the utility function that drives the content generation—thereby becoming creative in its own right. A static utility function that drives the generation is often referred to as *mere* generation within the computational creativity field [381].

While the line between autonomy and collaboration with designers is still an open research question, for the purposes of this book, we can safely claim that the PCG process is autonomous when the initiative of the designer is limited to algorithmic parameterizations before the generation starts.

4.4.3 Experience-Driven

As games offer one of the most representative examples of rich and diverse content creation applications and are elicitors of unique user experiences **experience-driven** PCG (EDPCG) [783, 784] views game content as the *building block* of games and the generated games as the potentiators of player experience. Based on the above, EDPCG is defined as a generic and effective approach for the optimization of user (player) experience via the adaptation of the experienced content. According to the experience-driven role of PCG in games *player experience* is the collection of affective patterns elicited, cognitive processes emerged and behavioral traits observed during gameplay [781].

By coupling player experience with procedural content generation, the experience-driven perspective offers a new, player-centered role to PCG. Since games are composed by game content that, when played by a particular player, elicits experience patterns, one needs to assess the quality of the content generated (linked to the experience of the player), search through the available content, and generate content that optimizes the experience for the player (see Fig. 4.14). In particular, the key components of EDPCG are:

- **Player experience modeling**: player experience is modeled as a function of game content and player.

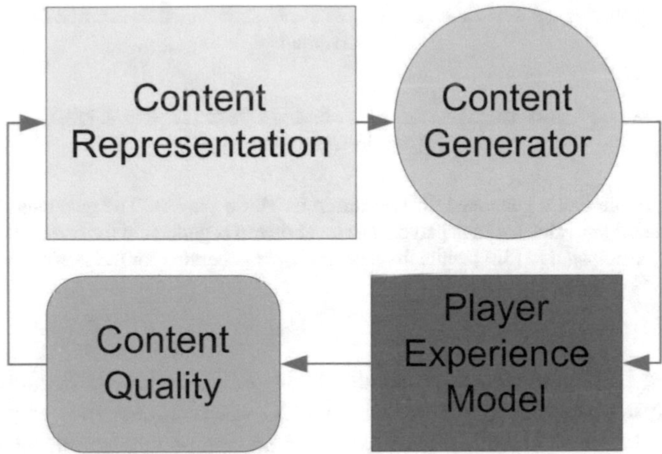

Fig. 4.14 The four key components of the experience-driven PCG framework.

- **Content quality**: the quality of the generated content is assessed and linked to the modeled experience.
- **Content representation**: content is represented accordingly to maximize search efficacy and robustness.
- **Content generator**: the generator searches through the generative space for content that optimizes the experience for the player according to the acquired model.

Each component of EDPCG has its own dedicated literature and the extensive review of each is covered in other parts of the book. In particular, player experience modeling is covered in Chapter 5 whereas the remaining three components of the framework have already been covered in this chapter. A detailed survey and discussion about EDPCG is available in [783].

4.4.3.1 Experience-Driven PCG in Practice

Left 4 Dead (Valve, 2008) is an example of the use of experience-driven PCG in a commercial game where an algorithm is used to adjust the pacing of the game on the fly based on the player's *emotional intensity*. In this case, adaptive PCG is used to adjust the difficulty of the game in order to keep the player engaged [60]. Adaptive content generation can also be used with another motive such as the generation of more content of the kind the player seems to like. This approach was followed, for instance, in the *Galactic Arms Race* [250] game where the weapons presented to the player are evolved based on her previous weapon use and preferences. In another EDPCG study, El-Nasr et al. implemented a direct fitness function—derived from visual attention theory—for the procedural generation of lighting [188, 185]. The procedural Zelda game engine [257], a game engine designed to emulate the

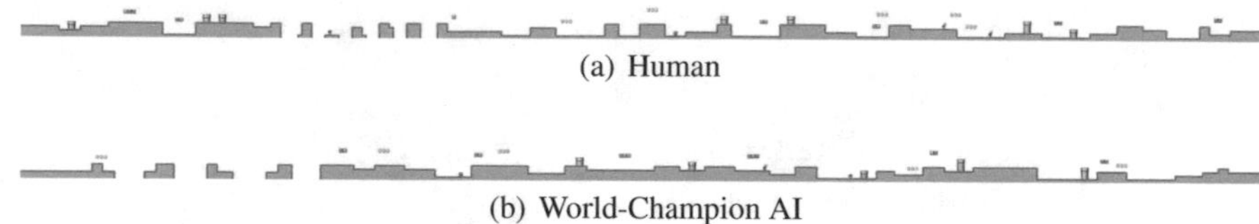

(a) Human

(b) World-Champion AI

Fig. 4.15 Example levels generated for two different Mario players. The generated levels maximize the modeled *fun* value for each player. The level on top is generated for one of the experiment subjects that participated in [521] while the level below is generated for the world champion agent of the Mario AI competition in 2009.

popular *The Legend of Zelda* (Nintendo, 1986–2017) action-RPG game series, is built mainly to support experience-driven PCG research. Another example is the work by Pedersen et al. [521], who modified an open-source clone of the classic platform game *Super Mario Bros* (Nintendo, 1985) to allow for personalized level generation. The realization of EDPCG in this example is illustrated in Fig. 4.16. The first step was to represent the levels in a format that would yield an easily searchable space. A level was represented as a short parameter vector describing the number, size and placement of gaps which the player can fall through, and the presence or absence of a switching mechanic. The next step was to create a model of player experience based on the level played and the player's playing style. Data was collected from hundreds of players, who played pairs of levels with different parameters and were asked to rank which of two levels best induced each of the following user states: fun, challenge, frustration, predictability, anxiety, boredom. While playing, the game also recorded a number of metrics of the players' playing styles, such as the frequency of jumping, running and shooting. This data was then used to train neural networks to predict the examined user states using evolutionary preference learning. Finally, these player experience models were utilized to optimize game levels for particular players [617]. Two examples of such levels can be seen in Fig. 4.15. It is worth noting—as discussed in Chapter 5—that one may wish to further improve the models of experience of Mario players by including information about the player beyond gameplay [29] such as her head pose [610] or her facial expressions [52].

4.4.4 Experience-Agnostic

With experience-agnostic PCG we refer to any PCG approach that does not consider the role of the player in the generation of content. But where should we set the boundary of involvement? When do we consider a player as part of the generation process and when don't we? While the borderline between experience-driven and experience-agnostic is not trivial to draw we define any PCG approach whose content quality function does not include a player (experience) model or it does not

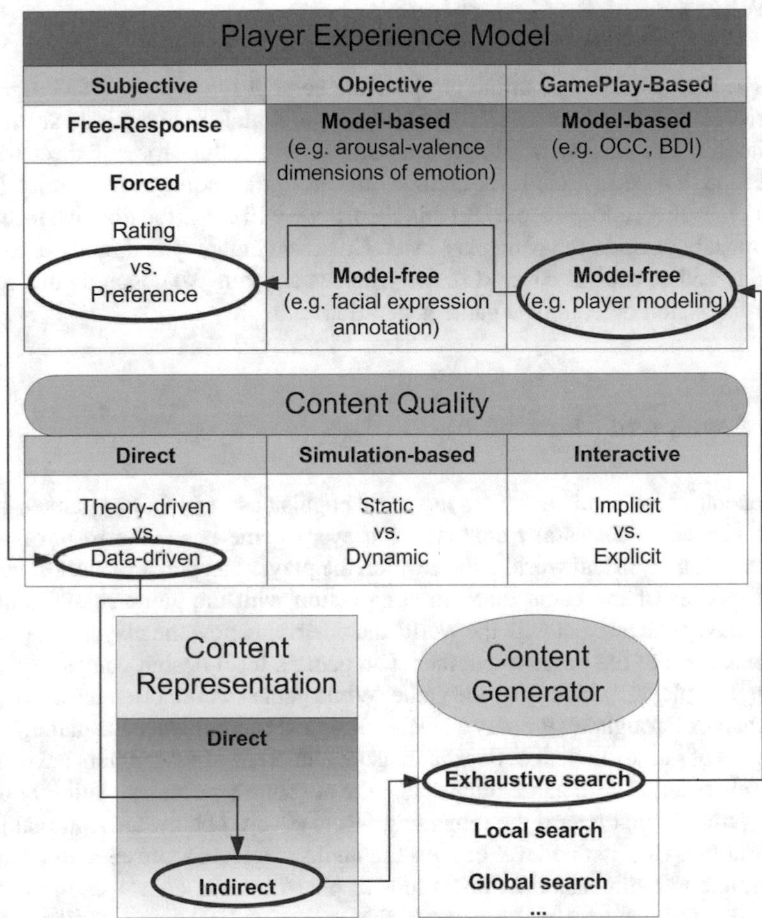

Fig. 4.16 The EDPCG framework in detail. The gradient grayscale-colored boxes represent a continuum of possibilities between the two ends of the box while white boxes represent discrete, exclusive options within the box. The blue arrows illustrate the EDPCG approach followed for the *Super Mario Bros* (Nintendo, 1985) example study [521, 617]: Content quality is assessed via a direct, data-driven evaluation function which is based on a combination of a gameplay-based (model-free) and a subjective (pairwise preference) player experience modeling approach; content is represented indirectly and exhaustive search is applied to generate better content.

interact with the player in any way during generation as experience-agnostic. As with the role of autonomous PCG, this chapter has already gone through several examples of content generation that do not involve a player or a player experience model. To avoid being repetitive we will refer the reader to the PCG studies covered already that are outside the definition of experience-driven PCG.

4.5 What Could Be Generated?

In this section we briefly outline the possible content types that a PCG algorithm can generate in a game. Generally speaking Liapis et al. [381] identified six creative domains (or else facets) within games that we will follow for our discussion in this section. These include level architecture (design), audio, visuals, rules (game design), narrative, and gameplay. In this chapter we will cover the first five facets and we purposely exclude the gameplay facet. Creative gameplay is directly associated with play and as such is covered in the previous chapter. We conclude this section with a discussion on complete game generation.

4.5.1 Levels and Maps

The generation of levels is by far the most popular use of PCG in games. Levels can be viewed as **necessary** content since every game has some form of spatial representation or virtual world within which the player can perform a set of actions. The properties of the game level, in conjunction with the game rules, frame the ways a player can interact with the world and determine how the player can progress from one point in the game to another. The game's level design contributes to the challenges a player faces during the game. While games would often have a fixed set of mechanics throughout, the way a level is designed can influence the gameplay and the degree of game challenge. For that reason, a number of researchers have argued that levels coupled with game rules define the absolutely necessary building blocks of any game; in that regard the remaining facets covered below are optional [371]. The variations of possible level designs are endless: a level representation can vary from simple two-dimensional illustrations of platforms and coins—as in the *Super Mario Bros* (Nintendo, 1985) series—to the constrained 2D space of *Candy Crush Saga* (King, 2012), to the three-dimensional and large urban spaces of *Assassin's Creed* (Ubisoft, 2007) and *Call of Duty* (Infinity Ward, 2003), to the 2D elaborated structures of *Angry Birds* (Rovio, 2009), to the voxel-based open gameworld of *Minecraft* (Mojang 2011).

Due to their several similarities we can view the procedural generation of game levels from the lens of procedural architecture. Similarly to architecture, level design needs to consider both the **aesthetic** properties and the **functional** requirements of whatever is designed within the game world. Depending on the game genre, functional requirements may vary from a reachable end-goal for platform games, to a challenging gameplay in driving games such as *Forza Motorsport* (Turn 10 Studios 2005), to waves of gameplay intensity as in *Pac-Man* (Namco, 1980), *Left 4 Dead* (Valve, 2008), *Resident Evil 4* (Capcom, 2005) and several other games. A procedural level generator also needs to consider the aesthetics of the content as the level's aesthetic appearance may have a significant impact not only on the visual stimuli it offers to the player but also on navigation. For example, a sequence of identical rooms can easily make the player disoriented—as was intended in the

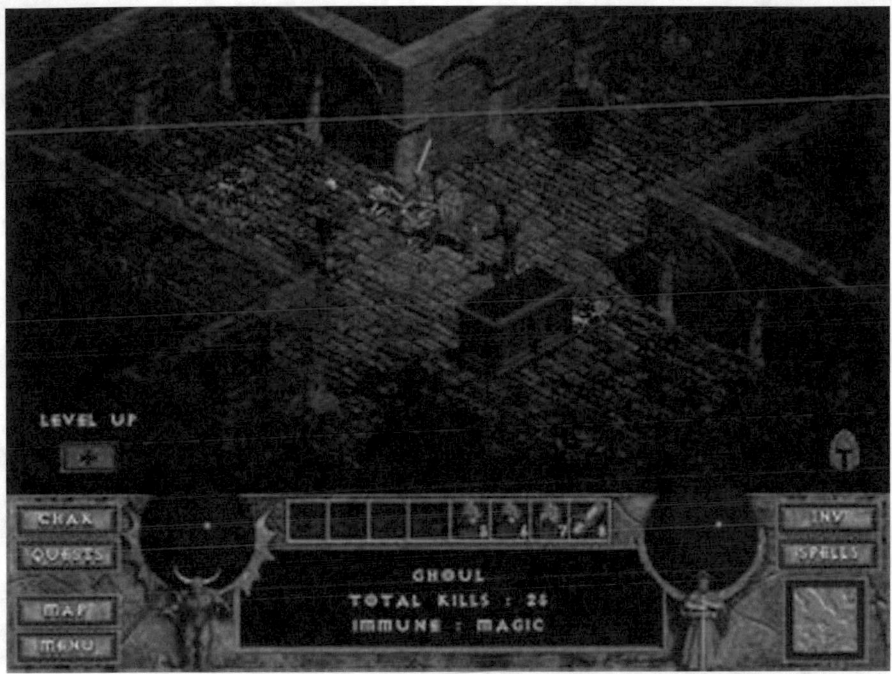

Fig. 4.17 The procedurally generated levels of *Diablo* (Blizzard Entertainment, 1996): one of the most characteristic examples of level generation in commercial games. Diablo is a relatively recent descendant of *Rogue* (Toy and Wichmann, 1980) (i.e., rogue-like) role-playing game characterized by dungeon-based procedurally generated game levels. Image obtained from Wikipedia (fair use).

dream sequences of *Max Payne* (Remedy, 2001)—while dark areas can add to the challenge of the game due to low visibility or augment the player's arousal as in the case of *Amnesia: The Dark Descent* (Frictional Games, 2010), *Nevermind* (Flying Mollusk, 2015) and *Sonancia* [394]. When the level generator considers larger, open levels or gameworlds then it draws inspiration from urban and city planning [410], with edges to constrain player freedom—landmarks to orient the player and motivate exploration [381]—as in the *Grand Theft Auto* (Rockstar Games, 1997) series and districts and areas to break the world's monotony—as in *World of Warcraft* (Blizzard Entertainment, 2004) which uses highly contrasting colors, vegetation and architectural styles to split the world into districts that are suitable for characters of different level ranges.

As already seen broadly in this chapter the generation of levels in a procedural manner is clearly the most popular and possibly the oldest form of PCG in the game industry. We already mentioned the early commercial use of PCG for automatic level design in games such as *Rogue* (Toy and Wichman, 1980) and the Rogue-inspired *Diablo* (see Fig. 4.17) series (Blizzard Entertainment, 1996), and the more recent world generation examples of *Civilization IV* (Firaxis, 2005) and *Minecraft* (Mojang, 2011). The level generation algorithms used in commercial games are usually

Fig. 4.18 The caricaturized and highly-emotive visuals of the game *Limbo* (Playdead, 2010). Image obtained from Wikipedia (fair use).

constructive, in particular, in games where players can interact with and change the game world via play. Players of *Spelunky* (Yu and Hull, 2009), for instance, are allowed to modify a level which is not playable (i.e., the exit cannot be reached) by blowing up the blocking tiles with a bomb provided by the game level.

The academic interest in procedural level generation is only recent [720, 783, 616] but it has produced a substantial volume of studies already. Most of the academic studies described in this chapter are focused on levels. The interested reader may refer to those for examples of level generation across various methodologies, PCG roles and game genres.

4.5.2 Visuals

Games are, by definition, visual media unless the game is designed explicitly to not have visuals—e.g., the *Real Sound: Kaze no Regret* (Sega, 1997) adventure audio game. The visual information displayed on the screen conveys messages to the player which are dependent on the graphical style, color palette and visual texture. Visuals in games can vary from simple abstract and pixelized representations as the 8-bit art of early arcade games, to caricaturized visuals as in *Limbo* (Playdead, 2010) (see Fig. 4.18), to photorealistic graphics as in the *FIFA* series (EA Sports, 1993) [299].

Within the game industry PCG has been used broadly for the generation of any of the above visual types. Arguably, the complexity of the visual generation task increases the more the resolution and the photorealism of the desired output increases. There are examples of popular generative tools such as *SpeedTree* (IDV, 2002) for vegetation and *FaceGen* (Singular Inversions, 2001) for faces, however, that can successfully output photorealistic 3D visuals. Within academia notable examples of visual generation are the games *Petalz* [565, 566] (flower generation), *Galactic Arms Race* [250] (weapon generation; see also Fig. 4.3) and *AudioInSpace*

[275] (weapon generation). In all three games visuals are represented by neural networks that are evolved via interactive evolution. Within the domain of weapon particle generation another notable example is the generation of surprising yet balanced weapons for the game *Unreal Tournament III* (Midway Games, 2007) using **constrained surprise search** [240]; an algorithm that maximizes the surprise score of a weapon but at the same time imposes designer constraints to it so that it is balanced. Other researchers have been inspired by theories about "universal" properties of beauty [18] to generate visuals of high appeal and appreciation [377]. The PCG algorithm in that study generates spaceships based on their size, simplicity, balance and symmetry, and adapts its visual output to the taste of the visual's designer via interactive evolution. The PCG-assisted design process referred to as *iterative refinement* [380] is another way of gradually increasing the resolution of the visuals a designer creates by being engaged in an iterative and creative dialog with the visuals generator. Beyond the generation of in-game entities a visuals PCG algorithm may focus on general properties of the visual output such as pixel shaders [282], lighting [188, 185], brightness and saturation, which can all influence the overall appearance of any game scene.

4.5.3 Audio

Even though audio can be seen as optional content it can affect the player directly and its impact on player experience is apparent in most games [219, 221, 129]. Audio in games has reached a great level of maturity as demonstrated by two BAFTA Game Awards and an MTV video music award for best video game soundtrack [381]. The audio types one meets in games may vary from fully orchestrated soundtrack (background) music, as in *Skyrim* (Bethesda, 2011), to sound effects, as the dying or pellet-eating sounds of *Pac-Man* (Namco, 1980), to the voice-acted sounds of *Fallout 3* (Bethesda, 2008). Most notably within indie game development, *Proteus* (Key and Kanaga, 2013) features a mapping between spatial positioning, visuals and player interaction, which collectively affect the sounds that are played. Professional tools such as the sound middleware of UDK (Epic Games, 2004) and the popular *sfxr* and *bfxr* sound generation tools provide procedural sound components to audio designers, demonstrating a commercial interest in and need of procedurally generated audio.

At a first glance, the generation of game audio, music and sounds might not seem to be particularly different from any other type of audio generation outside games. Games are interactive, however, and that particular feature makes the generation of audio a rather challenging task. When it comes to the definition of procedural audio in games, a progressive stance has been that its procedurality is caused by the very interaction with the game. (For instance, game actions that cause sound effects of music can be considered as procedural audio creation [220, 128].) Ideally, the game audio must be able to adapt to the current game state and the player behavior. As a result adaptive audio is a grand challenge for composers since the combinations of all possible game and player states could be largely unknown. Another difficulty for the autonomous generation of adaptive music, in particular, is that it requires real-time composition and production; both of which need to be embedded in a game engine. Aside from a few efforts in that direction[3] the current music generation models are not particularly designed to perform well in games. In the last decade, however, academic work on procedural game music and sound has seen substantial advancements in venues such as the Procedural Content Generation and the Musical Metacreation workshop series.

Generally speaking, sound can be either **diegetic** or **non-diegetic**. A sound is diegetic if its source is within the game's world. The source of the diegetic sound can be visible on the screen (on-screen) or can be implied to be present at the current game state (off-screen). Diegetic sounds include characters, voices, sounds made by items on-screen or interactions of the player, or even music represented as coming from instruments available in the game. A non-diegetic sound, on the other hand, is any sound whose source is outside the game's world. As such non-diegetic sounds cannot be visible on-screen or even implied to be off-screen. Examples include commentaries of a narrator, sound effects which are not linked to game actions, and background music.

A PCG algorithm can generate both diegetic and non-diegetic sounds including music, sound effects and commentaries. Examples of non-diegetic sound generation in games include the *Sonancia* horror sound generator that tailors the tension of the game to the desires of the designer based on the properties of the game level [394]. The mapping between tension and sounds in *Sonancia* has been derived through crowdsourcing [396]. Similarly to *Sonancia*—and towards exploring the creative space between audio and level design—*Audioverdrive* generates levels from audio and audio from levels [273]. Notably within diegetic audio examples, Scirea et al. [606] explores the relationship between procedurally generated music and narrative. Studies have also considered the presence of game characters on-display for the composition of game soundtracks [73] or the combination of short musical phrases that are driven by in-game events and, in turn, create responsive background audio for strategy games [280].

Finally, it is worth mentioning that there are games featuring PCG that use music as the input for the generation of other creative domains rather than music per se. For example, games such as *Audio Surf* (Fitterer, 2008) and *Vib Ribbon* (Sony Entertainment, 2000) do not procedurally generate music but they instead use music to drive the generation of levels. *AudioInSpace* [275] is another example of a side-scrolling space shooter game that does not generate music but uses the background music as the basis for weapon particle generation via artificial evolution.

[3] For instance, see the upcoming melodrive app at: http://melodrive.com.

4.5.4 Narrative

Many successful games are relying heavily on their narratives; the clear distinction however, between such narratives and traditional stories is the interactivity element that is offered by games. Now, whether games can tell stories [313] or games are instead a form of narrative [1] is still an open research question within game studies, and beyond the scope of this book. The study of computational (or procedural) narrative focuses on the representational and generational aspects of stories as those can be told via a game. Stories can play an essential part in creating the aesthetics of a game which, in turn, can impact affective and cognitive aspects of the playing experience [510].

By breaking the game narrative into subareas of game content we can find core game content elements such as the game's plotline [562, 229], but also the ways a story is represented in the game environment [730, 83]. The coupling of a game's representation and the **story of the game** is of vital importance for player experience. Stories and plots are taking place in an environment and are usually told via a **virtual camera** lens. The behavior of the virtual camera—viewed as a parameterized element of computational narrative—can drastically influence the player's experience. That can be achieved via an affect-based cinematographic representation of multiple cameras as those used in *Heavy Rain* (Quantic Dream, 2010) or through an affect-based automatic camera controller as that used in the Maze-Ball game [780]. Choosing the best camera angle to highlight an aspect of a story can be seen as a multi-level optimization problem, and approached with combinations of optimization algorithms [85]. Games such as *World of Warcraft* (Blizzard Entertainment, 2004) use cut scenes to raise the story's climax and lead the player to particular player experience states. The creation or semi-automatic generation of stories and narratives belongs to the area of interactive storytelling, which can be viewed as a form of story-based PCG. The story can adjust according to the actions of the player targeting personalized story generation (e.g., see [568, 106] among others). Ultimately, game worlds and plot point story representations can be co-generated as demonstrated in a few recent studies (e.g., see [248]).

Computational narrative methods for generating or adapting stories of expositions are typically build on planning algorithms, and **planning** is therefore essential for narrative [792]. The space of stories can be represented in various ways, and the representations in turn make use of dissimilar search/planning algorithms, including traditional optimization and reinforcement learning approaches [483, 117, 106]. Cavazza et al. [106], for instance, introduced an interactive storytelling system built with the Unreal game engine that uses Hierarchical Task Network planning to support story generation and anytime user intervention. Young et al. [792] introduced an architecture called *Mimesis*, primarily designed to generate intelligent, plan-based character and system behavior at runtime with direct uses in narrative generation. Finally the IDtension engine [682] dynamically generates story paths based on the player's choices; the engine was featured in *Nothing for Dinner*,[4] a 3D interactive story aiming to help teenagers living challenging daily life situations at home.

Similarly to dominant approaches of narrative and text generation, interactive storytelling in games relies heavily on stored knowledge about the (game) world. Games that rely on narratives—such as *Heavy Rain* (Quantic Dream, 2010)—may include thousands of lines of dialog which are manually authored by several writers. To enable interactive storytelling the game should be able to select responses (or paths in the narrative) based on what the player will do or say, as in *Façade* [441] (see Fig. 4.19). To alleviate, in part, the burden of manually representing world knowledge, data-driven approaches can be used. For instance, one may crowdsource actions and utterance data from thousand of players that interact with virtual agents of a game and then train virtual agents to respond in similar ways using *n*-grams [508]. Or instead, one might design a system in which designers collaborate with a computer by taking turns on adding sentences in a story; the computer is able to provide meaningful sentences by matching the current story with similar stories available on the cloud [673]. Alternatively, a designer could use the news of the day from sites, blogs or Wikipedia and generate games that tell the news implicitly via play [137].

Research on interactive narrative and story-based PCG benefits from and influences the use of **believable agents** that interact with the player and are interwoven in the story plot. The narrative can yield more (or less) believability to agents and thus the relationship between agents and the story they tell is important [801, 401, 531, 106]. In that sense, the computational narrative of a game may define the arena for believable agent design. Research on story-based PCG has also influenced the design and development of games. Starting from popular independent attempts like *Façade* [441] (see Fig. 4.19), *Prom Week* [448] and *Nothing for Dinner* to the commercial success of *The Elder Scrolls V: Skyrim* (Bethesda Softworks, 2011), *Heavy Rain* (Quantic Dream, 2010) and *Mass Effect* (Bioware, 2007) narrative has traditionally been amongst the key factors of player experience and immersion; particularly in narrative-heavy games as the ones aforementioned.

Examples of sophisticated computational narrative techniques crossing over from academia to commercial-standard products include the storytelling system *Versu* [197] which was used to produce the game *Blood & Laurels* (Emily Short, 2014). For the interested reader the *interactive fiction database*[5] contains a detailed list of games built on the principles of interactive narratives and fiction, and the stoygen.org[6] repository, by Chris Martens and Rogelio E. Cardona-Rivera, maintains existing openly-available computational story generation systems. Finally note that the various ways AI can be used to play text-based adventure games and interactive fiction are covered in Chapter 3.

[4] http://nothingfordinner.org

[5] http://ifdb.tads.org/

[6] http://storygen.org/

Fig. 4.19 A screenshot from the *Façade* [441] game featuring its main characters: Grace and Trip. The seminal design and AI technologies of *Façade* popularized the vision of interactive narrative and story-based PCG within games.

4.5.5 Rules and Mechanics

The game rules frame the playing experience by providing the conditions of play—for instance, winning and losing conditions—and the actions available to the player (game mechanics). Rules constitute necessary content as they are in a sense the core of any game, and a game's rules pervade it.

For most games, the design of their ruleset largely defines them and contributes to their success. It is common that the rule set follows some standard design patterns within its genre. For example, the genre of platform games is partly defined by running and jumping mechanics, whereas these are rare in puzzle games. Evidently, the genre constrains the possibility (design) space of the game rules. While this practice has been beneficial—as rule sets are built on earlier successful paradigms—it can also be detrimental to the creativity of the designer. It is often the case that the players themselves can create new successful game variants (or even sub-genres) by merely altering some rules of an existing game. A popular example is the modification of *Warcraft III* (Blizzard, 2002) which allowed the player to control a single "hero" unit and, in turn, gave rise to a new, popular game genre named Multiplayer Online Battle Arenas (MOBA).

Most existing approaches to rule generation take a search-based approach, and are thus dependent on some way of evaluating a set of rules [711, 355]. However, accurately estimating the quality of a set of game rules is very hard. Game rules differ from most other types of game content in that they are almost impossible to evaluate in isolation from the rest of the game. While levels, characters, textures, and many other types of content can to some extent be evaluated outside of the game, looking at a set of rules generally gives very little information of how they play. For a human, the only way to truly experience the rules of a game is to play the game. For a computer, this would translate to simulating gameplay in some way in order to evaluate the rules. (In this sense, rules can be said to be more similar to program code than they are to e.g., pictures or music.)

So how can simulated playthroughs be used to judge the quality of the rulesets? Several ideas about how to judge a game depending on how agents play it have been introduced. The first is **balance**; for symmetric two-player games in particular, balance between the winning chances of the two players is generally positive [274]. Another idea is **outcome uncertainty**, meaning that any particular game should be "decided" as late as possible [76]. Yet another idea is **learnability**: a good game, including its ruleset, is easy to learn and hard to master. In other words, it should have a long, smooth learning curve for the player, as learning to play the game is part of what makes it fun. This idea can be found expressed most clearly in Koster's "Theory of Fun" [351], but can also be said to be implicit in Schmidhuber's theory of artificial curiosity [602] and in theories in developmental psychology [204]. Within work in game rule generation, attempts have been made to capture this idea in different ways. One way is to use a reinforcement learning agent to try to learn to play the game; games where the agent improves the most over a long time score the best [716]. Another way of capturing this idea is to use several agents of different skill levels to try to play the game. Games score better when they maximize the performance difference between these agents [491]. This idea is also related to the idea of **depth** in games, which can be seen as the length of the chain of heuristics that can be acquired in a game [362].

Perhaps the most successful example of game rule generation within academic research is *Ludi* [76]. *Ludi* follows a search-based PCG approach and evolves grammars that represent the rules of board games (see Fig. 4.20). The fitness function that drives the rule generation is composed by several metrics that estimate good design patterns in board games, such as game depth and rule simplicity. A successfully designed game that came out of the *Ludi* generator is named *Yavalath* [75] (see Fig. 4.20). The game is played on a 5-by-5 hexagonal board by two or three players. *Yavalath*'s winning and losing conditions are very simple: you win if you make a line of four tokens, whereas you lose if you make a line of three tokens; the game is a draw if the board fills up.

One of the earliest examples of *video game* rule generation is Togelius and Schmidhuber's experiment with generating simple Pac-Man-like games [716]. In that study rules are evolved to maximize the learnability of player agents as measured via simulated playthroughs. Another example is the work by Nielsen et al. in which game rules are represented using the video game description language [492].

Using Answer Set Programming (a solver-based method) [69] rather than search-based methods, rules have been generated for simple 2D games that, however, respect constraints such as playability (i.e., the victory condition is attainable) [637]. It is fair to say that none of these attempts to generate video games have been able to produce *good* games, i.e., games that anyone (except the creator of the system that creates the games) would want to play. This points to the immense challenge in accurately estimating the quality of a video game rule set. One of the reasons this seems to be a more challenging problem than estimating the quality of a board game rule set is the time dimension, as human reaction time and ability to estimate and predict unfolding processes play an important role in the challenge of many video games.

For more examples and in-depth analysis on rule and mechanics generation the interested reader is referred to the "rules and mechanics" chapter of the PCG book [486].

4.5.6 Games

Game generation refers to the use of PCG algorithms for computationally designing new complete games. The vast majority of PCG studies so far, however, have been very *specific* to a particular game facet or domain. It is, for instance, either a level that is generated or the audio for some level but rarely both. Meanwhile it is surprising to think that the relationship between the different facets is naturally interwoven. A characteristic example of the interwoven nature among game facets is given in [381]: player actions—viewed as a manifestation of game rules—are usually accompanied by corresponding sound effects such as the sound of Mario jumping in *Super Mario Bros* (Nintendo, 1985). Now let us think of a PCG algorithm that introduces a new rule to the game—hence a new player action. The algorithm automatically constrains the sound effects that can be associated to this new action based on a number of factors such as the action's duration, purpose and overall contribution to the game plot. Actions and sounds appear to have a cause and effect (or hierarchical) relationship and a PCG algorithm would naturally prioritize the creation of the action before it generates its sound. Most relationships between facets, however, are not strictly hierarchical or unidirectional. For example, a game level can be successful because of a memorable landmark as much as the gameplay it affords [381]. Similarly, the narrative of a game relies on a multitude of factors including the camera placement as well as the visuals and the sounds.

The game generation topic has attracted a growing interest in recent years even though the relationship between the different games facets is not considered largely. Most game generation projects focus on a single facet of a game and do not investigate the interaction between different facets. The rule generator for Pac-Man-like games [716], for instance, evolves different rules for different colored agents but it does not evolve the agents' color to indicate different playing strategies. Similarly

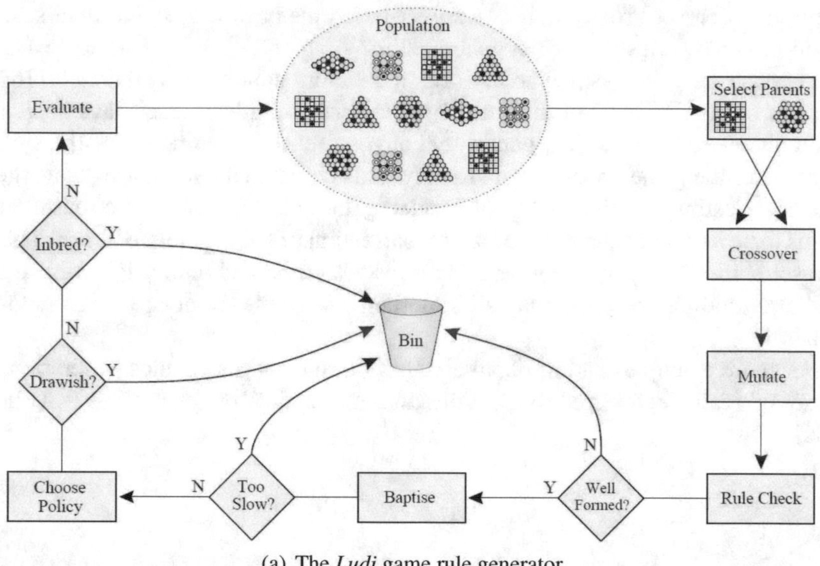

(a) The *Ludi* game rule generator

(b) A deluxe version of the *Yavalath* game

Fig. 4.20 The *Ludi* game rule generator (a) and its game *Yavalath* (b). Image (a) is adapted from [720]. Image (b) is published with permission by Cameron Browne and Néstor Romeral Andrés.

to the ghosts of *Pac-Man* (Namco, 1981) we could imagine that red represents an aggressive behavior whereas orange represents a passive behavior.

Among the few notable attempts of game generation, Game-O-Matic [723] is a game generation system that creates games representing ideas. More specifically, Game-O-Matic includes an authoring tool in which a user can enter entities and their interactions via concept maps. The entities and interactions are translated, respectively, into game visuals and game mechanics; the mechanics, however, do not take into account the visuals or semantics of the game objects they are applied on. One of the first preliminary discussions on how multi-facet integration might happen is offered by Nelson and Mateas [484]. In their paper they present a system for matching sprites (visuals) to very simple WarioWare-style mechanics. Their system is somewhat similar to Game-O-Matic, but works a bit differently: instead of the designer specifying verbs and nouns she wants a game to be about, she gives the system constraints on how verbs and nouns relate in the game (for example, a chasing game needs a "prey" sprite that is something that can do things like "flee" or "be hunted" and so on). The system then uses ConceptNet[7] and WordNet[8] to generate games that fit these constraints.

Arguably one of the most elaborate examples of game generation is ANGELINA [135, 137, 136]. ANGELINA[9] is a game generator that has seen several developments over the years and is currently capable of evolving the rules and levels of the game, collecting and placing pieces of visuals and music (that are relevant to the theme and the emotive mood of the game), giving names for the games it creates and even creating simple commentaries that characterize them. ANGELINA is able to generate games of different genres—including platformer games (see Fig. 4.21) and 3D adventure games—some of which have even participated in game design competitions [136].

The systems above make some initial, yet important, steps towards game generation and they attempt to interweave the different domains within games in a meaningful way, mostly in a hierarchical fashion. However, PCG eventually needs to rise to the challenge of tackling the compound generation of multiple facets in an **orchestrated** manner [711, 371]. An early study on the fusion of more than one generative facet (domain) in games is the one performed recently by Karavolos et al. [324]. The study employs machine learning-based PCG to derive the common generative space—or the common patterns—of game levels and weapons in first-person shooters. The aim of this orchestration process between level design and game design is the generation of level-weapon couplings that are balanced. The unknown mapping between level representations and weapon parameters is learned by a deep convolutional neural network which predicts if a given level with a particular set of weapons will be balanced or not. Balance is measured in terms of the win-lose ratio obtained by AI bots playing in a deathmatch scenario. Figure 4.22 illustrates the architecture used to fuse the two domains. For the interested reader in the or-

---

[7] http://conceptnet.io/

[8] https://wordnet.princeton.edu/

[9] http://www.gamesbyangelina.org/

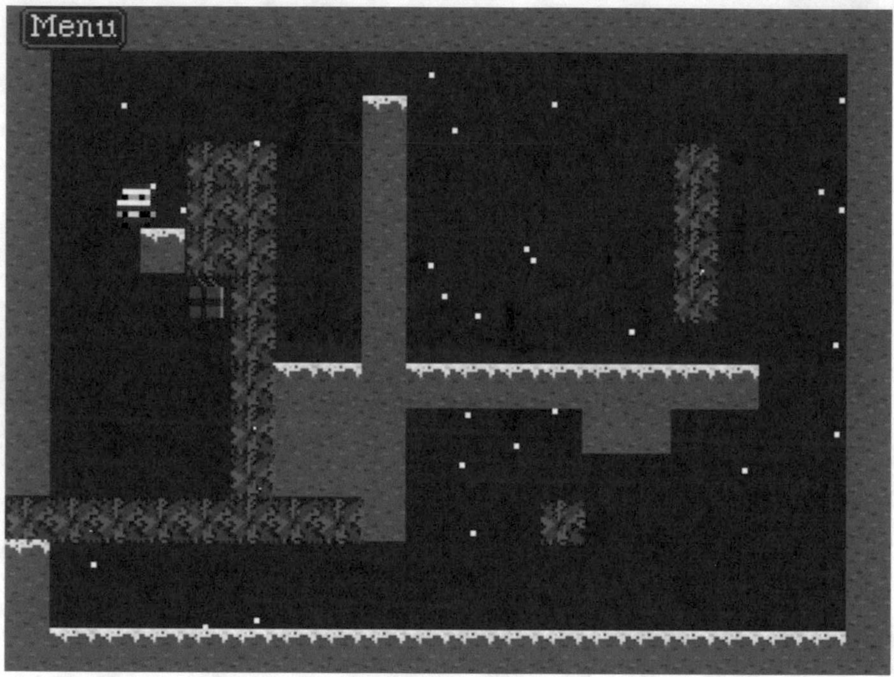

Fig. 4.21 ANGELINA's Puzzling Present game. The game features an invert gravity mechanic that allows a player to overcome the high obstacle on the left and complete the level. Image obtained with permission from http://www.gamesbyangelina.org/.

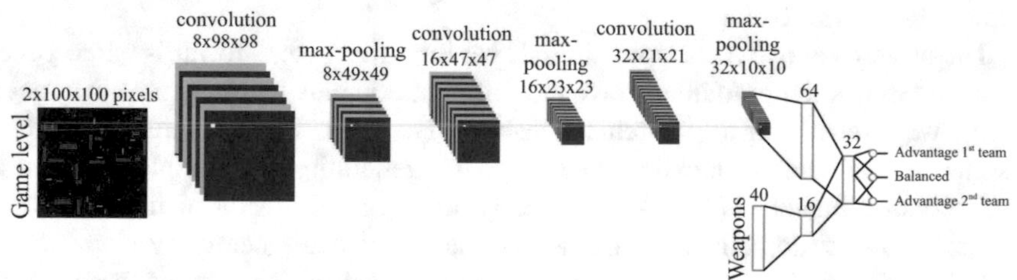

Fig. 4.22 A convolutional neural network (CNN) architecture used for fusing levels and weapons in first-person shooters. The network is trained to predict whether a combination of a level and a weapon would yield a balanced game or not. The CNN can be used to orchestrate the generation of a balanced level given a particular weapon and vice versa. Image adapted from [324].

chestration process we further elaborate on the topic in the last chapter of this book; some early discussions on this vision can also be found in [371].

4.6 Evaluating Content Generators

Creating a generator is one thing; evaluating it is another. Regardless of the method followed all generators shall be evaluated on their ability to achieve the desired goals of the designer. Arguably, the generation of any content is trivial; the generation of **valuable** content for the task at hand, on the other hand, is a rather challenging procedure. (One may claim, however, that the very process of generating valuable content is also, by itself, trivial as one can design a generator that returns a random sample of hand-crafted masterpieces.) Further, it is more challenging to generate content that is not only valuable but is also novel or even inspiring.

4.6.1 Why Is It Difficult?

But what makes the evaluation of content so difficult? First, it is the diverse, stochastic and subjective nature of the **users** that experience the content. Whether players or designers, content users have dissimilar personalities, gameplay aims and behavior, emotive responses, intents, styles and goals [378]. When designing a PCG system it is critical to remember that we can potentially generate massive amounts of content for designers to interact with and players to experience. It is thus of utmost importance to be able to evaluate how successful the outcomes of the generator might be across dissimilar users: players and designers. While content generation is a cheap process relying on algorithms, design and game-play are expensive tasks relying on humans who cannot afford the experience of bad content. Second, content quality might be affected by **algorithms** and their underlying stochasticity, for instance, in evolutionary search. Content generators often exhibit non-deterministic behavior, making it very hard to predict a priori what the outcomes of a particular generative system might be.

4.6.2 Function vs. Aesthetics

Particular properties of content can be **objectively** defined and tested whereas other properties of it can only be assessed **subjectively**. It is only natural to expect that functional properties of content quality can be objectively defined whereas a large part of its aesthetics can only be defined subjectively. For instance, playability of a level is a functional characteristic that can be objectively measured—e.g., an AI agent manages to complete the level; hence it is playable. Balance and symmetry can also be objectively defined to a degree through estimates of deviation from a norm—it may be a score (balance) or a distance from a central choke point in the map (symmetry). There are games, however, for which content balance, symmetry and other functional properties are not trivially measurable. And of course there are several aspects of content such as the comprehensibility of a narrative, the pleasantnesses of a color scheme, the preference for a room's architectural style or the graphics style, and the experience of sound effects and music that are not objectively measured.

Functional, objectively defined, content properties can be expressed either as metrics or as constraints that a generator needs to satisfy. **Constraints** can be specified by the content designers or imposed by other game content already available. For instance, let us assume that a well-designed generated strategy game level needs to be both balanced and playable. Playability can form a simple binary constraint: the level is playable when an AI agent manages to complete it; it is non-playable otherwise. Balance can form another constraint by which all items, bases and resources are accessible to similar degrees by all players; if equal accessibility is below a threshold value then the constraint is not satisfied. Next, let us suppose we wish to generate a new puzzle for the map we just generated. Naturally, the puzzle needs to be compatible with our level. A PCG algorithm needs to be able to satisfy these constraints as part of its quality evaluation. Constrained satisfaction algorithms such as the feasible-infeasible two-population genetic algorithm [379, 382], constrained divergent search rewarding content *value* but also content *novelty* [382] or *surprise* [240], and constraint solvers such as answer set programming [638] are able to handle this. The generated results are within constraints, thereby valuable for the designer. Value, however, may have varying degrees of success and this is where alternative methods or heuristics can help, as we cover in the section below.

4.6.3 How Can We Evaluate a Generator?

Generally speaking, a content generator can be evaluated in three ways: directly by the **designer** or indirectly by either **human players** or **AI agents**. Designers can directly observe properties of the content generator and take decisions based on data **visualization** methods. Human players can play and test the content and/or provide feedback about the content via subjective reporting. AI agents can do the same: play the content or measure something about the content and report it to us in the form of a quality metric, or metrics. Clearly, machines cannot *experience* the content but they can, instead, simulate it and provide us estimates of content experience. The overall evaluation process can very well combine and benefit from any of the above approaches. In the remainder of this section we cover the approaches of data visualization, AI automated playtesting and human playtesting in further detail.

4.6.3.1 Visualization

The visualization approach to content quality assessment is associated with a) the computation of meaningful metrics that can assign measurable characteristics to content and b) ways to visualize these metrics. The task of metric design can be viewed as the equivalent of fitness function design. As such, designing good con-

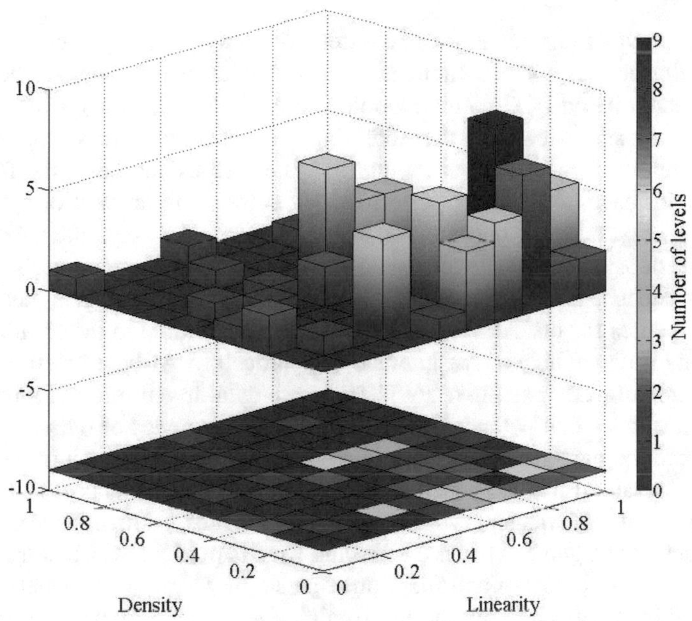

Fig. 4.23 The expressive range of the *Ropossum* level generator for the metrics of linearity and density. Adapted from [608]

tent generation quality metrics in an ad-hoc manner involves a degree of practical wisdom. Metrics, for instance, can be based on the **expressive range** of the generator under assessment, so-called expressivity metrics [640, 608]. The analysis of a generator's expressivity gives us access to the potential overall quality of the generator across its full range of generative space. The generative space can then be visualized as **heatmaps** or alternative graphical representations such as 2D or 3D scatter plots (see Fig. 4.23 for an example). It is one thing if a level generator is able to create only a few meaningful or playable levels and another if the generator is robust and consistent with respect to the playability of its generated levels. It is also one thing if our generator is only able to generate levels with very specific characteristics within a narrow space of its expressive range and another if our level generator is able to express a broad spectrum of level properties, yielding uniformly covered expressive ranges. Such information can be directly visible on the illustrated heatmaps or scatter plots. Alternatively, **data compression** methods can be used directly on the generated content and offer us 2D or 3D representations of the generative space, thereby, bypassing the limitations of ad-hoc metric design. An example of this approach is the use of autoencoders for compressing the images produced by the DeLeNoX autonomous content generator [373].

4.6.3.2 AI

Using AI to **playtest** our generator is a safe and relatively cheap way to rapidly retrieve quality metrics for it without relying on human playtesting. In the same way that a search-based PCG method would use AI to simulate the content before generating it, AI agents can test the potential of a generator across a number of metrics and return us values about its quality. The metrics can be in the form of classes—for instance, test checks performed for completing an area of a level—scalar values—e.g., a level's balance—or even ordinal values—e.g., the rank of the level in terms of asymmetry. The relevance of the metrics to the generator's quality is obviously dependent on the designer's ad-hoc decisions. Once again, designing appropriate metrics for our AI agent to compute is comparable to the challenge of designing any utility function. An interesting approach to AI-based testing is the use of **procedural personas** [267, 269]. These are data-driven inferred models of dissimilar play styles that potentially imitate the different styles of human play. In a sense, procedural personas provide a more human-realistic approach to AI-based testing of a generator. Finally, by visualizing particular game artifacts or simulating them through the use of AI agents we can have access to the information we might be able to extract from a game, we can understand what is possible within our content space, we can infer how rules and functions operate in whatever we generate, and we can possibly understand how the information we are able to extract relates to data we can extract from human playtesting [481].

4.6.3.3 Human Players

In addition to data visualization and AI-based simulation for the evaluation of a content generator a designer might wish to use complementary approaches that rely on quantitative user studies and playtesting. Playtesting is regarded to be an expensive way to test a generator but it can be of immense benefit for content quality assurance, in particular for those aspects of content that cannot be measured objectively—e.g., aesthetics and playing experience. The most obvious approach to evaluate the content experienced by players is to *explicitly* ask them about it. A game user study can involve a small number of dedicated players that will play through various amounts of content or, alternatively, a crowdsourcing approach can provide sufficient data to machine learn content evaluation functions (see [621, 370, 121] among others). Data obtained can be in any format including classes (e.g., a binary answer about the quality of a level), scores (e.g., the likeliness of a sound) or ranks (e.g., a preference about a particular level). It is important to note that the playtesting of content can be complemented by annotations coming from the designers of the game or other experts involved in content creation. In other words, our content generator may be labeled with both first-person (player) and third-person (designer) annotations. Further guidelines about which questionnaire type to use and advice about the design of user study protocols can be found in Chapter 5.

4.7 Further Reading

Extensive versions of most of the material covered in this chapter can be found in dedicated chapters of the PCG in Games Textbook [616]. In particular, all the methods and types of PCG in games are covered in further detail (see Chapters 1 to 9 of [616]). Considering the roles of PCG, the mixed-initiative and the experience-driven role of PCG are, respectively, detailed in Chapters 11 [374] and 10 [618] of that textbook [616]. In addition to Chapter 11 of [616] the framework named *mixed-initiative co-creativity* [774] provides a theoretical grounding for the impact of mixed-initiative interaction on the creativity of both designers and computational processes. Further, the original articles about the experience-driven PCG framework can be found in [783, 784]. Finally, the topic of PCG evaluation is covered also in the final chapter [615] of the PCG in Games textbook [616].

4.8 Exercises

PCG offers endless opportunities for generation and evaluation across the different creativity domains in games and across combinations of those. As an initial step we would recommend the reader to start experimenting with maze generation and platformer level generation (as outlined below). The website of the book contains details regarding both frameworks and potential exercises.

4.8.1 Maze Generation

Maze generation is a very popular type of level generation and relevant for several game genres. In the first exercise we recommend that you develop a maze generator using both a constructive and a search-based PCG approach and compare their performance according to a number of meaningful criteria that you will define. The reader may use the Unity 3D open-access maze generation framework which is available at: http://catlikecoding.com/unity/tutorials/maze/. Further guidelines and exercises for maze generation can be found at the book's website.

4.8.2 Platformer Level Generation

The platformer level generation framework is based on the *Infinite Mario Bros* (Persson, 2008) framework which has been used as the main framework of the Mario AI (and later Platformer AI) Competition since 2010. The competition featured several different tracks including gameplay, learning, Turing test and level generation. For the exercises of this chapter the reader is requested to download the level generation framework (https://sites.google.com/site/platformersai/) and apply constructive and generate-and-test methods for the generation of platformer levels. The levels need to be evaluated using one or more of the methods covered in this book. Further details and exercises with the platformer level generation framework can be found at the book's website.

4.9 Summary

This chapter viewed AI as a means for generating content in games. We defined procedural content generation as the algorithmic process of creating content in and for games and we explored the various benefits of this process. We then provided a general taxonomy about content and its generation and explored the various ways one can generate content including search-based, solver-based, grammar-based, machine learning-based, and constructive generation methods. The use of the PCG method is naturally dependent on the task at hand and the type of content one wishes to generate. It further depends on the potential role the generator might take within games. We outlined the four possible roles a generator can take in games which are determined by the degree to which they involve the designer (autonomous vs. mixed-initiative) and/or the player (experience-agnostic vs. experience-driven) in the process. The chapter ends with a discussion on the important and rather unexplored topic of evaluation, the challenges it brings, and a number of evaluation approaches one might consider.

We have so far covered the most traditional use of AI in games (Chapter 3) and the use of AI for generating parts of (or complete) games (this chapter). The next and final chapter of this part is dedicated to the player and the ways we can use AI to model aspects of her behavior and her experience.

Chapter 5
Modeling Players

This chapter is dedicated to *players* and the use of AI for modeling them. This area of research is often called **player modeling** [782, 636]. We take player modeling to mean the detection, prediction and expression of human player characteristics that are manifested through cognitive, affective and behavioral patterns while playing games. In the context of this book, player modeling studies primarily the use of AI methods for the construction of computational models of players. By **model** we refer to a mathematical representation—it may be a rule set, a vector of parameters, or a set of probabilities—that captures the underlying function between the characteristics of the player and her interaction with the game, and the player's response to that interaction. Given that every game features at least one player (with some notable exceptions [50]), and that player modeling affects work on game-playing and content generation, we consider the modeling of player behavior and experience as a very important use of AI in games [764, 785].

Psychology has studied human behavior, cognition and emotions for a long time. Branches of computer science and human-computer interaction that attempt to model and simulate human behavior, cognition, emotion or the feeling of emotion (**affect**) include the fields of **affective computing** and **user modeling**. Player modeling is related to these fields but focuses on the domain of games. Notably, games can yield dynamic and complex emotions in the player, the manifestations of which cannot be captured trivially by standard methods in empirical psychology, affective computing or cognitive modeling research. The high potential that games have in affecting players is mainly due to their ability to place the player in a continuous mode of interaction, which, in turn, elicits complex cognitive, affective and behavioral responses. Thus, the study of the player may not only contribute to the design of improved forms of human-computer interaction, but also advance our knowledge of human experiences.

As mentioned earlier, every game features at least one user—the player—who controls some aspect of the game environment. The player character could be visible in the game as an avatar or a group of entities [94], or could be invisible as in many puzzle games and casual games. Control may vary from the relatively simple (e.g., limited to movement in an orthogonal grid) to the highly complex (e.g., having

G. N. Yannakakis and J. Togelius, *Artificial Intelligence and Games*, https://doi.org/10.1007/978-3-319-63519-4_5

to decide several times per second between hundreds of different possibilities in a highly complex 3D world). Given these intricacies, understanding and modeling the interaction between the player and the game can be seen as a **holy grail** of game design and development. Designing the interaction and the emergent experience *right* results in a successful game that manages to elicit unique experiences.

The interaction between the player(s) and the game is dynamic, real-time and in many cases highly complex. The interaction is also **rich** in the sense that many modalities of interaction may be involved and that the information exchange between the game and the player may both be fast and entail large amounts of data for us to process. If the game is well-designed, the interaction is also highly engaging for the player. Given the great amount of information that can be extracted through this interaction and used for creating models of the player, the game should be able to learn much about the person playing it, as a player and perhaps as a human in general. In fact, there is no reason why the model should not know more about how you play than you do.

In the remainder of this chapter we first attempt to define the core ingredients of player modeling (Section 5.1) and then we discuss reasons why AI should be used to model players (Section 5.2). In Section 5.3 we provide a high-level taxonomy of player modeling focusing on two core approaches for constructing a player model: top-down and bottom-up. We then detail the available types of data for the model's **input** (Section 5.4), a classification for the model's **output** (Section 5.5) and the various AI methods that are appropriate for the player modeling task (Section 5.6). The key components of player modeling as discussed in this chapter (input, output and model) are depicted in Fig. 5.1. We conclude, in Section 5.7, with a number of concrete examples of AI being used for modeling players.

5.1 What Player Modeling Is and What It Is Not

One could arguably detect behavioral, emotional or cognitive aspects of both human players and non-human players, or non-player characters (notwithstanding the actual existence of emotions in the latter). However, in this book we focus on aspects that can be detected from, modeled from, and expressed in games with human players [782]. We explicitly exclude the modeling of NPCs from our discussion in this chapter, as in our definition, player modeling is modeling of a **human player**. Modeling the experience of an NPC would seem to be a futile exercise, as one can hardly say that an NPC possesses actual emotions or cognition. Modeling the behavior of an NPC is also of little interest, at least if one has access to the game's code: a perfect model for the NPC already exists. NPC modeling, however, can be a useful testbed for player modeling techniques, for instance, by comparing the model derived from human players with the hand-crafted one. More interestingly, it can be an integral component of AI that adapts its behavior in response to the dynamics of the NPCs—as in [28]. Nevertheless, while the challenges faced in modeling NPCs

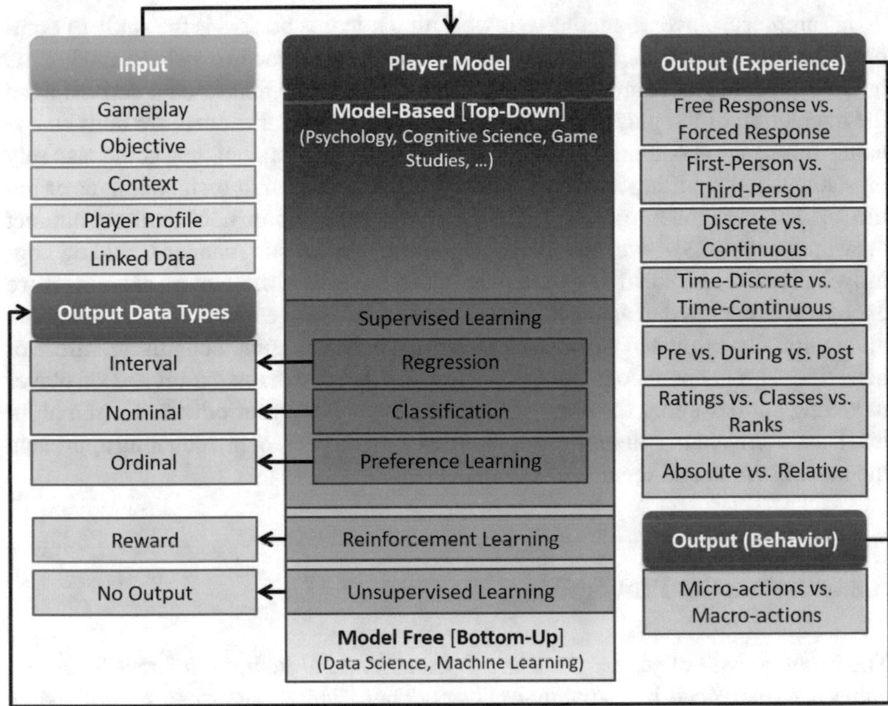

Fig. 5.1 The key components of player modeling as discussed in this chapter. The distinction between model-based and model-free approaches is outlined in Section 5.3. The various options for the input of the model are discussed in Section 5.4. The taxonomy for the model's output is discussed in Section 5.5—each box represents a dedicated subsection. Finally, the various AI methods (supervised learning, reinforcement learning and unsupervised learning) used for modeling corresponding output data types are discussed thoroughly in Section 5.6.

are substantial, the issues raised from the modeling of human players define a far more complex and important problem for the understanding of player experience.

Sometimes the terms player modeling and **opponent modeling** [214, 592, 48] are used interchangeably when a human player is modeled. However, opponent modeling is a more narrow concept referring to predicting behavior of an adversarial player when playing to win in an imperfect information game like Poker [48] or *StarCraft* (Blizzard Entertainment, 1988) [504]. Some aspects of modeling NPCs or simulated playthroughs for winning in a game are discussed in Chapter 3.

We also make a distinction between player modeling [116, 281] and **player profiling** [782]. The former refers to modeling complex **dynamic** phenomena during gameplay interaction, whereas the latter refers to the categorization of players based on **static** information that does not alter during gameplay. Information of static nature includes personality, cultural background, gender and age. We put an emphasis on the former, but will not ignore the latter, as the availability of a good player profile may contribute to the construction of reliable player models.

In summary, player modeling—as we define it in this book—is the study of computational means for the modeling of a player's experience or behavior which is based on theoretical frameworks about player experience and/or data derived from the interaction of the player with a game [782, 764]. Player models are built on dynamic information obtained during game-player interaction, but they could also rely on static player profiling information. Unlike studies focusing on taxonomies of **behavioral** player modeling—e.g., via a number of dimensions [636] or direct/indirect measurements [623]—we view player modeling in a holistic manner including cognitive, affective, personality and demographic aspects of the player. Moreover, we exclude approaches that are not directly based on human-generated data or not based on empirically-evaluated theories of player experience, human cognition, affect or behavior. The chapter does not intend to provide an exhaustive review of player modeling studies under the above definition, but rather an introduction and a high-level taxonomy that explores the possibilities with respect to the modeling approach, the model's input and the model's output.

5.2 Why Model Players?

The *primary goal* of player modeling is to *understand* how the interaction with a game is experienced by individual players. Thus, while games can be utilized as an arena for eliciting, evaluating, expressing and even synthesizing experience, we argue that the main aim of the study of players in games is the understanding of players' cognitive, affective and behavioral patterns. Indeed, by the very nature of games, one cannot dissociate games from player experience.

There are two core reasons that drive the use of AI for modeling game players and their play, thereby serving the primary goal of player modeling as stated above. The *first* is for understanding something about their players' **experience** during play. Models of player experience are often built using machine learning methods, typically supervised learning methods like support vector machines or neural networks. The training data here consists of some aspect of the game or player-game interaction, and the targets are labels derived from some assessment of player experience, gathered for example from physiological measurements or questionnaires [781]. Once predictors of player experience are derived they can be taken into account for designing the in-game experience. That can be achieved by adjusting the behavior of non-player characters (see Chapter 3) or by adjusting the game environment (see Chapter 4).

The *second* reason why one would want to use AI to model players is for understanding players' **behavior** in the game. This area of player modeling is concerned with structuring observed player behavior even when no measures of experience are available—for instance, by identifying player types or predicting player behavior via game and player **analytics** [178, 186]. A popular distinction in data derived from games [186] is the one between **player metrics** and **game metrics**. The latter is a superset of the former as it also includes metrics about the game software (system metrics) and the game development process as a whole (process metrics). System metrics and process metrics are important aspects of modern game development that influence decision making with respect to procedures, business models, and marketing. In this book, however, we focus on player metrics. The interested reader may refer to [186] for alternative uses of metrics in games and the application of analytics to game development and research—i.e., **game analytics**.

Once aspects of player behavior are identified a number of actions can be taken to improve the game such as the personalization of content, the adjustment of NPCs or, ultimately, the redesign of (parts of) the game. Derived knowledge about the in-game behavior of the player can lead to improved game testing and game design procedures, and better monetization and marketing strategies [186]. Within behavior modeling we identify four main player modeling subtasks that are particularly relevant for game AI: **imitation** and **prediction**—achieved via supervised learning or reinforcement learning—and **clustering** and **association mining**—achieved via unsupervised learning. The two main purposes of player **imitation** is the development of non-player characters with believable, human-like behavioral characteristics, and the understanding of human play per se through creating generative models of it. The **prediction** of aspects of player behavior, instead, may provide answers to questions such as "when will this player stop playing?" or "how often will that player get stuck in that area of the level?" or "which item type will this player pick in the next room?". The aim of **clustering** is the classification of player behaviors within a number of clusters depending of their behavioral attributes. Clustering is important for both the personalization of the game and the understanding of playing behavior in association with the game design [178]. Finally, **association mining** is useful in instances where frequent patterns or sequences of actions (or in-game events) are important for determining how a player behaves in a game.

While player behavior and player experience are interwoven notions there is a subtle difference between them. Player behavior points to **what a player does** in a game whereas player experience refers to **how a player feels** during play. The feeling of one's gameplay experience is clearly associated with what one does in the game; player experience, however, is primarily concerned with affective and cognitive aspects of play as opposed to mere reactions of gameplay which refer to player behavior.

Given the above aims, core tasks and sub-tasks of player modeling in the next section we discuss the various available options for constructing a player model.

5.3 A High-Level Taxonomy of Approaches

Irrespective of the application domain, computational models are characterized by three core components: the input the model will consider, the computational model per se, and the output of the model (see Fig. 5.1). The model itself is a mapping between the input and the output. The mapping is either hand-crafted or derived from data, or a mix of the two. In this section we will first go through the most common approaches for constructing a computational model of players, then we will go through a taxonomy of possible inputs for a player model (Section 5.4) and finally we will examine aspects of player experience and behavior that a player model can represent as its output (Section 5.5).

A high-level classification of the available approaches for player modeling can be made between **model-based** (or top-down) and **model-free** (or bottom-up) approaches [782, 783]. The above definitions are inspired by the analogous classification in RL by which a world model is available (i.e., model-based) or not (i.e., model-free). Given the two ends of this continuum **hybrid** approaches between them can naturally exist. The gradient red color of the player model box in Fig. 5.1 illustrates the continuum between top-down and bottom-up approaches. The remainder of this section presents the key elements of and core differences among the various approaches for modeling of players.

5.3.1 Model-Based (Top-Down) Approaches

In a **model-based** or top-down [782] approach a player model is built on a theoretical framework. As such, researchers follow the modus operandi of the humanities and social sciences, which hypothesize models to explain phenomena. Such hypotheses are usually followed by an empirical phase in which it is experimentally determined to what extent the hypothesized models fit observations; however, such a practice is not the norm within player experience research. While user experience has been studied extensively across several disciplines, in this book we identify three main disciplines we can borrow theoretical frameworks from and build models of player experience: **psychology and affective sciences**, **neuroscience**, and finally, **game studies and game research**.

5.3.1.1 Psychology and Affective Sciences

Top-down approaches to player modeling may refer to models derived from popular theories about emotion [364] such as the cognitive appraisal theory [212, 601]. Further, the player model may rely on well established affect representations such as the emotional dimensions of arousal and valence [200] that define the circumplex model of affect of Russell [539] (see Fig. 5.2(a)). Valence refers to how pleasurable (positive) or unpleasurable (negative) the emotion is whereas arousal refers to how intense (active) or lethargic (inactive) that emotion is. Following a theoretical model, emotional manifestations of players are often mapped directly to specific player states. For instance, by viewing player experience as a psychophysiological phenomenon [779] a player's increased heart rate may correspond to high arousal and, in turn, to high levels of excitement or frustration.

Beyond established theories of emotion, model-based approaches can also be inspired by a general cognitive-behavioral theoretical framework such as the *theory of mind* [540] for modeling aspects of social interactions in games. Popular example frameworks for deriving user models in games include the usability theory [489, 290], the belief-desire-intention (BDI) model [66, 224], the cognitive model by Ortony, Clore, and Collins [512] and Skinner's behavioristic approach [633] with its links to reward systems in games. Further we can draw inspiration from social sciences and linguistics in order to model lexical aspects of gameplay interaction (e.g., chatting). Natural language processing, opinion mining and sentiment analysis are normally relying on theoretical models that build on affective and sociological aspects of textual communication [517, 514].

Of particular importance is the concept of **flow** by Csikszentmihalyi [151, 149, 150] which has been a popular psychological construct for modeling player experience in a top-down fashion. When in a state of flow (or else, state of "happiness") during an activity we tend to concentrate on the present moment, we lose our ability of reflective self-consciousness, we feel a sense of personal control over the situation or activity, our perception of time is altered, and we experience the activity as intrinsically rewarding. Analogously the optimal experience during play has been associated with a fine balance between boredom and anxiety, also known as the **flow channel** (see Fig. 5.2(b)). Given its direct relevance to player experience, flow has been adapted and incorporated for use in game design and for the understanding of player experience [678, 675, 473].

5.3.1.2 Neuroscience

A number of studies have relied on the working hypothesis of an underlying mapping between the brain, its neural activity and player experience. However, this relationship is not well explored and the presumptive mapping is largely unknown. For example, interest has been associated with activity in the visual cortex and the release of endomorphin whereas the sense of achievement has been linked to dopamine levels [35]. According to [35], neuroscientific evidence suggests that the reward systems of games are directly associated with the dopamine-based reward structures in the brain and that dopamine is released during gameplay [346]. Further, pleasure has been associated with areas in the brain responsible for decision making, thereby revealing the direct links between gameplay experience and decision making [575]. Pleasure has also been associated with uncertain outcomes or uncertain rewards [625] as well as with interest and curiosity [43], which are all key elements of successful game design. Stress is also tightly coupled with player experience given its clear association with anxiety and fear; stress can be both monitored via physiology and regulated via game design. The testosterone levels of players have also been measured in association to digital game activities [444] and findings reveal particular patters of competition in games as testosterone factors. Finally, it appears that trust between players in a social gaming setup could be measured indirectly via oxytocin levels [350].

The degree to which current findings from neuroscience are applicable to player experience research is largely unknown since access to neural activity and brain hor-

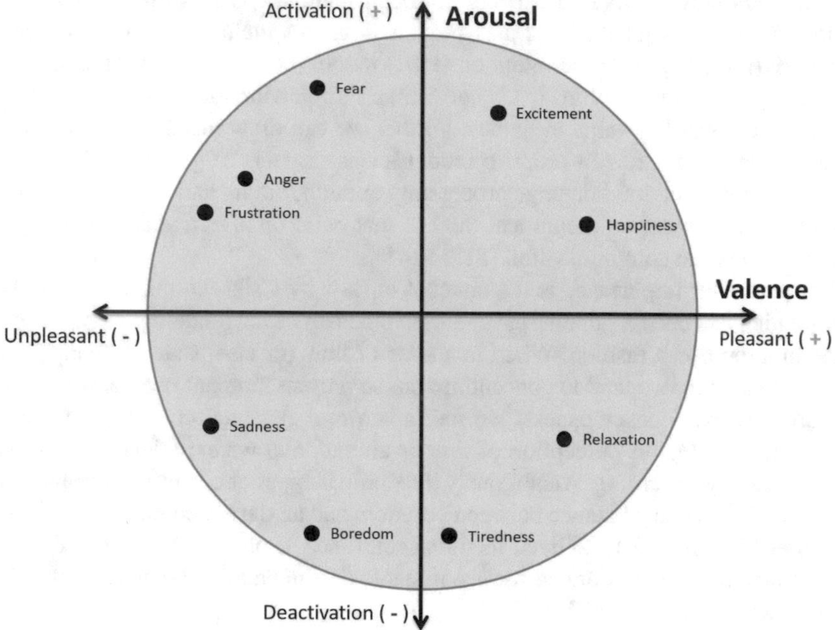

(a) Russell's two-dimensional circumplex model of affect. The figure contains a small number of representative affective states (black circles).

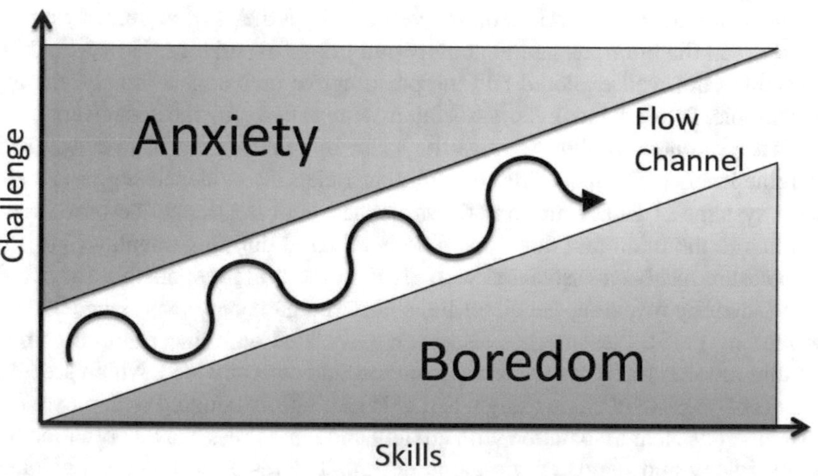

(b) An illustration of the *flow channel*.

Fig. 5.2 Two popular frameworks used for modeling users and their experience in games: (a) the arousal-valence circumplex model of affect and (b) the flow channel concept.

mone levels remains a rather intrusive process at the time of writing. Manifestations of brain activity such as the brain's electrical waves—measured through electroencephalography on our scalp—or more indirect manifestations such as stress and anxiety—measured through skin conductance—can give us access to approximates of brain activity. These approximates can be used for modeling the experience of play as discussed later in this chapter.

5.3.1.3 Game Studies and Game Research

Theoretical models of user experience in games are often driven by work in game studies and game research. Examples of models that have been used extensively in the literature include Malone's core design dimensions that collectively contribute to 'fun' games [419] defined as **challenge**, **curiosity** and **fantasy**. In particular, challenge refers to the uncertainty of achieving a goal due to e.g., variable difficulty level, multiple level goals, hidden information, and randomness. Curiosity refers to the player's feeling of uncertainty with respect to what will happen next. Finally, fantasy is the ability of the game to show (or evoke) situations or contexts that are not actually present. These three dimensions have been quantified, operationalized and successfully evaluated in prey-predator games [766], physical games [769, 775], preschooler games [320] and racing games [703].

Bartle's [33] classification of player types within games as a form of *general* player profiles can be used indirectly for modeling players. Bartle identifies four archetypes of players he names **killers** (i.e., players that focus on winning and are engaged by ranks and leaderboards), **achievers** (i.e., players that focus on achieving goals quickly and are engaged by achievements), **socializers** (i.e., players that focus on social aspects of games such as developing a network of friends) and **explorers** (i.e., players who focus on the exploration of the unknown). Various other methodologies have also been followed to derive *specific* player experience archetypes for particular classes of games [34, 787].

Other popular and interconnected views of player experience from a *game design* perspective include the theory of 'fun' by Koster [351], the notion of the 'magic circle' in games [587] and the four "fun" factor model of Lazzaro [365]. Indicatively, Koster's theory relates the concept of **fun** with learning in games: the more you learn the more you tend to play a game. According to his theory you stop playing a game that is way too easy (no learning of new skills) or way too hard (no learning either). Lazzaro's four fun factors are named **hard fun** (e.g., playing to win and see how good I am at it), **easy fun** (e.g., playing to explore new worlds and game spaces), **serious fun** (e.g., playing to feel better about myself or get better at something that matters to me) and **people fun** (e.g., playing as an excuse to invite friends over, or having fun merely by watching them play). Within *game studies*, the theoretical model of incorporation [94] is a notable multifaceted approach for capturing player immersion. The model is composed of six types of player involvement: affective, kinaesthetic, spatial, shared, ludic, and narrative.

With a careful analysis of the models proposed and their subcomponents one could coherently argue that there is one underlying theoretical model of player experience after all. While it is not the intention of this book to thoroughly discuss the interconnections between the aforementioned models it is worth pointing out a number of indicative examples of our envisaged overarching player experience model. An explorer (Bartle), for instance, can be associated with the easy fun factor of Lazzaro and the curiosity dimension of Malone. Further, the achiever archetype (Bartle) can be linked to the serious fun factor (Lazzaro). Accordingly, a killer archetype (Bartle) maps to the hard fun factor (Lazzaro), the challenge dimension of Malone's model, and a number of flow aspects. Finally, a socializer player profile (Bartle) could be associated to people fun (Lazzaro) and, in turn, to the shared involvement facet of Calleja [94].

Even though the literature on theoretical models of experience is rather rich, one needs to be cautious with the application of such theories to games (and game players) as the majority of the models have not been derived from or tested on interactive media such as games. Calleja [94], for instance, reflects on the inappropriateness of the concepts of 'fun' and 'magic circle' (among others) for games. At this point it is worth noting that while ad-hoc designed models can be an extremely powerful and expressive they need to be cross-validated empirically to be of practical use for computational player modeling; however, such practices are not as common within the broader area of game studies and game design.

5.3.2 Model-Free (Bottom-Up) Approaches

Model-free approaches refer to the data-driven construction of an unknown mapping (model) between a player **input** and a **player state**. Any manifestation of player affect or behavioral pattern could define the input of the model (see more in Section 5.4 below). A player state, on the other hand, is any representation of the player's experience or current emotional, cognitive, or behavioral state; this is essentially the output of the computational model (see more in Section 5.5). Evidently, model-free approaches follow the modus operandi of the exact sciences, in which observations are collected and analyzed to generate models without a strong initial assumption on what the model looks like or even what it captures. Player data and labels of player states are collected and used to derive the model.

Classification, regression and preference learning techniques adopted from machine learning—see Chapter 2—or statistical approaches are commonly used for the construction of the mapping between the input and the output. Examples include studies in which player actions, goals and intentions are modeled and predicted for the purpose of believable gameplay, adaptive gameplay, interactive storytelling or even the improvement of a game's monetization strategy [511, 800, 414, 693, 592]. In contrast to supervised learning, reinforcement learning can be applied when a reward function, instead, can characterize aspects of playing behavior or experience. Unsupervised learning is applicable when target outputs are not available for predictive purposes but, alternatively, data is used for the analysis of playing behavior (see Fig. 5.1).

We meet bottom-up player modeling attempts since the early years of the game AI field in first-person shooters [695, 696], racing games [703] and variants of *Pac-Man* (Namco, 1980) [776]. Recently, the availability of large sets of game and player data has opened up the horizons of behavioral data mining in games—i.e., **game data mining** [178]. Studies that attempt to identify different behavioral, playing and action patterns within a game are well summarized in [186] and include [36, 176, 687, 690, 750], among many others.

5.3.3 *Hybrids*

The space between a completely model-based and a completely model-free approach can be viewed as a continuum along which any player modeling approach might be placed. While a completely model-based approach relies solely on a theoretical framework that maps a player's responses to game stimuli, a completely model-free approach assumes there is an unknown function between modalities of user input and player states that a machine learner (or a statistical model) may discover, but does not assume anything about the structure of this function. Relative to these extremes, the vast majority of studies in player modeling may be viewed as **hybrids** that synergistically combine elements of the two approaches. The continuum between top-down and bottom-up player modeling approaches is illustrated with a gradient color in Fig. 5.1.

5.4 What Is the Model's Input Like?

By now we have covered the various approaches available for modeling players and we will, in this section, focus on what the input of such a model might be like. The model's input can be of three main types: (1) anything that a player is doing in a game environment gathered from **gameplay** data—i.e., behavioral data of any type such as user interface selections, preferences, or in-game actions; (2) **objective** data collected as responses to game stimuli such as physiology, speech and body movements; and (3) the **game context** which comprises of any player-agent interactions but also any type of game content viewed, played, and/or created. The three input types are detailed in the remainder of this section. At the end of the section we also discuss *static* profile information on the player (such as personality) as well as web data beyond games that could feed and enhance the capacity of a player model.

5.4.1 Gameplay

Given that games may affect the player's cognitive processing patterns and cognitive focus we assume that a player's actions and preferences are linked directly to her experience. Consequently, one may infer the player's current experience by analyzing patterns of her interaction with the game, and by associating her experience with game context variables [132, 239]. Any element derived from the direct interaction between the player and the game can be classified as **gameplay** input. These interpretable measures of gameplay have also been defined as **player metrics** [186]. Player metrics include detailed attributes of the player's behavior derived from responses to game elements such as NPCs, game levels, user menus, or embodied conversational agents. Popular examples of data attributes include detailed spatial locations of players viewed as **heatmaps** [177], statistics on the use of in-game menus, as well as descriptive statistics about gameplay, and communication with other players. Figure 5.3 shows examples of heatmaps in the *MiniDungeons*[1] puzzle game. Both general measures (such as performance and time spent on a task) and game-specific measures (such as the weapons selected in a shooter game [250]) are relevant and appropriate player metrics.

A major limitation with the gameplay input is that the actual player experience is only **indirectly** observed. For instance, a player who has little interaction with a game might be thoughtful and captivated, or just bored and busy doing something else. Gameplay metrics can only be used to approach the likelihood of the presence of certain player experiences. Such statistics may hold for player populations, but may provide little information for individual players. Therefore, when one attempts to use pure player metrics to make estimates of player experiences and make the game respond in an appropriate manner to these perceived experiences, it is advisable to keep track of the feedback of the player to the game responses, and adapt when the feedback indicates that the player experience was gauged incorrectly.

5.4.2 Objective

Computer game players are presented with a wide palette of affective stimuli during game play. Those stimuli vary from simple auditory and visual events (such as sound effects and textures) to complex narrative structures, virtual cinematographic views of the game world and emotively expressive game agents. Player emotional responses may, in turn, cause changes in the player's physiology, reflect on the player's facial expression, posture and speech, and alter the player's attention and focus level. Monitoring such bodily alterations may assist in recognizing and constructing the player's model. As such, the **objective** approach to player modeling incorporates access to multiple modalities of player input.

[1] http://minidungeons.com/

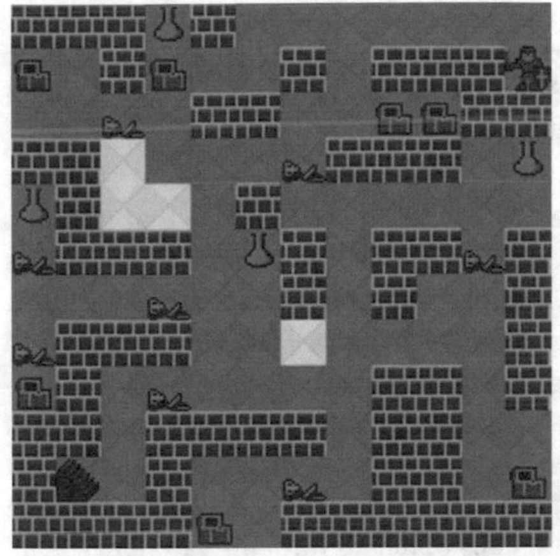

(a) In this example the player acts as a completionist: succeeding in killing all monsters, drinking all potions and collecting all treasure.

(b) In this example the player prioritizes reaching the exit, avoiding any monsters and only collecting potions and treasures that are near the path to the exit.

Fig. 5.3 Two example heatmaps (human playtraces) in the *MiniDungeons* game. *MiniDungeons* is a simple turn-based rogue-like puzzle game, implemented as a benchmark problem for modeling the decision-making styles of human players [267].

The relationship between psychology and its physiological manifestations has been studied extensively ([17, 95, 779, 558] among many others). What is widely evidenced is that the sympathetic and the parasympathetic components of the autonomic nervous system are involuntary affected by affective stimuli. In general, arousal-intense events cause dynamic changes in both nervous systems: an increase and a decrease of activity, respectively, in the sympathetic and the parasympathetic nervous system. Alternatively, activity at the parasympathetic nervous system is high during relaxing or resting states. As mentioned above, such nervous system activities cause alterations in one's facial expression, head pose, electrodermal activity, heart rate variability, blood pressure, pupil dilation [91, 624] and so on.

Recent years have seen a significant volume of studies that explore the interplay between **physiology** and gameplay by investigating the impact of different gameplay stimuli on dissimilar physiological signals ([697, 473, 421, 420, 556, 721, 175, 451] among others). Such signals are usually obtained through electrocardiography (ECG) [780], photoplethysmography [780, 721], galvanic skin response (GSR) [421, 271, 270, 272], respiration [721], electroencephalography (EEG) [493] and electromyography (EMG).

In addition to physiology one may track the player's **bodily expressions** (motion tracking) at different levels of detail and infer the real-time affective responses from the gameplay stimuli. The core assumption of such input modalities is that particular bodily expressions are linked to expressed emotions and cognitive processes. Objective input modalities, beyond physiology, that have been explored extensively include facial expressions [321, 19, 236, 88, 794], muscle activation (typically face) [133, 164], body movement and posture [23, 731, 321, 172, 47], speech [741, 319, 308, 306, 30], text [517, 137, 391], haptics [509], gestures [283], brain waves [559, 13], and eye movement [23, 469].

While objective measurements can be a very informative way of assessing the player's state during the game a major limitation with most of them is that they can be invasive, thus affecting the player's experience with the game. In fact, some types of objective measures appear to be **implausible** within commercial-standard game development. **Pupillometry** and **gaze tracking**, for instance, are very sensitive to distance from screen, and variations in light and screen luminance, which collectively make them rather impractical for use in a game application. The recent rebirth of virtual reality (VR), however, gives eye gaze sensing technologies entirely new opportunities and use within games [628]; a notable example of a VR headset that features eye-tracking is FOVE.[2] Other **visual cues** obtained through a camera (facial expressions, body posture and eye movement) require a well-lit environment which is often not present in home settings (e.g., when playing video-games) and they can be seen by some players as privacy hazards (as the user is continuously recorded). Even though highly unobtrusive, the majority of the vision-based affect-detection systems currently available have additional limitations when asked to operate in real-time [794]. We argue that an exception to this rule is body posture, which can both be effectively detected nowadays and provide us with meaningful estimates of player experience [343]. Aside from the potential they might have, however, the appropriateness of camera-based input modalities for games is questionable since experienced players tend to stay still while playing games [22].

As a response to the limitations of camera-based measurements, **speech** and **text** (e.g., chat) offer two highly accessible, real-time efficient and unobtrusive modalities with great potential for gaming applications; however, they are only applicable to games where speech (or text) forms a control modality (as e.g., in conversational games for children [320, 789]), collaborative games that naturally rely on speech or text for communication across players (e.g., in collaborative first-person shooters), or games that rely on natural language processing such as text-based adventure games or interactive fiction (see discussion of Chapter 4).

Within players' **physiology**, existing hardware for EEG, respiration and EMG require the placement of body parts such as the head, the chest or parts of the face on the sensors, making those physiological signals rather impractical and highly intrusive for most games. On the contrary, recent sensor technology advancements for the measurement of electrodermal activity (skin conductivity), photoplethysmography (blood volume pulse), heart rate variability and skin temperature have made those physiological signals more attractive for the study of affect in games. Real-time recordings of these can nowadays be obtained via comfortable wristbands and stored in a personal computer or a mobile device via a wireless connection [779].

At the moment of writing there are a few examples of **commercial games** that utilize physiological input from players. One particularly interesting example is

[2] https://www.getfove.com/

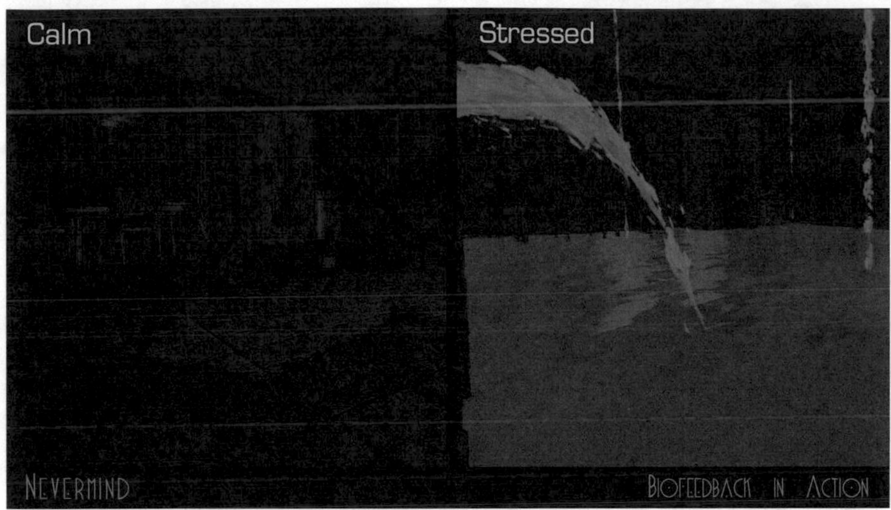

Fig. 5.4 A screenshot from *Nevermind* (Flying Mollusk, 2015). The game supports several off-the-shelf sensors that allow the audiovisual content of the game to adapt to the stress levels of the player. Image obtained from Erin Reynolds with permission.

Nevermind (Flying Mollusk, 2015), a biofeedback-enhanced adventure horror game that adapts to the player's stress levels by increasing the level of challenge it provides: the higher the stress the more the challenge for the player (see Fig. 5.4). A number of sensors which detect heart activity are available for affective interaction with *Nevermind*. *The Journey of Wild Divine* (Wild Divine, 2001) is another biofeedback-based game designed to teach relaxation exercises via the player's blood volume pulse and skin conductance. It is also worth noting that AAA game developers such as Valve have experimented with the player's physiological input for the personalization of games such as *Left 4 Dead* (Valve, 2008) [14].

5.4.3 Game Context

In addition to gameplay and objective input, the game's context is a necessary input for player modeling. **Game context** refers to the momentanous state of the game during play and excludes any gameplay elements; those are already discussed in the gameplay input section. Clearly, our gameplay affects some aspects of the game context and vice versa but the two can be viewed as separate entities. Viewing this relationship from an analytics lens, the game context can be seen as a form of game metrics, opposed to gameplay which is a form of player metrics.

The importance of the game context for modeling players is obvious. In fact, we could argue that the context of the game during the interaction is a necessary input for detecting reliably any cognitive and affective responses of players. It could also be argued that the game context is necessary as a guide during the annotation of the player experience; but more of that we will discuss in Section 5.5. The same way that we require the current social and cultural context to better detect the underlying emotional state of a particular facial expression of our discussant any player reactions cannot be dissociated from the stimulus (or the game context) that elicited them. Naturally, player states are always linked to game context. As a result, player models that do not take context into account run a risk of inferring erroneous states for the player. For example, an increase in galvanic skin response can be linked to different high-arousal affective states such as frustration and excitement. It is very hard to tell however, what the heightened galvanic skin response "means" without knowing what is happening in the game at the moment. In another example, a particular facial expression of the player, recorded though a camera, could be associated with either an achievement in the game or a challenging moment, and needs to be triangulated with the current game state to be understood. Evidently, such dualities of the underlying player state may be detrimental for the design of the player model.

While a few studies have investigated physiological reactions of players in isolation, good practice in player modeling commands that any reactions of the players is triangulated with information about the current game state. For instance, the model needs to know if the GSR increases because the player died or completed the level. The game context—naturally combined (or **fused**) with other input modalities from the player—has been used extensively in the literature for the prediction of different affective and cognitive states relevant to playing experience [451, 434, 521, 617, 572, 133, 254, 558, 452, 433].

5.4.4 Player Profile

A **player profile** includes all the information about the player which is *static* and it is not directly (nor necessarily) linked to gameplay. This may include information on player personality (such as expressed by the Five Factor Model of personality [140, 449]), culture dependent factors, and general demographics such as gender and age. A player's profile may be used as input to the player model to complement the captured in-game behavior with general attributes about the player. Such information may lead to more precise predictive models about players.

While gender, age [787, 686], nationality [46] and player expertise level [96] have already proven important factors for profiling players the role of personality remains somewhat contentious. On the one hand, the findings of van Lankveld et al. [736, 737], for instance, reveal that gameplay behavior does not necessarily correspond to a player's behavior beyond the game. On the other hand, Yee et al. have identified strong correlations between player choices in *World of Warcraft* (Blizzard Entertainment, 2004) and the personalities of its players [788]. Strong correlations have also been found between the playing style and personality in the first-person shooter *Battlefield 3* (Electronic Arts, 2011) [687]. In general, we need to acknowledge that there is no guaranteed one-to-one mapping between a player's in-game behavior and personality, and that a player's personality profile does not necessarily indicate what the player would prefer or like in a game [782].

5.4.5 Linked Data

Somewhere between the highly dynamic in-game behavior and the static profile information about the player we may also consider **linked data** retrieved from web services that are not associated with gameplay per se. This data, for instance, may include our social media posts, emoticons, emojis [199], tags used, places visited, game reviews written, or any relevant semantic information extracted from diverse Web content. The benefit of adding such information to player models is manyfold but it has so far seen limited use in games [32]. In contrast to current player modeling approaches the use of massive amounts and dissimilar types of content across linked online sources would enable the design of player models which are based on user information stored across various online datasets, thereby realizing semantically-enriched game experiences. For example, both scores and sentiment-analyzed textual reviews [517, 514] from game review sites such as Metacritic[3] or GameRankings[4] can be used as input to a model. This model can then be used to create game content which is expected to appeal to the specific parts of the community, based, for instance, on demographics, skill or interests collected from the user's in-game achievements or favored games [585].

5.5 What Is the Model's Output Like?

The model's **output**, i.e., that which we wish to model, is usually a representation of the player's state. In this section we explore three options for the output of the model that serve different purposes in player modeling. If we wish to model the **experience** of the player the output is provided predominately through manual annotation. If instead we wish to model aspects of player **behavior** the output is predominately based on in-game actions (see Fig. 5.1). Finally, it may very well be that the model has **no output**. Section 5.5.1 and Section 5.5.2 discuss the particularities of the output, respectively, for the purpose of behavioral modeling and experience modeling whereas Section 5.5.3 explores the condition where the model has no outputs.

[3] http://www.metacritic.com

[4] http://www.gamerankings.com/

5.5.1 Modeling Behavior

The task of modeling player behavior refers to the prediction or imitation of a particular behavioral state or a set of states. Note that if no target outputs are available then we are faced with either an unsupervised learning problem or a reinforcement learning problem which we discuss in Section 5.5.3. The output we must learn to predict (or imitate) in a supervised learning manner can be of two major types of **gameplay** data: either **micro-actions** or **macro-actions** (see Fig. 5.1). The first machine learning problem considers the moment-to-moment game state and player action space that are available at a frequency of frame rates. For example, we can learn to imitate the moves of a player on a frame-to-frame basis by comparing the play traces of an AI agent and a human as e.g., done for *Super Mario Bros* (Nintendo, 1985) [511, 469]. When macro-actions are considered instead, the target output is normally an aggregated feature of player behavior over time, or a behavioral pattern. Examples of such outputs include game completion times, win rates, churn, trajectories, and game balance.

5.5.2 Modeling Experience

To model the experience of the player one needs to have access to labels of that experience. Those labels ideally need to be as close to the **ground truth** of experience as possible. The ground truth (or gold standard) in affective sciences refers to a hypothesized and unknown label, value, or function, that best characterizes and represents an affective construct or an experience. Labels are normally provided through manual annotation which is a rather laborious process. Manual annotation is however necessary given that we require some estimate of the ground truth for subjective notions such as the emotional states of the player. The **accuracy** of that estimation is regularly questioned as there are numerous factors contributing to a deviation between a label and the actual underlying player experience.

Manually annotating players and their gameplay is a challenge in its own right with respect to both the human annotators involved and the annotation protocol chosen [455, 777]. On one hand, the annotators need to be skilled enough to be able to approximate the actual experience well. On the other hand, there are still many open questions left for us to address when it comes to the annotation tools and protocols used. Such questions include: Who will do the labeling: the person experiencing the gameplay or others? Will the labeling of player experience involve states (discrete representation) or instead involve the use of intensity or experience dimensions (continuous representation)? When it comes to time, should it be done in real-time or offline, in discrete time periods or continuously? Should the annotators be asked to rate the affect in an absolute fashion or, instead, rank it in a relative fashion? Answers to the above questions yield different data annotation protocols and, inevitably, varying degrees of data quality, validity and reliability. In the following sections we attempt to address a number of such critical questions that are usually raised in subjective annotations of player states.

5.5.2.1 Free Response or Forced Response?

Subjective player state annotations can be based either on a player's **free response**—retrieved via e.g., a think-aloud protocol [555]—or on **forced responses** retrieved through questionnaires or annotation tools. Free response naturally contains richer information about the player's state, but it is often unstructured, even chaotic, and thus hard to analyze appropriately. On the other hand, forcing players to self-report their experiences using directed questions or tasks constrains them to specific questionnaire items which could vary from simple tick boxes to multiple choice items. Both the questions and the answers we provide to annotators may vary from single words to sentences. Questionnaires can contain elements of player experience (e.g., the Game Experience Questionnaire [286]), demographic data and/or personality traits (e.g., a validated psychological profiling questionnaire such as the NEO-PI-R [140]). In the remainder of this section we will focus on forced responses as these are easier to analyze and are far more appropriate for data analysis and player modeling (as defined in this book).

5.5.2.2 Who Annotates?

Given the subjective nature of player experience the first natural question that comes in mind is *who annotates players*? In other words, who has the right authority and the best capacity to provide us with reliable tags of player experience? We distinguish two main categories: annotations can either be self-reports or reports expressed indirectly by experts or external observers [783].

In the first category the player states are provided by the players themselves and we call that **first-person** annotation. For example, a player is asked to rate the level of engagement while watching her playthrough video. First-person is clearly the most direct way to annotate a player state and build a model based on the solicited annotations. We can only assume there is disparity between the true (inner) experience of each player and the experience as felt by herself or perceived by others. Based on this assumption the player's annotations should normally be closer to her inner experience (**ground truth**) compared to third-person annotation. First-person annotation, however, may suffer from self-deception and memory limitations [778]. These limitations have been attributed mainly to the discrepancies between "the experiencing self" and "the remembering self" of a person [318] which is also known as the **memory-experience gap** [462].

Expert annotators—as a response to the above limitations—may instead be able to surpass the perception of experience and reach out to the inner experience of the player. In this second annotation category, named **third-person** annotation, an expert—such as a game designer—or an external observer provides the player state in a more objective manner, thereby reducing the subjective biases of first-person perceptions. For instance, a user experience analyst may provide particular player state tags while observing a first-person shooter deathmatch game. The benefit of third-person annotation is that multiple annotators can be used for a particular gameplay experience. In fact, the availability of several such subjective perceptions of experience may allow us to approximate the ground truth better as the agreement between many annotators enhances the validity of our data directly. A potential disagreement, on the other hand, might suggest that the gameplay experience we examine is non-trivial or may indicate that some of our annotators are untrained or inexperienced.

5.5.2.3 How Is Player Experience Represented?

Another key question is how player experience is best represented: as a number of different states (**discrete**) or, alternatively, as a set of dimensions (**continuous**)? On one hand, discrete labeling is practical as a means of representing player experience since the labels can easily form individual items (e.g., "excited", "annoyed" etc.) in an experience questionnaire, making it easy to ask the annotator/player to pick one (e.g., in [621]). Continuous labeling, on the other hand, appears to be advantageous for two key reasons. First, experiential states such as *immersion* are hard to capture with words or linguistic expressions that have fuzzy boundaries. Second, states do not allow for variations in *experience intensity* over time since they are binary: either the state is present or not. For example, the complex notions of *fun*, or even *engagement*, cannot be easily captured by their corresponding linguistic representation in a questionnaire or define well a particular *state* of a playing experience. Instead it seems natural to represent them as a continuum of experience intensity that may vary over time. For these reasons we often observe low agreement among the annotators [143] when we represent playing experience via discrete states.

As discussed earlier, the dominant approach in continuous annotation is the use of Russell's two-dimensional (arousal-valence) circumplex model of affect [581] (see Fig. 5.2(a)). Figure 5.5 illustrates two different annotation tools (FeelTrace and *AffectRank*) that are based on the arousal-valence circumplex model of affect. Figure 5.6 depicts the *RankTrace* continuous annotation tool which can be used for the annotation of a single dimension of affect (i.e., tension in this example). All three tools are accessible and of direct use for annotating playing experience.

5.5.2.4 How Often to Annotate?

Annotation can happen either within particular time intervals or continuously. **Time-continuous** annotation has been popularized due to the existence of freely available tools such as FeelTrace [144] (see Fig. 5.5(c)) and GTrace [145], which allows for continuous annotation of content (mostly videos and speech) across the dimensions of arousal and valence. In addition to FeelTrace there are annotation tools like the continuous measurement system [454] and EmuJoy [474], where the latter is designed for the annotation of music content. User interfaces such as wheels and knobs linked to the above annotation tools show further promise for the continuous annotation of experience in games [125, 397, 97] (see Fig. 5.6). The continuous annotation process, however, appears to require a higher amount of cognitive load compared to a time-discrete annotation protocol. Higher cognitive loads often result in lower levels of agreement between different annotators and yield unreliable data for modeling player experience [166, 418].

As a response to the above limitations, **time-discrete** annotation provides data at particular intervals when the annotator feels there is a *change* in the player's state. And changes are best indicated *relatively* rather than *absolutely*. *AffectRank*, for instance (see Fig. 5.5(b)), is a discrete, rank-based annotation tool that can be used for the annotation of any type of content including images, video, text or speech and it provides annotations that are significantly more reliable (with respect to inter-rater agreement) than the annotations obtained from continuous annotation tools such as FeelTrace [777]. The rank-based design of *AffectRank* is motivated by observations of recent studies in third-person video annotation indicating that "...humans are better at rating emotions in *relative* rather than absolute terms." [455, 777]. Further, *AffectRank* is grounded in numerous findings showcasing the supremacy of ranks over ratings for obtaining annotations of lower inconsistency and order effects [777, 773, 778, 436, 455, 761].

A recent tool that builds on the relative-based annotation of *AffectRank* and allows for the annotation of affect in a continuous yet unbounded fashion is *RankTrace* (see Fig. 5.6). The core idea behind *RankTrace* is introduced in [125]: the tool asks participants to watch the recorded playthrough of a play session and annotate in real-time the perceived intensity of a single emotional dimension. The annotation process in *RankTrace* is controlled through a "wheel-like" hardware, allowing participants to meticulously increase or decrease emotional intensity by turning the wheel, similarly to how volume is controlled in a stereo system. Further, the general interfacing design of *RankTrace* builds on the one-dimensional GTrace annotation tool [145]. Unlike other continuous annotation tools, however, annotation in *RankTrace* is *unbounded*: participants can continuously increase or decrease the intensity as desired without constraining themselves to an absolute scale. This design decision is built on the *anchor* [607] and *adaptation level* [258] psychology theories by which affect is a temporal notion based on earlier experience that is best expressed in relative terms [765, 777, 397]. The use of *RankTrace* has revealed the benefits of relative and unbounded annotation for modeling affect more reliably [397] and has also showed promise for the construction of general models of player emotion across games [97].

5.5.2.5 When to Annotate?

When is it best to annotate experience: *before*, *during* or *after* play (see Fig. 5.1)? In a **pre**-experience questionnaire we usually ask annotators to set the baseline of

(a) Annotating facial expressions of players during play. The annotation is context-dependent as the video includes the gameplay of the player (see top left corner).

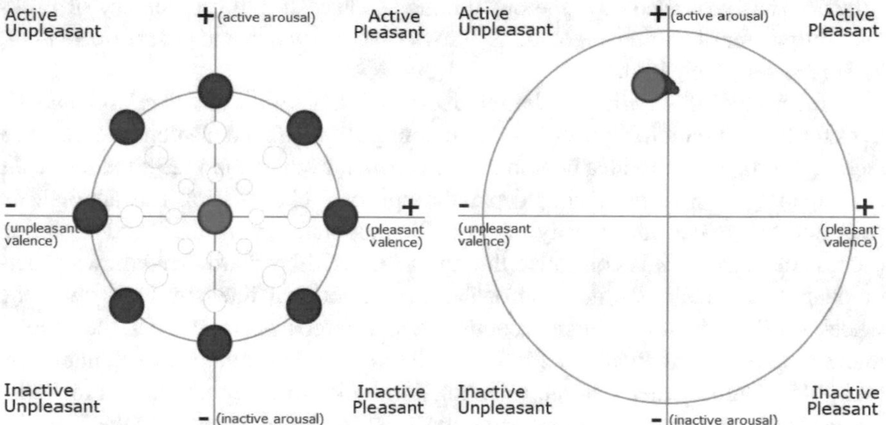

(b) *AffectRank*: A time-discrete annotation tool for arousal and valence.

(c) FeelTrace. A time-continuous annotation tool for arousal and valence.

Fig. 5.5 An example of third-person annotation based on videos of players and their gameplay using either (a) the *AffectRank* (b) or the FeelTrace (c) annotation tool. *AffectRank* is freely available at: https://github.com/TAPeri/AffectRank. FeelTrace is freely available at: http://emotion-research.net/download/Feeltrace%20Package.zip.

a player's state prior to playing a game. This state can be influenced by a number of factors such as the mood of the day, the social network activity, the caffeine consumption, earlier playing activity and so on. This is a wealth of information that can be used to enrich our models. Again, what is worth detecting is the relative *change* [765] from the baseline state of the user prior playing to the game.

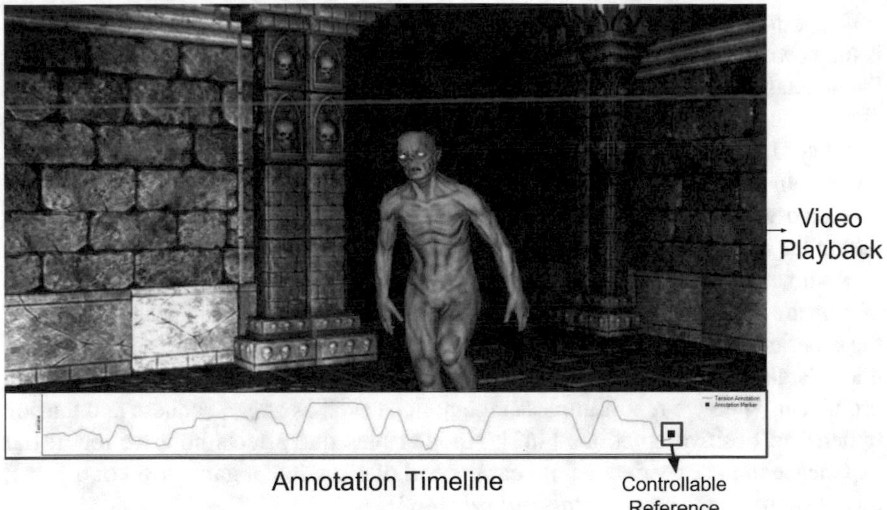

Fig. 5.6 The *RankTrace* annotation tool. In this example the tool is used for the annotation of tension in horror games. Participants play a game and then they annotate the level of tension by watching a video-recorded playthrough of their game session (top of image). The annotation trace is controlled via a wheel-like user interface. The entire annotation trace is shown for the participant's own reference (bottom of image). Image adapted from [397]. The *RankTrace* tool is available at: http://www.autogamedesign.eu/software.

A **during**-experience protocol, on the other hand, may involve the player in a first-person think-aloud setup [555] or a third-person annotation design. For the latter protocol you may think of user experience experts that observe and annotate player experience during the beta release of a game, for example. As mentioned earlier, first-person annotation during play is a rather intrusive process that disrupts the gameplay and risks adding experimental noise to annotation data. In contrast, third-person annotation is not intrusive; however, there are expected deviations from the actual first-person experience, which is inaccessible to the observer.

The most popular approach for the annotation of player experience is after a game (or a series of games) has been played, in a **post**-experience fashion. Post-experience annotation is unobtrusive for the player and it is usually performed by the players themselves. Self-reports, however, are memory-dependent by nature and memory is, in turn, time-dependent. Thus, one needs to consider carefully the **time window** between the experience and its corresponding report. For the reported post-experience to be a good approximation of the actual experience the playing time window needs to be small in order to minimize memory biases, yet sufficiently large to elicit particular experiences to the player. The higher the cognitive load required to retrieve the gameplay context is, the more the reports are memory-biased and not relevant to the actual experience. Further, the longer the time window between the real experience and the self-report the more the annotator activates aspects of *episodic* memory associated with the gameplay [571]. Episodic memory traces that form the basis of self-reports fade over time, but the precise rate at which this memory decay occurs is unknown and most likely individual [571]. Ideally, memory decay is so slow that the annotator will have a clear feeling of the gameplay session when annotating it. Now, if the time window becomes substantial—on the scale of hours and days—the annotator has to activate aspects of *semantic* memory such as general beliefs about a game. In summary, the more the episodic memory, and even more so the semantic memory, are activated during annotation, the more systematic errors are induced within the annotation data.

As a general rule of thumb the longer it takes for us to evaluate an experience of ours the larger the discrepancy between the true experience and the evaluation of the experience, which is usually more intense than the true experience. It also seems that this gap between our memory of experience and our real experience is more prominent when we report unpleasant emotions such as anger, sadness and tension rather than positive emotions [462]. Another bias that affects how we report our experience is the experience felt near the end of a session, a game level or a game; this effect has been named *peak-end rule* [462].

An effective way to assist episodic memory and minimize post-experience cognitive load is to show annotators replay videos of their gameplay (or the gameplay of others) and ask them to annotate those. This can be achieved via crowdsourcing [96] in a third-person manner or on a first-person annotation basis [125, 271, 397, 97]. Another notable approach in this direction is the *data-driven retrospective interviewing* method [187]. According to that method player behavioral data is collected and is analyzed to drive the construction of interview questions. These questions are then used in retrospect (post-experience) to reflect on the annotator's behavior.

5.5.2.6 Which Annotation Type?

We often are uncertain about the type of labels we wish to assign to a player state or a player experience. In particular, we can select from three data types for our annotation: ratings, classes, and ranks (see Fig. 5.1). The **rating**-based format represents a player's state with a scalar value or a vector of values. Ratings are arguably the dominant practice for quantitatively assessing aspects of a user's behavior, experience, opinion or emotion. In fact, the vast majority of user and psychometric studies have adopted rating questionnaires to capture the opinions, preferences and perceived experiences of experiment participants—see [78, 442, 119] among many. The most popular rating-based questionnaire follows the principles of a Likert scale [384] in which users are asked to specify their level of agreement with (or disagreement against) a given statement—see Fig. 5.7(a) for an example. Other popular rating-based questionnaires for user and player experience annotation include the Geneva Wheel model [600], the Self-Assessment Manikin [468], the Positive and Negative Affect Schedule [646], the Game Experience Questionnaire [286], the Flow State Scale [293] and the Player Experience of Need Satisfaction (PENS) survey [583], which was developed based on self-determination theory [162].

Rating-based reporting has notable inherent **limitations** that are often overlooked, resulting in fundamentally flawed analyses [778, 298]. First, ratings are analyzed traditionally by comparing their values across participants; see [233, 427] among many. While this is a generally accepted and dominant practice it neglects the existence of **inter-personal differences** as the meaning of each level on a rating scale may differ across experiment participants. For example, two participants assessing the difficulty of a level may assess it as exactly the same difficulty, but then one rates it as "very easy to play" and the other as "extremely easy to play". It turns out that there are numerous factors that contribute to the different internal rating scales existent across participants [455] such as differences in personality, culture [643], temperament and interests [740]. Further, a large volume of studies has also identified the presence of primacy and recency order effects in rating-based questionnaires (e.g., [113, 773]), systematic biases towards parts of the scale [388] (e.g., right-handed participants may tend to use the right side of the scale) or a fixed tendency over time (e.g., on a series of experimental conditions, the last ones are rated higher). Indicatively, the comparative study of [773] between ratings and ranks showcases higher inconsistency effects and significant order (recency) effects existent in ratings.

In addition to inter-personal differences, a critical limitation arises when ratings are treated as interval values since ratings are by nature **ordinal values** [657, 298]. Strictly speaking, any approach or method that treats ratings as numbers by, for instance, averaging their ordinal labels is *fundamentally flawed*. In most questionnaires Likert items are represented as pictures (e.g., different representations of arousal in the Self-Assessment Manikin [468]) or as adjectives (e.g., "moderately", "fairly" and "extremely"). These labels (images or adjectives) are often erroneously converted to integer numbers, violating basic axioms of statistics which suggest that ordinal values cannot be treated as interval values [657] since the underlying numerical scale is unknown. Note that even when a questionnaire features ratings as numbers (e.g., see Fig. 5.7(a)), the scale is still ordinal as the numbers in the instrument represent labels. Thus, the underlying numerical scale is still unknown and dependent on the participant [657, 515, 361]. Treating ratings as interval values is grounded in the assumption that the difference between consecutive ratings is fixed and equal. However, there is no valid assumption suggesting that a subjective rating scale is linear [298]. For instance, the difference between "fairly (4) " and "extremely (5)" may be larger than the distance between "moderately (3)" and "fairly (4)" as some experiment participants rarely use the extremes of the scale or tend to use one extreme more than the other [361]. If, instead, ratings are treated naturally as ordinal data no assumptions are made about the distance between rating labels, which eliminates introducing data noise to the analysis.

The second data type for the annotation of players is the **class**-based format. Classes allow annotators to select from a finite and non-structured set of options and, thus, a class-based questionnaire provides nominal data among two (binary) or more options. The questionnaire asks subjects to pick a player state from a particular representation which could vary from a simple boolean question (*was that game level frustrating or not? is this a sad facial expression? which level was the most* *stressful?*) to a player state selection from, for instance, the circumplex model of affect (*is this a high- or a low-arousal game state for the player?*). The limitations of ratings are mitigated, in part, via the use of class-based questionnaires. By not providing information about the intensity of each player state, however, classes do not have the level of granularity ratings naturally have. A class-based questionnaire might also yield annotations with an unbalanced number of samples per class. A common practice in psychometrics consists of transforming sets of consecutive ratings into separate classes (e.g., see [226, 260] among many). In an example study [255], arousal ratings on a 7-point scale are transformed into *high*, *neutral* and *low* arousal classes using 7-5, 4 and 3-1 ratings, respectively. While doing so might seem appropriate, the ordinal relation among classes is not being taken into account. More importantly, the transformation process adds a new set of bias to the subjectivity bias of ratings, namely **class splitting criteria** [436].

Finally, **rank**-based questionnaires ask the annotator to rank a preference among options such as two or more sessions of the game [763]. In its simplest form, the annotator compares two options and specifies which one is preferred under a given statement (**pairwise preference**). With more than two options, the participants are asked to provide a ranking of some or all the options. Examples of rank-based questions include: *was that level more engaging than this level? which facial expression looks happier?*). Another example of a rank-based questionnaire (4-alternative forced choice) is illustrated in Fig. 5.7(b). Being a form of subjective reporting, rank-based questionnaires (as much as rating-based and class-based questionnaires) are associated with the well known limitations of memory biases and self-deception. Reporting about subjective constructs such as experience, preference or emotion via rank-based questionnaires, however, has recently attracted the interest of researchers in marketing [167], psychology [72], user modeling [761, 37] and affective computing [765, 721, 436, 455, 773] among other fields. This gradual paradigm shift is driven by both the reported benefits of ranks minimizing the effects of self-reporting subjectivity biases and recent findings demonstrating the advantages of ordinal annotation [765, 773, 455].

5.5.2.7 What Is the Value of Player Experience?

Describing, labeling and assigning values to subjective notions, such as player experience, is a non-trivial task as evidenced by a number of disciplines including neuroscience [607], psychology [258], economics [630], and artificial intelligence [315]. Annotators can attempt to assign numbers to such notions in an **absolute** manner, using for instance a rating scale. Annotators can alternatively assign values in a **relative** fashion, using for instance a ranking. There are, however, a multitude of theoretical and practical reasons to doubt that subjective notions can be encoded as numbers in the first place [765]. For instance, according to Kahneman [317], co-founder of behavioral economics, "...it is safe to assume that changes are more accessible than absolute values"; his theory about judgment heuristics is built on Herbert Simon's *psychology of bounded rationality* [630]. Further, an important

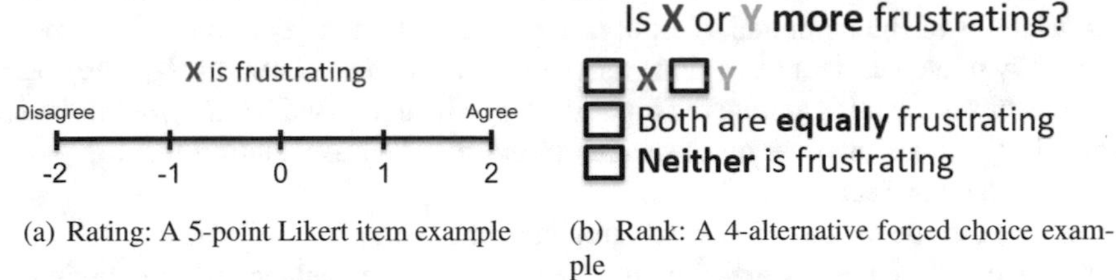

(a) Rating: A 5-point Likert item example

(b) Rank: A 4-alternative forced choice example

Fig. 5.7 Examples of rating-based (a) vs. rank-based (b) questionnaires.

thesis in psychology, named **adaptation level theory** [258], suggests that humans lack the ability to maintain a constant value about subjective notions and their preferences about options are, instead, made on a pairwise comparison basis using an internal ordinal scale [460]. The thesis claims that while we are efficient at discriminating among options, we are not good at assigning accurate absolute values for the intensity of what we perceive. For example, we are particularly bad at assigning absolute values to tension, frequency and loudness of sounds, the brightness of an image, or the arousal level of a video. The above theories have also been supported by neuroscientific evidence suggesting that experience with stimuli gradually creates our own internal context, or *anchor* [607], against which we rank any forthcoming stimulus or perceived experience. Thus, our choice about an option is driven by our internal ordinal representation of that particular option within a sample of options; not by any absolute value of that option [658].

As a remote observation, one may argue that the relative assessment provides less information than the absolute assessment since it does not express a quantity explicitly and only provides ordinal relations. As argued earlier, however, any additional information obtained in an absolute fashion (e.g., when ratings are treated as numbers) violates basic axioms of applied statistics. Thus the value of the additional information obtained (if any) is questioned directly [765].

In summary, results across different domains investigating subjective assessment suggest that relative (rank-based) annotations minimize the assumptions made about experiment participants' notions of highly subjective constructs such as player experience. Further, annotating experience in a relative fashion, instead of an absolute fashion, leads to the construction of more generalizable and accurate computational models of experience [765, 436].

5.5.3 No Output

Very often we are faced with datasets where target outputs about player behavioral or experience states are not available. In such instances modeling of players must rely on unsupervised learning [176, 244, 178] (see Fig. 5.1). Unsupervised learning, as discussed in Chapter 2, focuses on fitting a model to observations by discovering associations of the input and without having access to a target output. The input is generally treated as a set of random variables and a model is built through the observations of associations among the input vectors. Unsupervised learning as applied to modeling players involves tasks such as **clustering** and **association mining** which are described in Section 5.6.3.

It may also be the case that we do not have target outputs available but, nevertheless, we can design a **reward** function that characterizes behavioral or experiential patterns of play. In such instances we can use reinforcement learning approaches to discover policies about player behavior or player experience based on in-game play traces or other state-action representations (see Section 5.6.2). In the following section we detail the approaches used for modeling players in a supervised learning, reinforcement learning and unsupervised learning fashion.

5.6 How Can We Model Players?

In this section we build upon the data-driven approach of player modeling and discuss the application of **supervised**, **reinforcement** and **unsupervised** learning to model players, their behavior and their experience. To showcase the difference between the three learning approaches let us suppose we wish to classify player behavior. We can only use unsupervised learning if no behavioral classes have been defined a priori [176]. We can instead use supervised learning if, for example, we have already obtained an initial classification of players (either manually or via clustering) and we wish to fit new players into these predefined classes [178]. Finally, we can use reinforcement learning to derive policies that imitate different types of playing behavior or style. In Section 5.6.1 we focus on the supervised learning paradigm whereas in Section 5.6.2 and Section 5.6.3 we outline, respectively, the reinforcement learning and the unsupervised learning approach for modeling players. All three machine learning approaches are discussed in Chapter 2.

5.6.1 Supervised Learning

Player modeling consists of finding a function that maps a set of measurable attributes of the player to a particular player state. Following the supervised learning approach this is achieved by machine learning, or automatically adjusting, the parameters of a model to fit a dataset that contains a set of input samples, each one paired with target outputs. The input samples correspond to the list of measurable attributes (or features) while the target outputs correspond to the annotations of the player's states for each of the input samples that we are interested to learn to predict. As mentioned already, the annotations may vary from behavioral characteristics, such as completion times of a level or player archetypes, to estimates of player experience, such as player frustration.

As we saw in Chapter 2 popular supervised learning techniques, including artificial neural networks (shallow or deep architectures), decision trees, and support vector machines, can be used in games for the analysis, the imitation and the prediction of player behavior, and the modeling of playing experience. The data type of the annotation determines the output of the model and, in turn, the type of the machine learning approach that can be applied. The three supervised learning alternatives for learning from numerical (or interval), nominal and ordinal annotations—respectively, **regression**, **classification** and **preference learning**—are discussed in this section.

5.6.1.1 Regression

When the outputs that a player model needs to approximate are interval values, the modeling problem is known as metric or standard **regression**. Any regression algorithm is applicable to the task, including linear or polynomial regression, artificial neural networks and support vector machines. We refer the reader to Chapter 2 for details on a number of popular regression algorithms.

Regression algorithms are appropriate for **imitation** and **prediction** tasks of player *behavior*. When the task, however, is modeling of player *experience* caution needs to be put on the data analysis. While it is possible, for instance, to use regression algorithms to learn the exact numeric ratings of experience, in general it should be avoided because regression methods assume that the target values follow an interval (numerical) scale. Ratings naturally define an ordinal scale [765, 773] instead. As mentioned already, ordinal scales such as ratings should not be converted to numerical values due to the subjectivity inherent to reports, which imposes a non-uniform and varying distance among questionnaire items [778, 657]. Prediction models trained to approximate a real-value representation of a rating—even though they may achieve high prediction accuracies—do not necessarily capture the true reported playing experience because the ground truth used for training and validation of the model has been undermined by the numerous effects discussed above. We argue that the known fundamental pitfalls of self-reporting outlined above provide sufficient evidence against the use of regression for player experience modeling [765, 515, 455, 361]. Thus, we leave the evaluation of regression methods on experience annotations outside the scope of this book.

5.6.1.2 Classification

Classification is the appropriate form of supervised learning for player modeling when the annotation values represent a finite and non-structured set of classes. Classification methods can infer a mapping between those classes and player attributes. Available algorithms include artificial neural networks, decision trees, random forests, support vector machines, K-nearest neighbors, and ensemble learning among many others. Further details about some of these algorithms can be found in Chapter 2.

Classes can represent playing behavior which needs to be **imitated** or **predicted**, such as completion times (e.g., expressed as low, average or high completion time) or user retention in a free-to-play game (e.g., expressed as weak, mild or strong retention). Classes can alternatively represent player experience such as an *excited* versus a *frustrated* player as manifested from facial expressions or *low*, *neutral* and *high* arousal states for a player.

Classification is perfectly suited for the task of modeling player experience if discrete annotations of experience are selected from a list of possibilities and provided as target outputs [153, 344]. In other words, annotations of player experience need to be nominal for classification to be applied. A common practice, however, as already mentioned in Section 5.5.2.6, is to treat ratings of experience as classes and transform the ordinal scale—that defines ratings—into a nominal scale of separate classes. For example, ratings of arousal that lie between -1 and 1 are transformed into low, neutral and high arousal classes. By classifying ratings not only the ordinal relation among the introduced classes is ignored but, most importantly, the transformation process induces several biases to the data (see Section 5.5.2.6). These biases appear to be detrimental and mislead the search towards the ground truth of player experience [765, 436].

5.6.1.3 Preference Learning

As an alternative to regression and classification methods, **preference learning** [215] methods are designed to learn from ordinal data such as ranks or preferences. It is important to note that the training signal in the preference learning paradigm merely provides information for the **relative** relation between instances of the phenomenon we attempt to approximate. Target outputs that follow an ordinal scale do not provide information about the intensity (regression) or the clusters (classification) of the phenomenon.

Generally we could construct a player model based on in-game *behavioral preferences*. The information that this player, for example, prefers the mini-gun over a number of other weapons could form a set of pairwise preferences we could learn from. Alternatively we can build a model based on *experience preferences*. A player, for instance, reported that area X of the level is more challenging than area Y of the same level. Based on a set of such pairwise preferences we can derive a global function of challenge for that player.

As outlined in Chapter 2 a large palette of algorithms is available for the task of preference learning. Many popular classification and regression techniques have been adapted to tackle preference learning tasks, including linear statistical models such as linear discriminant analysis and large margins, and non-linear approaches such as Gaussian processes [122], deep and shallow artificial neural networks [430], and support vector machines [302].

Preference learning has already been extensively applied to modeling aspects of players. For example, Martínez et al. [430, 431] and Yannakakis et al. [780, 771] have explored several artificial neural network approaches to learn to predict affective and cognitive states of players reported as pairwise preferences. Similarly, Garbarino et al. [217] have used linear discriminant analysis to learn pairwise enjoyment predictors in racing games. To facilitate the use of proper machine learning methods on preference learning problems, a number of such preference learning methods as well as data preprocessing and feature selection algorithms have been made available as part of the Preference Learning Toolbox (PLT) [198]. PLT is an open-access, user-friendly and accessible toolkit[5] built and constantly updated for the purpose of easing the processing of (and promoting the use of) ranks.

As ratings, by definition, express ordinal scales they can directly be transposed to any ordinal representation (e.g., pairwise preferences). For instance, given an annotator's rating indicating that a condition A felt 'slightly frustrating' and a condition B felt 'very frustrating', a preference learning method can train a model that predicts a higher level of frustration for B than for A. In this way the modeling approach avoids introducing artifacts of what is the actual difference between 'very' and 'slightly' or the usage of the scale for this particular annotator. Further, the limitation of different subjective scales across users can be safely bypassed by transforming rating reports into ordinal relations on a per-annotator basis. Finally, the problem of the scale varying across time due to episodic memory still persists but can be minimized by transforming only consecutive reports, i.e., given a report for three conditions A, B and C, the player model can be trained using only the relation between A and B, and B and C (dropping the comparison between A and C).

5.6.1.4 Summary: The Good, the Bad and the Ugly

The last section on supervised learning is dedicated to the comparison among the three methods—regression, classification and preference learning—for modeling players. Arguably the discussion is limited when the in-game *behavior* of players is imitated or predicted. If behavioral data about players follows interval, nominal or ordinal scales then naturally, regression, classification and preference learning should be applied, respectively. Behavioral data have an objective nature which makes the task of learning less challenging. Given the subjective notion of player *experience*, however, there are a number of caveats and limitations of each algorithm that need to be taken into account. Below we discuss the comparative advantages of each and we summarize the key outcomes of supervised learning as applied to modeling the experience of players.

Regression vs. Preference Learning: Motivated by psychological studies suggesting that interval ratings misrepresent experience [515, 455, 361], we will not dedicate ourselves an extensive comparison between preference learning and regression methods. The performance comparison between a regression and preference-

[5] http://sourceforge.net/projects/pl-toolbox/

learned model is also irrelevant as the former is arguably *a priori* incapable of capturing the underlying experience phenomenon as precisely as the latter. Such deviations from the ground truth, however, are not trivial to illustrate through a data modeling approach and thus the comparison is not straightforward. The main reason is that the objective ground truth is fundamentally ill-defined when numbers are used to characterize subjective notions such as player experience.

Regression vs. Classification: Classes are easy to analyze and create player models from. Further, their use eliminates part of the inter-personal biases introduced with ratings. For these reasons classification should be preferred to regression for player experience modeling. We already saw that classification, instead of regression, is applied when ratings are available to overcome part of the limitations inherent in rating-based annotation. For instance, this can be achieved by transforming arousal ratings to *high*, *neutral* and *low* arousal classes [255]. While this common practice in psychometrics eliminates part of the rating subjectivity it adds new forms of data biases inherent in the ad-hoc decisions to split the classes. Further, the analysis of player models across several case studies in the literature has already shown that transforming ratings into classes creates a more complicated machine learning problem [765, 436].

Classification vs. Preference Learning: Preference learning is the supreme method for modeling experience when ranks or pairwise preferences are available. Even when ratings or classes are available comparisons between classification and preference learning player models in the literature suggest that preference learning methods lead to more efficient, generic and robust models which capture more information about the ground truth [765]. Indicatively, Crammer and Signer [147] compare classification, regression and preference learning training algorithms in a task to learn ratings. They report the supremacy of preference learning over the other methods based on several synthetic datasets and a movie-ratings dataset. In addition, extensive evidence already shows that preference learning better approximates the underlying function between input (e.g., experience manifestations such as gameplay) and output (e.g., annotations) [436]. Figure 5.8 showcases how much closer a preference learned model can reach a hypothesized (artificial) ground truth, compared to a classification model trained on an artificial dataset. In summary, preference learning via rank-based annotation controls for reporting memory effects, eliminates subjectivity biases and builds models that are closer to the ground truth of player experience [778, 777].

Grounded in extensive evidence our final note for the selection of a supervised learning approach for modeling player experience is clear: Independently of how experience is annotated we argue that preference learning (*the good*) is a superior supervised learning method for the task at hand, classification (*the ugly*) provides a good balance between simplicity and approximation of the ground truth of player experience whereas regression (*the bad*) is based on rating annotations which are of questionable quality with respect to their relevance to the true experience.

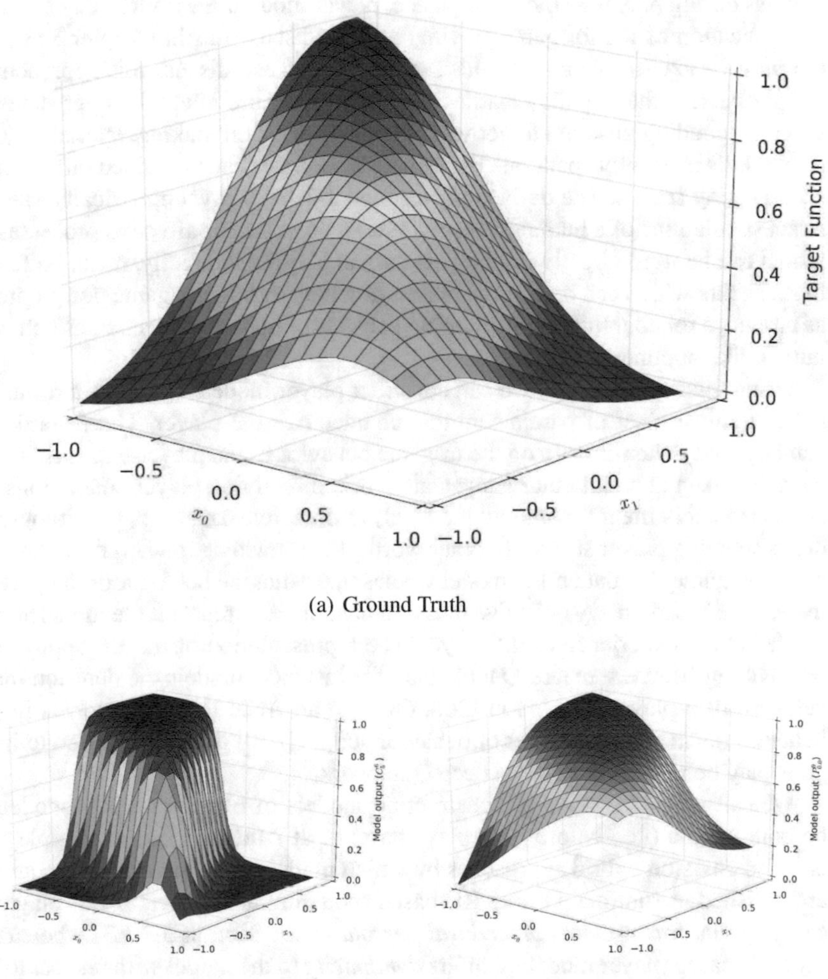

(a) Ground Truth

(b) Classification

(c) Preference learning

Fig. 5.8 A hypothesized (artificial) ground truth function (z-axis) which is dependent on two player attributes, x_1 and x_2 (Fig. 5.8(a)), the best classification model (Fig. 5.8(b)) and the best preference learned model (Fig. 5.8(c)). All images are retrieved from [436].

5.6.2 *Reinforcement Learning*

While it is possible to use **reinforcement learning** to model aspects of users during their interaction the RL approach for modeling players has been tried mostly in comparatively simplistic and abstract games [195] and has not seen much application in computer games. The key motivation for the use of RL for modeling players is that it can capture the relative valuation of game states as encoded internally by humans during play [685]. At first glance, player modeling with RL may seem to be an application of RL for game playing, and we discuss this in Chapter 3 as part of the *play for experience* aim. In this section, we instead discuss this approach from the perspective that a policy learned via RL can capture internal player states with no corresponding absolute target values such as decision making, learnability, cognitive skills or emotive patterns. Further, those policies can be trained on player data such as **play traces**. The derived player model depicts psychometrically-valid, abstract simulations of a human player's internal cognitive or affective processes. The model can be used directly to interpret human play, or indirectly, it can be featured in AI agents which can be used as playtesting bots during the game design process, as baselines for adapting agents to mimic classes of human players, or as believable, human-like opponents [268].

Using the RL paradigm, we can construct player models via RL if a reward signal can adjust a set of parameters that characterize the player. The reward signal can be based either directly on the in-game behavior of the player—for instance, the decision taken at a particular game state—or indirectly on player annotations (e.g., annotated excitement throughout the level) or objective data (e.g., physiological indexes showing player stress). In other words, the immediate reward function can be based on gameplay data if the model wishes to predict the behavior of the player or, instead, be based on any objective measure or subjective report if the model attempts to predict the experience of the player. The representation of the RL approach can be anything from a standard Q table that, for instance, models the decision making behavior of a player (e.g., as in [268, 685]) to an ANN that e.g., models in-game behavior (as in [267]), to a set of behavior scripts [650] that are adjusted to imitate gameplay behavior via RL, to a deep Q network.

We can view two ways of constructing models of players via RL: models can be built **offline** (i.e., before gameplay starts) or at **runtime** (i.e., during play). We can also envision hybrid approaches by which models are first built offline and then are polished at runtime. Offline RL-based modeling adds value to our playtesting capacity via, for instance, *procedural personas* (see Section 5.7.1.3) whereas runtime RL-based player modeling offers *dynamicity* to the model with respect to time. Runtime player modeling further adds on the capacity of the model to adapt to the particular characteristics of the player, thereby increasing the degree of personalization. We can think of models of players, for instance, that are continuously tailored to the current player by using the player's in-game annotations, behavioral decisions, or even physiological responses during the game.

This way of modeling players is still in its infancy with only a few studies existent on player behavioral modeling in educational games [488], in roguelike adventure games [268, 267] via TD learning or evolutionary reinforcement learning, and in first-person shooter games [685] via inverse RL. However, the application of RL for modeling users beyond games has been quite active for the purposes of modeling web-usage data and interactions on the web [684] or modeling user simulations in dialog systems [114, 225, 595]. Normally in such systems a statistical model is first trained on a corpus of human-computer interaction data for simulating (imitating) user behavior. Then reinforcement learning is used to tailor the model towards an optimal dialog strategy which can be found through trial and error interactions between the user and the simulated user.

5.6.3 Unsupervised Learning

The aim of **unsupervised learning** (see Chapter 2) is to derive a model given a number of observations. Unlike in supervised learning, there is no specified target output. Unlike in reinforcement learning there is no reward signal. (In short, there is no training signal of any kind.) In unsupervised learning, the signal is hidden internally in the interconnections among data attributes. So far, unsupervised learning has mainly been applied to two core player modeling tasks: **clustering** behaviors and **mining associations** between player attributes. While Chapter 2 provides the general description of these unsupervised learning algorithms, in this section we focus on their specific application for modeling players.

5.6.3.1 Clustering

As discussed in Chapter 2, **clustering** is a form unsupervised learning aiming to find clusters in datasets so that data within a cluster is similar to each other and dissimilar to data in other clusters. When it comes to the analysis of user behavior in games, clustering offers a means for reducing the dimensionality of a dataset, thereby yielding a manageable number of critical features that represent user behavior. Relevant data for clustering in games include player behavior, navigation patterns, assets bought, items used, game genres played and so on. Clustering can be used to group players into archetypical playing patterns in an effort to evaluate how people play a particular game and as part of a user-oriented testing process [176]. Further, one of the key questions of user testing in games is *whether people play the game as intended*. Clustering can be utilized to derive a number of different playing or behavioral styles directly addressing this question. Arguably, the key challenge in successfully applying clustering in games is that the derived clusters should have an intelligible meaning with respect to the game in question. Thus, clusters should be clearly interpretable and labeled in a language that is meaningful to the involved stakeholders (such as designers, artists and managers) [176, 178]. In the case studies of Section 5.7.1 we meet the above challenges and demonstrate the use of clustering in the popular *Tomb Raider: Underworld* (EIDOS interactive, 2008) game.

5.6.3.2 Frequent Pattern Mining

In Chapter 2 we defined frequent pattern mining as the set of problems and techniques related to finding patterns and structures in data. Patterns include sequences and itemsets. Both **frequent itemset mining** (e.g., Apriori [6]) and **frequent sequence mining** (e.g., GSP [652]) are relevant and useful for player modeling. The key motivation for applying frequent pattern mining on game data is to find inherent and hidden regularities in the data. In that regard, key player modeling problems, such as player type identification and detection of player behavior patterns, can be viewed as frequent pattern mining problems. Frequent pattern mining can for example be used to to discover what game content is often purchased together—e.g., players that buy X tend to buy Y too—or what are the subsequent actions after dying in a level—e.g., players that die often in the tutorial level pick up more health packs in level 1 [120, 621].

5.7 What Can We Model?

As already outlined at the beginning of this chapter, modeling of users in games can be classified into two main tasks: modeling of the **behavior** of players and modeling of their **experience**. It is important to remember that modeling of player behavior is (mostly) a task of **objective** nature whereas the modeling of player experience is **subjective** given the idiosyncratic nature of playing experience. The examples we present in the remainder of this chapter highlight the various uses of AI for modeling players.

5.7.1 Player Behavior

In this section, we exemplify player behavior modeling via three representative use cases. The two first examples are based on a series of studies on player modeling by Drachen et al. in 2009 [176] and later on by Mahlmann et al. in 2010 [414] in the *Tomb Raider: Underworld* (EIDOS interactive, 2008) game. The analysis includes both the **clustering** of players [176] and the **prediction** [414] of their behavior, which make it an ideal case study for the purposes of this book. The third study presented in this section focuses on the use of play traces for the procedural creation of player models. That case study explores the creation of **procedural personas** in the *MiniDungeons* puzzle roguelike game.

Fig. 5.9 A screenshot from the *Tomb Raider: Underworld* (EIDOS interactive, 2008) game featuring the player character, Lara Croft. The game is used as one of the player behavior modeling case studies presented in this book. Image obtained from Wikipedia (fair use).

5.7.1.1 Clustering Players in *Tomb Raider: Underworld*

Tomb Raider: Underworld (TRU) is a third-person perspective, advanced platform-puzzle game, where the player has to combine strategic thinking in planning the 3D-movements of Lara Croft (the game's player character) and problem solving in order to go through a series of puzzles and navigate through a number of levels (see Fig. 5.9).

The **dataset** used for this study includes entries from $25,240$ players. The $1,365$ of those that completed the game were selected and used for the analysis presented below. Note that TRU consists of seven main levels plus a tutorial level. Six features of gameplay behavior were extracted from the data and are as follows: number of deaths by opponents, number of deaths by the environment (e.g., fire, traps, etc.), number of deaths by falling (e.g., from ledges), total number of deaths, game completion time, and the times help was requested. All six features were calculated on the basis of completed TRU games. The selection of these particular features was based on the core game design of the TRU game and their potential impact on the process of distinguishing among dissimilar patterns of play.

Three different **clustering** techniques were applied to the task of identifying the number of meaningful and interpretable clusters of players in the data: k-means, hierarchical clustering and self-organizing maps. While the first two have been covered in Chapter 2 we will briefly outline the third method here.

A **self-organizing map** (SOM) [347] creates and iteratively adjusts a low-dimensional projection of the input space via vector quantization. In particular, a type of large SOM called an emergent self-organizing map [727] was used in conjunction with reliable visualization techniques to help us identify clusters. A SOM consists of neurons organized in a low-dimensional grid. Each neuron in the grid (map) is connected to the input vector through a connection weight vector. In addition to the input vector, the neurons are connected to neighbor neurons of the map through neighborhood interconnections which generate the structure of the map: rectangular and hexagonal lattices organized in a two-dimensional sheet or a three-dimensional toroid shape are some of the most popular topologies used. SOM training can be viewed as a **vector quantization** algorithm which resembles the k-means algorithm. What differentiates SOM, however, is the update of the topological neighbors of the best-matching neuron—a best-matching neuron is a neuron for which there exists at least one input vector for which the Euclidean distance to the weight vector of this neuron is minimal. As a result, the whole neuron neighborhood is stretched towards the presented input vector. The outcome of SOM training is that neighboring neurons have similar weight vectors which can be used for projecting the input data to the two-dimensional space and thereby clustering a set of data through observation on a 2D plane. For a more detailed description of SOMs, the reader is referred to [347].

To get some insight into the possible number of clusters existent in the data, **k-means** was applied for all k values less than or equal to 20. The sum of the Euclidean distances between each player instance and its corresponding cluster centroid (i.e., quantization error) is calculated for all 20 trials of k-means. The analysis reveals that the percent decrease of the mean quantization error due to the increase of k is notably high when $k = 3$ and $k = 4$. For $k = 3$ and $k = 4$ this value equals 19% and 13% respectively while it lies between 7% and 2% for $k > 4$. Thus, the k-means clustering analysis provides the first indication of the existence of three or four main player behavioral clusters within the data.

As an alternative approach to k-means, **hierarchical clustering** is also applied to the dataset. This approach seeks to build a hierarchy of clusters existent in the data. The Ward's clustering method [747] is used to specify the clusters in the data by which the squared Euclidean distance is used as a measure of dissimilarity between data vector pairs. The resulting **dendrogram** is depicted in Fig. 5.10(a). As noted in Chapter 2 a dendrogram is a treelike diagram that illustrates the merging of data into clusters as a result of hierarchical clustering. It consists of many U-shaped lines connecting the clusters. The height of each U represents the squared Euclidean distance between the two clusters being connected. Depending on where the data analyst sets the squared Euclidean distance threshold a dissimilar number of clusters can be observed.

Both k-means and hierarchical clustering already demonstrate that the 1,365 players can be clustered in a low number of different player types. K-means indicates there are for three or four clusters while the Ward's dendrogram reveals the existence of two populated and two smaller clusters, respectively, in the middle and at the edges of the tree.

Applying SOM, as the third alternative approach, allows us to cluster the TRU data by observation on a two-dimensional plane. The U-matrix depicted in Fig. 5.10(b) is a visualization of the local distance structure in the data placed onto a two-dimensional map. The average distance value between each neuron's weight vector and the weight vectors of its immediate neighbors corresponds to the height

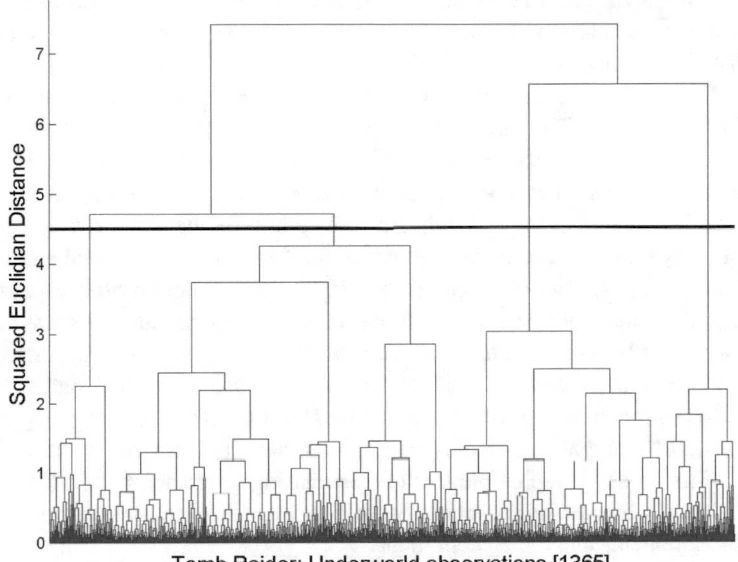

(a) Dendrogram of TRU data using Ward hierarchical clustering. A squared Euclidean distance of 4.5 (illustrated with a horizontal black line) reveals four clusters. Image adapted from [176].

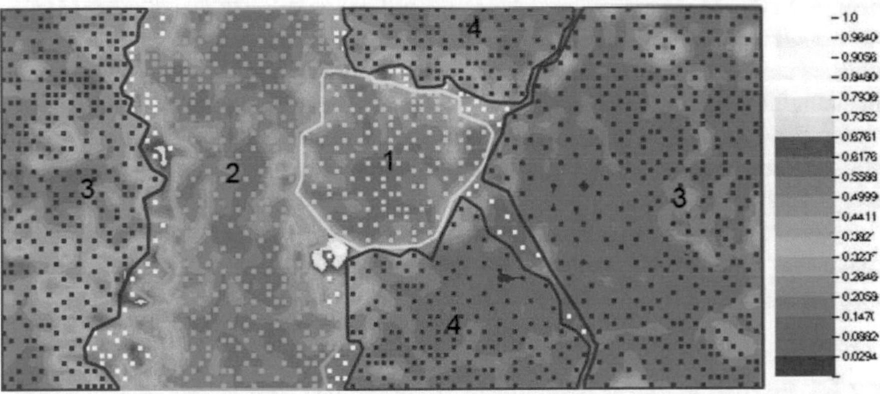

(b) U-matrix visualization of a self-organizing map depicting the four player clusters identified in a population of 1,365 TRU players (shown as small colored squares). Different square colors depict different player clusters. Valleys represent clusters whereas mountains represent cluster borders. Image adopted from [176].

Fig. 5.10 Detecting player types from TRU data using hierarchical (a) and SOM (b) clustering methods.

of that neuron in the U-matrix (positioned at the map coordinates of the neuron).

Thus, U-matrix values are large in areas where no or few data points reside, creating mountain ranges for cluster boundaries. On the other hand, visualized valleys indicate clusters of data as small U-matrix values are observed in areas where the data space distances of neurons are small.

The SOM analysis reveals four main classes of behavior (player types) as depicted in Fig. 5.10(b). The different colors of the U-matrix correspond to the four different clusters of players. In particular, cluster 1 (8.68% of the TRU players) corresponds to players that die very few times, their death is caused mainly by the environment, they do not request help from the game frequently and they complete the game very quickly. Given such game skills these players were labeled as *Veterans*. Cluster 2 (22.12%) corresponds to players that die quite often (mainly due to falling), they take a long time to complete the game—indicating a slow-moving, careful style of play—and prefer to solve most puzzles in the game by themselves. Players of this cluster were labeled as *Solvers*, because they excel particularly at solving the puzzles of the game. Players of cluster 3 form the largest group of TRU players (46.18%) and are labeled as *Pacifists* as they die primarily from active opponents. Finally, the group of players corresponding to cluster 4 (16.56% of TRU players), namely the *Runners*, is characterized by low completion times and frequent deaths by opponents and the environment.

The results showcase how clustering of player behavior can be useful to evaluate game designs. Specifically, TRU players seem to not merely follow a specific strategy to complete the game but rather they fully explore the affordances provided by the game in dissimilar ways. The findings are directly applicable to TRU game design as clustering provides answers to the critical question of *whether people play the game as intended.* The main limitation of clustering, however, is that the derived clusters are not intuitively interpretable and that clusters need to be represented into meaningful behavioral patterns to be useful for game design. Collaborations between the data analyst and the designers of the game—as was performed in this study—is essential for meaningful interpretations of the derived clusters. The benefit of such a collaboration is both the enhancement of game design features and the effective **phenomenological debugging** of the game [176]. In other words, we make sure both that no feature of the game is underused or misused and that the playing experience and the game balance are debugged.

5.7.1.2 Predicting Player Behavior in *Tomb Raider: Underworld*

Building on the same set of TRU player data, a second study examined the possibilities of **predicting** particular aspects of playing behavior via supervised learning [414]. An aspect of player behavior that is particularly important for game design is to predict *when a player will stop playing.* As one of the perennial challenges of game design is to ensure that as many different types of players are facilitated in the design, being able to predict when players will stop playing a game is of interest because it assists with locating potentially problematic aspects of game design. Further, such information helps toward the redesign of a game's monetization strategy for maximizing user retention.

Data was drawn from the Square Enix Europe Metrics Suite and was collected during a two month period (1st Dec 2008 – 31st Jan 2009), providing records from approximately 203,000 players. For the player behavior prediction task it was decided to extract a subsample of 10,000 players which provides a large enough and representative dataset for the aims of the study, while at the same time is manageable in terms of computational effort. A careful data-preprocessing approach yielded 6,430 players that were considered for the prediction task—these players had completed the first level of the game.

As in the TRU cluster analysis the features extracted from the data relate to the core mechanics of the game. In addition to the six features investigated in the clustering study of TRU the extracted features for this study include the number of times the adrenalin feature of the game was used, the number of rewards collected, the number of treasures found, and the number of times the player changes settings in the game (including player ammunition, enemy hit points, player hit points, and recovery time when performing platform jumps). Further details about these features can be found in [414].

To test the possibility of predicting the TRU level the player completed last a number of **classification** algorithms are tested on the data using the Weka machine learning software [243]. The approach followed was to experiment with at least one algorithm from each of the algorithm families existent in Weka and to put additional effort on those classification algorithms that were included in a recent list of the most important algorithms in data mining: decision tree induction, backpropagation and simple regression [759]. The resulting set of algorithms chosen for classification are as follows: logistic regression, multi-layer perceptron backpropagation, variants of decision trees, Bayesian networks, and support vector machines. In the following section, we only outline the most interesting results from those reported in [414]. For all tested algorithms, the reported classification prediction accuracy was achieved through 10-fold cross validation.

Most of the tested algorithms had similar levels of performance, and were able to predict when a player will stop playing substantially better than the baseline. In particular, considering only gameplay of level 1 classification algorithms reach an accuracy between 45% and 48%, which is substantially higher than the baseline performance (39.8%). When using additional features from level 2, the predictions are much more accurate—between 50% and 78%—compared to the baseline (45.3%). In particular, decision trees and logistic regression manage to reach accuracies of almost 78% in predicting on what level a player will stop playing. The difference in the predictive strength of using level 1 and 2 data as compared to only level 1 data is partly due to increased amount of features used in the latter case.

Beyond accuracy an important feature of machine learning algorithms is their transparency and their expressiveness. The models are more useful to a data analyst and a game designer if they can be expressed in a form which is easy to visualize and comprehend. Decision trees—of the form constructed by the ID3 algorithm [544] and its many derivatives—are excellent from this perspective, especially if pruned to

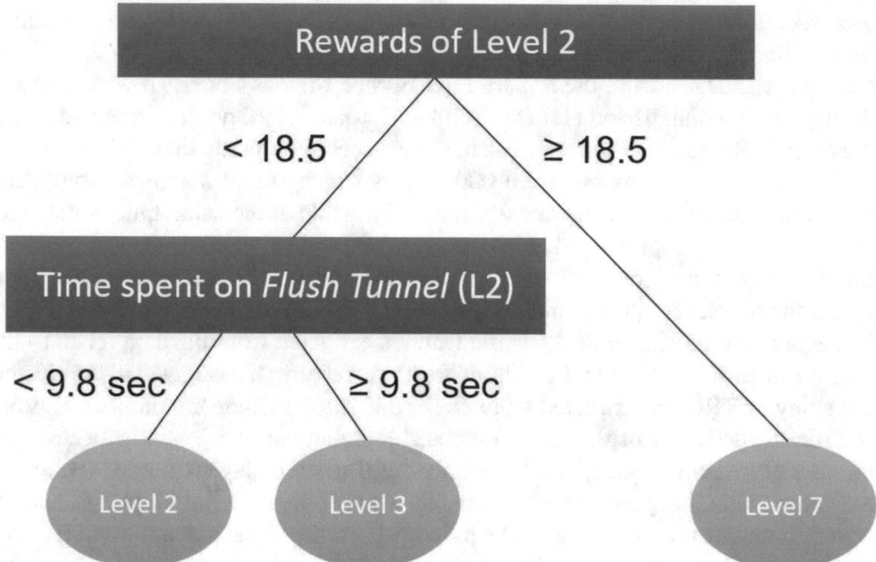

Fig. 5.11 A decision tree trained by the ID3 algorithm [544] to predict when TRU players will stop playing the game. The leaves of the tree (ovals) indicate the number of the level (2, 3 or 7) the player is expected to complete. Note that the TRU game has seven levels in total. The tree is constrained to tree depth 2 and achieves a classification accuracy of 76.7%.

a small size. For instance, the extremely small decision tree depicted in Fig. 5.11 is constrained to tree depth 2 and was derived on the set of players who completed both levels 1 and 2 with a classification accuracy of 76.7%. The predictive capacity of the decision tree illustrated in Fig. 5.11 is impressive given how extremely simple it is. The fact that we can predict the final played level—with a high accuracy—based only on the amount of time spent in the room named *Flush Tunnel* of level 2 and the total rewards collected in level 2 is very appealing for game design. What this decision tree indicates is that the amount of time players spend within a given area early in the game and how well they perform are important for determining if they continue playing the game. Time spent on a task or in an area of the game can indeed be indicative of challenges with progressing through the game, which can result in a frustrating experience.

5.7.1.3 Procedural Personas in *MiniDungeons*

Procedural personas are generative models of player behavior, meaning that they can replicate in-game behavior and be used for playing games in the same role as players; additionally, procedural personas are meant to represent archetypical players rather than individual players [268, 267, 269]. A procedural persona can be defined as the parameters of a utility vector that describe the preferences of a player. For example, a player might allocate different weight to finishing a game fast, exploring dialog options, getting a high score, etc.; these preferences can be numerically encoded in a utility vector where each parameter corresponds to the persona's interest in a particular activity or outcome. Once these utilities are defined, reinforcement learning via TD learning or neuroevolution can be used to find a policy that reflects these utilities, or a tree search algorithm such as MCTS can be used with these utilities as evaluation functions. Approaches similar to the procedural persona concept have also been used for modeling the learning process of the player in educational games via reinforcement learning [488].

As outlined in Section 5.4.1, *MiniDungeons* is a simple rogue-like game which features turn-based discrete movement, deterministic mechanics and full information. The player avatar must reach the exit of each level to win it. Monsters block the way and can be destroyed, at the cost of decreasing the player character's health; health can be restored by collecting potions. Additionally, treasures are distributed throughout levels. In many levels, potions and treasures are placed behind monsters, and monsters block the shortest path from the entrance to the exit. Like many games, it is therefore possible to play *MiniDungeons* with different goals, such as reaching the exit in the shortest possible time, collecting as much treasure as possible or killing all the monsters (see Figs. 5.3 and 5.12).

These different playing styles can be formalized as procedural personas by attaching differing utilities to measures such as the number of treasures collected, the number of monsters killed or the number of turns taken to reach the exit. Q-learning can be used to learn policies that implement the appropriate persona in single levels [268], and evolutionary algorithms can be used to train neural networks that implement a procedural persona across multiple levels [267]. These personas can be compared with the play traces of human players by placing the persona in every situation that the human encountered in the play trace and comparing the action chosen by the procedural persona with the action chosen by the human (as you are comparing human actions with those of a Q function, you might say that you are asking "what would Q do?"). It is also possible to learn the utility values for a procedural persona clone of a particular human player by evolutionary search for those values that make the persona best match a particular play trace (see Fig. 5.12). However, it appears that these "clones" of human players generalize less well than designer-specified personas [269].

5.7.2 Player Experience

The modeling of player experience involves learning a set of target outputs that approximate the experience (as opposed of the behavior) of the player. By definition, that which is being modeled (experience) is of **subjective nature** and the modeling therefore requires target outputs that somehow approximate the ground truth of experience. A model of player experience predicts some aspect of the experience a player would have in some game situation, and learning such models is naturally a

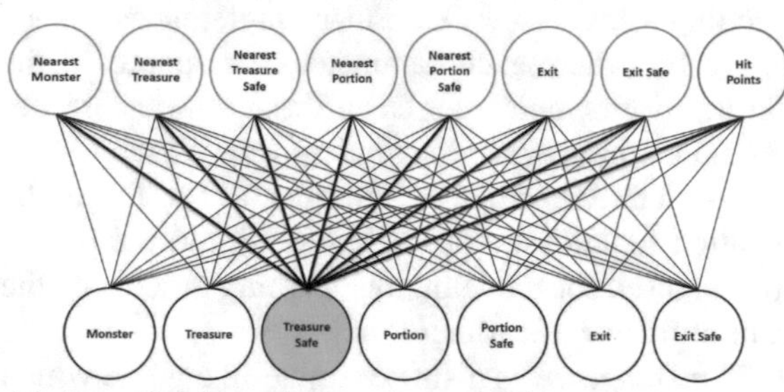

(a) An artificial neural network mapping between a game state (input) and plans (output). The input contains blue circles representing distances to various important elements of the game and a red circle representing hit points of the player character.

(b) A level of *MiniDungeons 2* depicting the current state of the game.

Fig. 5.12 A example of a *procedural persona*: In this example we evolve the weights of artificial neural networks—an ANN per persona. The ANN takes observations of the player character and its environment and uses these to choose from a selection of possible plans. During evolution the utility function of each persona is used as the fitness function for adjusting its network weights. Each individual of each generation is evaluated by simulating a full game. A utility function allows us to evolve a network to pursue multiple goals across a range of different situations. The method depends on the designer providing carefully chosen observations, appropriate planning algorithms, and well-constructed utility functions. In this example the player opts to move towards a safe treasure. This is illustrated with a green output neuron and highlighted corresponding weights (a) and a green path on the game level (b).

supervised learning problems. As mentioned, there are many ways this can be done, with approaches to player experience modeling varying regarding the inputs (from what the experience is predicted, e.g., physiology, level design parameters, playing style or game speed), the outputs (what sort of experience is predicted, e.g., fun, frustration, attention or immersion) and the modeling methodology.

In this section we will outline a few examples of supervised learning for modeling the experience of players. To best cover the material (methods, algorithms, uses of models) we rely on studies which have been thoroughly examined in the literature. In particular, in the remainder of this section we outline the various approaches and extensive methodology for modeling experience in two games: a variant of the popular *Super Mario Bros* (Nintendo, 1985) game and a 3D prey-predator game named *Maze-Ball*.

5.7.2.1 Modeling Player Experience in *Super Mario Bros*

Our first example builds upon the work of Pedersen et al. [521, 520] who modified an open-source clone of the classic platform game *Super Mario Bros* (Nintendo, 1985) to allow for personalized level generation. That work is important in that it set the foundations for the development of the *experience-driven procedural content generation* framework [783] which constitutes a core research trend within procedural content generation (see also Chapter 4).

The game used in this example is a modified version of Markus Persson's *Infinite Mario Bros* which is a public domain clone of Nintendo's classic platform game *Super Mario Bros* (Nintendo, 1985). All experiments reported in this example rely on a **model-free** approach for the modeling of player experience. Models of player experience are based on both the level played (game context) and the player's playing style. While playing, the game recorded a number of behavioral metrics of the players, such as the frequency of jumping, running and shooting, that are taken into consideration for modeling player experience. Further, in a follow-up experiment [610], the videos of players playing the game were also recorded and used to extract a number of useful visual cues such as the average head movement during play. The output (ground truth of experience) for all experiments is provided as first-person, rank-based reports obtained via **crowdsourcing**. Data was crowdsourced from hundreds of players, who played pairs of levels with different level parameters (e.g., dissimilar numbers, sizes and placements of gaps) and were asked to rank which of two levels best induced a number of player states. Across the several studies of player experience modeling for this variant of *Super Mario Bros* (Nintendo, 1985) collectively players have been asked to annotate fun, engagement, challenge, frustration, predictability, anxiety, and boredom. These are the target outputs the player model needs to predict based on the input parameters discussed above.

Given the rank-based nature of the annotations the use of preference learning is necessary for the construction of the player model. The collected data is used to train artificial neural networks that predict the players' experiential states, given a player's behavior (and/or affective manifestations) and a particular game context, using evolutionary preference learning. In **neuroevolutionary preference learning** [763], a genetic algorithm evolves an artificial neural network so that its output matches the pairwise preferences in the data set. The input of the artificial neural network is a set of features that have been extracted from the data set—as mentioned earlier, the input may include gameplay and/or objective data in this example. It is worth noting that automatic feature selection is applied to pick the set of features (model input) that are relevant for predicting variant aspects of player experience. The genetic algorithm implemented uses a fitness function that measures the difference between the reported preferences and the relative magnitude of the model output. Neuroevolutionary preference learning has been used broadly in the player modeling literature and the interested reader may refer to the following studies (among many): [432, 610, 763, 521, 520, 772].

The crowdsourcing experiment of Pedersen et al. [521, 520] resulted in data (gameplay and subjective reports of experience) from **181 players**. The best predictor of reported *fun* reached an accuracy of around 70% on unseen subjective reports of fun. The input of the neural network *fun*-model is obtained through automatic feature selection consists of the time Mario spent moving left, the number of opponents Mario killed from stomping, and the percentage of level played in the left direction. All three playing features appear to contribute positively for the prediction of reported *fun* in the game. The best-performing model for *challenge*

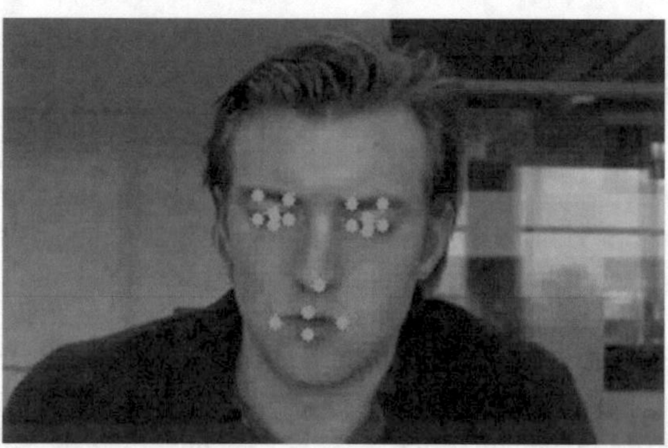

Fig. 5.13 Facial feature tracking for head movement. Image adapted from [610].

prediction had an accuracy of approximately 78%. It is more complex than the best fun predictor, using five features: time Mario spends standing still, jump difficulty, coin blocks Mario pressed, number of cannonballs Mario killed, and Mario kills by stomping. Finally, the best predictor for *frustration* reaches an accuracy of 89%. It is indeed an impressive finding that a player experience model can predict (with near-certainty) whether the player is frustrated by the current game by merely calculating the time Mario spent standing still, the time Mario spent on its last life, the jump difficulty, and the deaths Mario had from falling in gaps. The general findings of Pedersen et al. [520] suggest that good predictors for experience can be found if a preference learning approach is applied on crowdsourced reports of experience and gameplay data. The prediction accuracies, however, depend on the complexity of the reported state—arguably *fun* is a much more complicated and fuzzier notion to report compared to *challenge* or *frustration*. In a follow up study by Pedersen et al. [521] the additional player states of *predictability*, *anxiety* and *boredom* were predicted with accuracies of approximately 77%, 70% and 61%, respectively. The same player experience methodology was tested on an even larger scale, soliciting data from a total number of **780 players** of the game [621]. Frequent pattern mining algorithms were applied to the data to derive frequent sequences of player actions. Using sequential features of gameplay the models reach accuracies of up to 84% for *engagement*, 83% for *frustration* and 79% for *challenge*.

In addition to the behavioral characteristics the **visual cues** of the player can be taken into account as objective input to the player model. In [610] visual features were extracted from videos of **58 players**, both throughout whole game sessions and during small periods of critical events such as when a player completes a level or when the player loses a life (see Figs. 5.13 and 5.14). The visual cues enhance the quality of the information we have about a player's affective state which, in turn, allows us to better approximate player experience. Specifically, fusing the gameplay and the visual reaction features as inputs to the artificial neural network we achieve average accuracies of up to 84%, 86% and 84% for predicting reported *engagement*, *frustration* and *challenge*, respectively. The key findings of [610] suggest

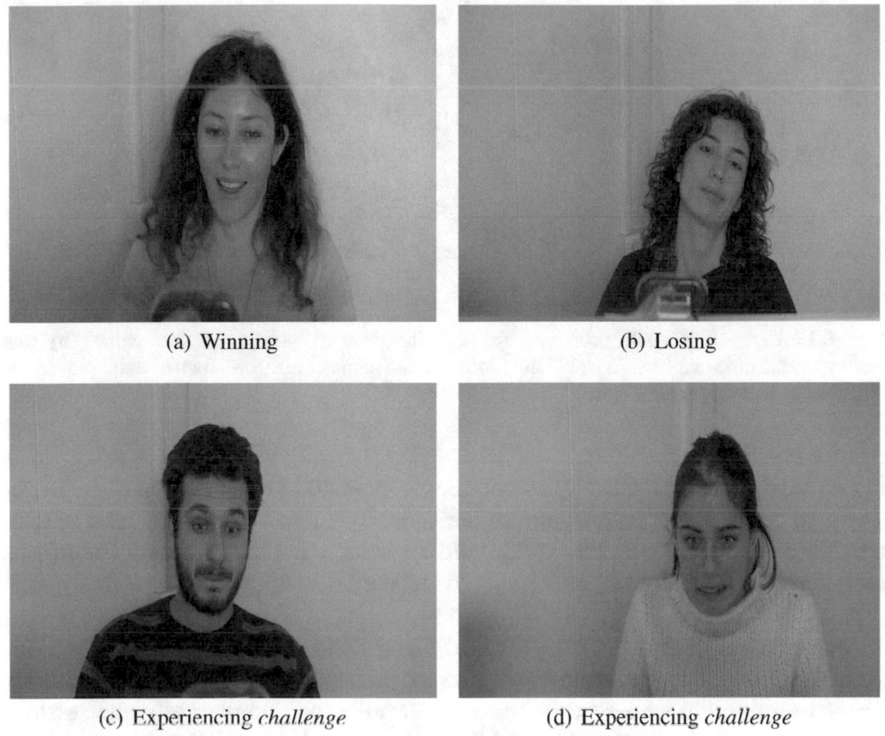

(a) Winning (b) Losing

(c) Experiencing *challenge* (d) Experiencing *challenge*

Fig. 5.14 Examples of facial expressions of *Super Mario Bros* (Nintendo, 1985) players for different game states. All images are retrieved from the Platformer Experience Dataset [326].

that players' visual reactions can provide a rich source of information for modeling experience preferences and lead to more accurate models of player experience.

5.7.2.2 Modeling Player Experience in *Maze-Ball*

Our second example for modeling player experience builds largely upon the extensive studies of Martínez et al. [434, 430, 435] who analyzed player experience using a simple 3D prey-predator game named *Maze-Ball* towards achieving affective-driven camera control in games. While the game is rather simple, the work on *Maze-Ball* offers a thorough analysis of player experience via a set of sophisticated techniques for capturing the psychophysiological patterns of players including preference learning, frequent pattern mining and deep convolutional neural networks. In addition the dataset that resulted from these studies is publicly available for further experimentation and forms a number of suggested exercises for this book.

Maze-Ball is a three-dimensional prey-predator game (see Fig. 5.15); similar to a 3D version of *Pac-Man* (Namco,1981). The player (prey) controls a ball which

(a) Maze-Ball (b) Space-Maze

Fig. 5.15 Early Maze-Ball prototype (a) and a polished variant of the game (b) that features real-time camera adaptation [345]. The games can be found and played at http://www.hectorpmartinez.com/.

moves inside a maze hunted by 10 opponents (predators) moving around the maze. The goal of the player is to maximize her score by gathering as many tokens, scattered in the maze, as possible while avoiding being touched by the opponents in a predefined time window of 90 seconds. A detailed description of *Maze-Ball* gameplay can be found in [780].

Gameplay attributes and physiological signals (skin conductance and heart rate variability) were acquired from **36 players** of *Maze-Ball*. Each subject played a predefined set of eight games for 90 seconds, thus the total number of game sessions available is 288. Gameplay and physiological signals define the **input** of the player experience model. To obtain the ground truth of experience players self-reported their preference about game pairs they played using a rank-based questionnaire, in particular a 4-alternative forced choice (4-AFC) protocol [780]. They were asked to rank the two games of each game pair with respect to *fun*, *challenge*, *boredom*, *frustration*, *excitement*, *anxiety* and *relaxation*. These annotations are the target **outputs** the player model will attempt to predict based on the input parameters discussed above.

Several features were extracted from the gameplay and physiological signals obtained. These included features related to kills and deaths in the game as well as features associated with the coverage of the level. For extracting features from the physiological signals the study considered their average, standard deviation, and maximum and minimum values. The complete feature list and the experimental protocol followed can be found in [780].

As in the example with the game variant of *Super Mario Bros* (Nintendo, 1985) the rank-based nature of the annotations requires the use of **preference learning** for the construction of the player experience model. Thus, the collected data is used to train neural networks that predict the player states, given a player's gameplay behavior and its physiological manifestations, using evolutionary preference learning. The architecture of the neural network can be either shallow using a simple multi-layered perceptron or deep (using convolutional neural networks [430, 435]). Figure 5.16 shows the different ways information from gameplay and physiology can been

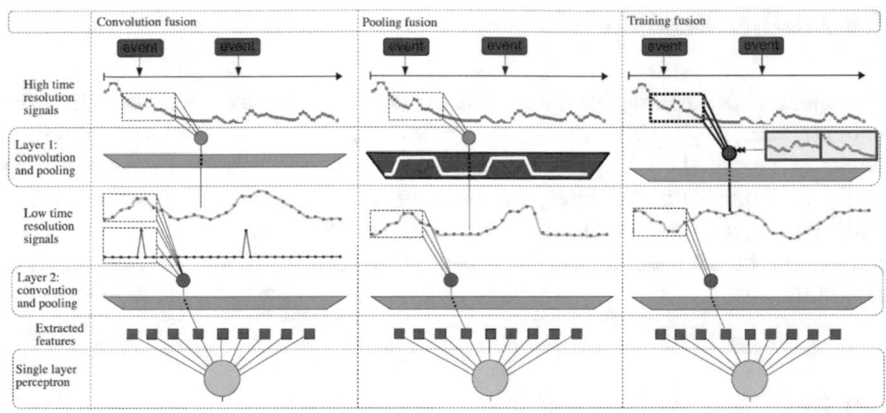

Fig. 5.16 Three dissimilar approaches to deep multimodal fusion via convolutional neural networks. Gameplay events are fused with skin conductance in this example. The networks illustrated present two layers with one neuron each. The first convolutional layer receives as input a continuous signal at a high time resolution, which is further reduced by a pooling layer. The resulting signal (feature map) presents a lower time resolution. The second convolutional layer can combine this feature map with additional modalities at the same low resolution. In the convolution fusion network (left figure), the two events are introduced at this level as a pulse signal. In the pooling fusion network (middle figure), the events are introduced as part of the first pooling layer, resulting in a filtered feature map. Finally, in the training fusion network (right figure), the events affect the training process of the first convolutional layer, leading to an alternative feature map. Image adapted from [435].

fused on a **deep convolutional neural network** which is trained via preference learning to predict player experience in any game experience dataset that contains discrete in-game events and continuous signals (e.g., the player's skin conductance).

Predictors of the player experience states can reach accuracies that vary from 72% for *challenge* up to 86% for *frustration* using a shallow multi-layer perceptron player model [780]. Significant improvements are observed in those accuracies when the input space of the model is augmented with frequent sequences of in-game and physiological events (i.e., fusion on the input space). As in [621], Martínez et al. used GSP to extract those frequent patterns that were subsequently used as inputs of the player model [434]. Further accuracy improvements can be observed when physiology is fused with physiological signals on deep architectures of convolutional neural networks [435]. Using **deep fusion** (see Fig. 5.16) accuracies of predicting player experience may surpass 82% for all player experience states considered. Further information about the results obtained in the *Maze-Ball* game can be found in the following studies: [780, 434, 430, 435, 436].

5.8 Further Reading

For an extensive reading on game and player analytics (including visualization, data preprocessing, data modeling and game domain-dependent tasks) we refer the reader to the edited book by El-Nasr et al. [186]. When it comes to player modeling two papers offer complementary perspectives and taxonomies of player modeling and a thorough discussion on what aspects of players can be modeled and the ways players can be modeled: the survey papers of Smith et al. [636] and Yannakakis et al. [782].

5.9 Exercises

In this section we propose a set of exercises for modeling both the behavior and the experience of game players. For that purpose, we outline a number of datasets that can be used directly for analysis. Please note, however, that the book's website will remain up to date with more datasets and corresponding exercises beyond the ones covered below.

5.9.1 Player Behavior

A proposed semester-long game data mining project is as follows. You have to choose a dataset containing player behavioral attributes to apply the necessary preprocessing on the data such as extracting features and selecting features. Then you must apply a relevant **unsupervised learning** technique for compressing, analyzing, or reducing the dimensionality of your dataset. Based on the outcome of unsupervised learning you will need to implement a number of appropriate **supervised learning** techniques that learn to predict a data attribute (or a set of attributes). We leave the selection of algorithms to the reader or the course instructor. Below we discuss a number of example datasets one might wish to start from; the reader, however, may refer to the book's website for more options on game data mining projects.

5.9.1.1 *SteamSpy* Dataset

SteamSpy (http://steamspy.com/) is a rich dataset of thousands of games released on Steam[6] containing several attributes each. While strictly not a dataset focused on player modeling, *SteamSpy* offers an accessible and large dataset for game analytics. The data attributes of each game include the game's name, the developer, the publisher, the score rank of the game based on user reviews, the number of owners

[6] http://store.steampowered.com/

of the game on Steam, the people that have played this game since 2009, the people that have played this game in the last two weeks, the average and median playtime, the game's price and the game's tags. The reader may use an API[7] to download all data attributes from all games contained in the dataset. Then one might wish to apply supervised learning to be able to predict an attribute (e.g., the game's price) based on other game features such as the game's score, release date and tags. Or alternatively, one might wish to construct a score predictor of a new game. The selection of the modeling task and the AI methods is left to the reader.

5.9.1.2 *StarCraft: Brood War* Repository

The *StarCraft: Brood War* repository contains a number of datasets that include thousands of professional *StarCraft* replays. The various data mining papers, datasets as well as replay websites, crawlers, packages and analyzers have been compiled by Alberto Uriarte at Drexel University.[8] In this exercise you are faced with the challenge of mining game replays with the aim to predict a player's strategy. Some results on the *StarCraft: Brood War* datasets can be found in [750, 728, 570] among others.

5.9.2 *Player Experience*

As a semester project on player experience modeling it is suggested you choose a game, one or more affective or cognitive states to model (model's output) and one or more input modalities. You are expected to collect empirical data using your selected game and build models for the selected psychological state of the players that rely on the chosen input modalities.

As a smaller project that does not involve data collection you may opt to choose one of the following datasets and implement a number of AI methods that will derive accurate player experience models. The models should be compared in terms of a performance measure. The two datasets accompanying this book and outlined below are the **platformer experience dataset** and the ***Maze-Ball* dataset**. The book's website will be up to date with more datasets and exercises beyond the ones covered below.

5.9.2.1 Platformer Experience Dataset

The extensive analysis of player experience in *Super Mario Bros* (Nintendo, 1985) and our wish to further advance our knowledge and understanding on player expe-

[7] http://steamspy.com/api.php

[8] Available at: http://nova.wolfwork.com/dataMining.html

rience had led to the construction of the Platformer Experience Dataset [326]. This is the first open-access game experience corpus that contains multiple modalities of data from players of *Infinite Mario Bros*, a variant of *Super Mario Bros* (Nintendo, 1985). The open-access database can be used to capture aspects of player experience based on **behavioral** and **visual** recordings of platform game players. In addition, the database contains aspects of the **game context**—such as level attributes—demographic data of the players and self-reported annotations of experience in two forms: **ratings** and **ranks**.

Here are a number of questions you might wish to consider when attempting to build player experience models that are as accurate as possible: Which AI methods should I use? How should I treat my output values? Which feature extraction and selection mechanism should I consider? The detailed description of the dataset can be found here: http://www.game.edu.mt/PED/. The book website contains further details and a set of exercises based on this dataset.

5.9.2.2 *Maze-Ball* Dataset

As in the case of the Platformer Experience Dataset the *Maze-Ball* dataset is also publicly available for further experimentation. This open-access game experience corpus contains two modalities of data obtained from *Maze-Ball* players: their **gameplay** attributes and three **physiological signals**: blood volume pulse, heart rate and skin conductance. In addition, the database contains aspects of the game such as features of the virtual camera placement. Finally the dataset contains demographic data of the players and self-reported annotations of experience in two forms: **ratings** and **ranks**.

The aim, once again, is to construct the most accurate models of experience for the players of *Maze-Ball*. So, which modalities of input will you consider? Which annotations are more reliable for predicting player experience? How will your signals be processed? These are only a few of the possible questions you will encounter during your efforts. The detailed description of the dataset can be found here: http://www.hectorpmartinez.com/. The book website contains further details about this dataset.

5.10 Summary

This chapter focused on the use of AI for modeling players. The core reasons why AI should be used for that purpose is either to derive something about the players' experience (how they feel in a game) or for us to understand something about their behavior (what they do in a game). In general we can model player behavior and player experience by following a top-down or a bottom-up approach (or a mix of the two). Top-down (or model-based) approaches have the advantage of solid theoretical frameworks usually derived from other disciplines or other domains than games. Bottom-up (or model-free) instead rely on data from players and have the advantage of not assuming anything about players other than that player experience and behavior are associated with data traces left by the player and that these data traces are representative of the phenomenon we wish to explain. While a hybrid between model-based and model-free approaches is in many ways a desirable approach to player modeling, we focus on bottom-up approaches, where we provide a detailed taxonomy for the options available regarding the input and the output of the model, and the modeling mechanism per se. The chapter ends with a number of player modeling examples, for modeling both the behavior of players and their experience.

The player modeling chapter is the last chapter of the second part of this book, which covered the core uses of AI in games. The next chapter introduces the third and last part of the book, which focuses on the holistic synthesis of the various AI areas, the various methods and the various users of games under a common game AI framework.

Part III
The Road Ahead

Chapter 6
Game AI Panorama

This chapter attempts to give a high-level overview of the field of game AI, with particular reference to how the different core research areas within this field inform and interact with each other, both actually and potentially. For that purpose we first identify the main research areas and their sub-areas within the game AI field. We then view and analyze the areas from three key perspectives: (1) the dominant AI method(s) used under each area; (2) the relation of each area with respect to the end (human) user; and (3) the placement of each area within a human-computer (player-game) interaction perspective. In addition, for each of these areas we consider how it could inform or interact with each of the other areas; in those cases where we find that meaningful interaction either exists or is possible, we describe the character of that interaction and provide references to published studies, if any.

The main motivations for us writing this chapter is to help the reader understand how a particular area relates to other areas within this increasingly growing field, how the reader can benefit from knowledge created in other areas and how the reader can make her own research more relevant to other areas. To facilitate and foster synergies across active research areas we place all key studies into a taxonomy with the hope of developing a common understanding and vocabulary within the field of AI and games. The structure of this chapter is based on the first holistic overview of the game AI field presented in [785]. The book takes a new perspective on the key game AI areas given its educational and research focus.

The main game AI areas and core subareas already identified in this book and covered in this chapter are as follows:

- **Play Games** (see Chapter 3) which includes the subareas of *Playing to Win* and *Playing for the Experience*. Independently of the purpose (winning or experience) AI can control either the player character or the non-player character.
- **Generate Content** (see Chapter 4) which includes the subareas of *autonomous (procedural) content generation* and *assisted content generation*. Please note that the terms *assisted (procedural) content generation* and *mixed-initiative (procedural) content generation* (as defined in Chapter 4) are used interchangeably in this chapter.

G. N. Yannakakis and J. Togelius, *Artificial Intelligence and Games*, https://doi.org/10.1007/978-3-319-63519-4_6

- **Model Players** (see Chapter 5) which includes the subareas of *player experience modeling* and *player behavior modeling*, or else, *game data mining* [178].

The scope of this chapter is not to provide an inclusive survey of all game AI areas—the details of each area have been covered in preceding chapters of the book—but rather a roadmap of interconnections between them via representative examples. As research progresses in this field, new research questions will pop up and new methods be invented, and other questions and methods recede in importance. We believe that all taxonomies of research fields are by necessity tentative. Consequently, the list of areas defined in this chapter should not be regarded as fixed and final.

The structure of the chapter is as follows: In Section 6.1, we start by holistically analyzing the game AI areas within the game AI field and we provide three alternative views over game AI: one with respect to the methods used, one with respect to the end users within game research and development and one where we outline how each of the research areas fits within the player-game interaction loop of digital games. Then, Section 6.2, digs deeper into the research areas and describes each one of them in detail. With the subsection describing each area, there is a short description of the area and a paragraph on the possible interactions with each of the other areas for which we have been able to identify strong or weak influences . The chapter ends with a section containing our key conclusions and vision for the future of the field.

6.1 Panoramic Views of Game AI

Analyzing any research field as a composition of various subareas with interconnections and interdependencies can be achieved in several different ways. In this section we view game AI research from three high-level perspectives that focus on the computer (i.e., the AI methods), the human (i.e., the potential end user of game AI) and the interaction between the key end user (i.e., player) and the game. Instead in Section 6.2 we outline each game AI area and present the interconnections between the areas.

Game AI is composed of (a set of) methods, processes and algorithms in artificial intelligence as those are applied to, or inform the development of, games. Naturally, game AI can be analyzed through the **method** used by identifying the dominant AI approaches under each game AI area (see Section 6.1.1). Alternatively, game AI can be viewed from the game domain perspective with a focus on the **end users** of each game AI area (see Section 6.1.2). Finally game AI is, by nature, realized through systems that entail rich human-computer interaction (i.e., games) and, thus, the different areas can be mapped to the interaction framework between the player and the game (see Section 6.1.3).

Table 6.1 Dominant (•) and secondary (○) AI methods for each of the core AI areas we cover in this book. The total number of methods used for each area appears at the bottom row of the table.

	Play Games		Generate Content		Model Players	
	Winning	Experience	Autonomously	Assisted	Experience	Behavior
Behavior Authoring	•	•				
Tree Search	•	○	○	○		
Evolutionary Computation	•	○	•	•	•	
Supervised Learning	○	•			•	•
Reinforcement Learning	•	○				
Unsupervised Learning				○	○	•
Total (Dominant)	5 (4)	5 (2)	2 (1)	3 (1)	3 (2)	2 (2)

6.1.1 Methods (Computer) Perspective

The first panoramic view of game AI we present is centered around the AI methods used in the field. As the basis of this analysis we first list the core AI methods mostly used in the game AI field. The key methodology areas identified in Chapter 2 include ad-hoc behavior authoring, tree search, evolutionary computation, reinforcement learning, supervised learning, and unsupervised learning. For each of the game AI areas investigated we have identified the AI methods that are **dominant** or **secondary** in the area. While the dominant methods represent the most popular techniques used in the literature, secondary methods represent techniques that have been considered from a substantial volume of studies but are not dominant.

We have chosen to group methods according to what we perceive as a received taxonomy and following the structure of Chapter 2. While it would certainly be possible to classify the various methods differently, we argue that the proposed classification is compact (containing solely key methodology areas) and it follows standard method classifications in AI. While this taxonomy is commonly accepted, the lines can be blurred. In particular evolutionary computation, being a very general optimization method, can be used to perform supervised, unsupervised or reinforcement learning (more or less proficiently). The model-building aspect of reinforcement learning can be seen as a supervised learning problem (mapping from action sequences to rewards), and the commonly used tree search method Monte Carlo tree search can be seen as a form of TD learning. The result of any tree search algorithm can be seen as a plan, though it is often not guaranteed to lead to the desired end state. That the various methods have important commonalities and some overlap does not detract from the fact that each of them is clearly defined.

Table 6.1 illustrates the relationship between game AI areas and corresponding methods. It is evident that evolutionary computation and supervised learning appear to be of dominant or secondary use in most game AI areas. Evolutionary computation is a dominant method for playing to win, for generating content (in an assisted/mixed-initiative fashion or autonomously), and for modeling players; it has also been considered for the design of believable play (play for experience) research. Supervised learning is of substantial use across the game AI areas and appears to be dominant in player experience and behavioral modeling, as well as in the area of AI that plays for experience. Behavior authoring, on the other hand, is useful solely for game-playing. Reinforcement learning and unsupervised learning find limited use across the game AI areas, respectively, being dominant only on AI that plays to win and player behavior modeling. Finally, tree search finds use primarily in playing to win and it is also considered—as a form of planning—for controlling play for experience and in computational narrative (as part of autonomous or assisted PCG).

Viewing Table 6.1 from the game AI areas' perspective (columns) it seems that AI that plays games (either for wining or for the experience) defines the game AI area with the most diverse and richest palette of AI methods. On the contrary, procedural content generation is solely dominated by evolutionary computation and tree search to a secondary degree. It is important to state that the popularity of any AI method within a particular area is closely tied to the task performed or the goal in mind. For example, evolutionary computation is largely regarded as a computationally heavy process which is mostly used in tasks associated with offline training. As PCG so far mainly relies on content that is generated offline, evolutionary computation offers a good candidate method and the core approach behind search-based PCG [720]. If online learning is a requirement for the task at hand, however, other methods (such as reinforcement learning or pruned tree-search) tend to be preferred.

Clearly the possibility space for future implementations of AI methods under particular game AI areas seems rather large. While particular methods have been traditionally dominant in specific areas for good reasons (e.g., planning in computational narrative) there are equally good reasons to believe that the research in a game AI area itself has been heavily influenced by (and limited to) its corresponding dominant AI methods. The empty cells of Table 6.1 indicate potential areas for exploration and offer us an alternative view of promising new intersections between game AI areas and methods.

6.1.2 End User (Human) Perspective

The second panoramic view of the game AI field puts an emphasis on the end user of the AI technology or general outcome (product or solution). Towards that aim we investigate three core dimensions of the game AI field and classify all game AI areas with respect to the **process** AI follows, the game **context** under which algorithms operate and, finally, the **end user** that benefits most from the resulting outcome. The classes identified under the above dimensions are used as the basis of the taxonomy we propose.

The first dimension (phrased as a question) refers to the AI process: *In general, what can AI **do** within games?* We identify two potential classes in this dimension: AI can **model** or **generate**. For instance, an artificial neural network can model a playing pattern, or a genetic algorithm can generate game assets. Given that AI can model or generate the second dimension refers to the context: ***What** can AI methods model or generate in a game?* The two possible classes here are **content**

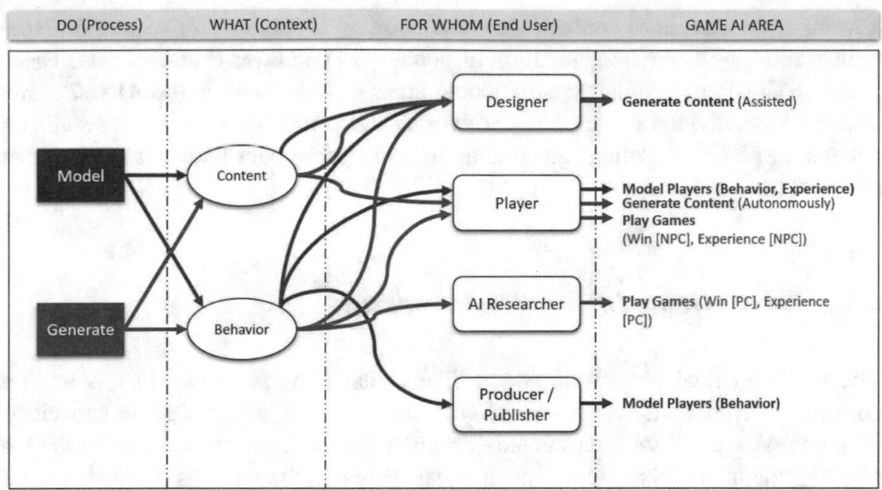

Fig. 6.1 The end user perspective of the identified game AI areas. Each AI area follows a **process** (model or generate) under a **context** (content or behavior) for a particular **end user** (designer, player, AI researcher or game producer/publisher). Blue and red arrows represent the processes of modeling and generation, respectively. Modified graph from [785].

and **behavior**. For example, AI can model a players' affective state, or generate a level. Finally, the third dimension is the end user: *AI can model, or generate, either content or behavior; but,* ***for whom****?* The classes under the third dimension are the **designer**, the **player**, the **AI researcher**, and the **producer/publisher**.

Note that the above taxonomy serves as a framework for classifying the game AI areas according to the end user and is, by no means, inclusive of all potential processes, contexts, and end users. For instance, one could claim that the producer's role should be distinct from the publisher's role and that a developer should also be included in that class. Moreover, game content could be further split into smaller sub-classes such as narrative, levels, etc. Nevertheless, the proposed taxonomy provides distinct roles for the AI process (model vs. generate vs. evaluate), clear-cut classification for the context (content vs. behavior) and a high-level classification of the available stakeholders in game research and development (designer vs. player vs. AI researcher vs. producer/publisher). The taxonomy presented here is a modified version of the one introduced in [785] and it does not consider **evaluation** as a process for AI since it is out of the primary scope of this book.

Figure 6.1 depicts the relationship between the game AI core areas, the subareas and the end users in game research and development. Assisted, or mixed-initiative, content generation is useful for the designer and entails all possible combinations of processes and context as both content and behavior can be either modeled or generated for the designer. Compared to the other stakeholders the player benefits directly from more game AI research areas. In particular the player and her experience are affected by research on player modeling, which results from the modeling of experience and behavior; research on autonomous procedural content generation, as a result of generation of content; and studies on NPC playing (for wining or experience) resulting from the generation of behavior. The player character (PC)-based game playing (for winning or experience) areas provide input to the AI researcher primarily. Finally, the game producer/publisher is primarily affected by results on behavioral player modeling, game analytics and game data mining as a result of behavior modeling.

6.1.3 Player-Game Interaction Perspective

The third and final panoramic perspective of game AI presented in this section couples the computational processes with the end user within a game and views all game AI areas through a human-computer interaction—or, more accurately, a player-game interaction—lens. The analysis builds on the findings of Section 6.1.2 and places the five game AI areas that concern the player as an end user on a player-game interaction framework as depicted in Fig. 6.2. Putting an emphasis on player experience and behavior, player modeling directly focuses on the interaction between a player and the game context. Game content is influenced primarily by research on autonomous procedural content generation. In addition to other types of content, most games feature NPCs, the behavior of which is controlled by some form of AI. NPC behavior is informed by research in NPCs that play the game to win or any other playing-experience purpose such as believability.

Looking at the player-game interaction perspective of game AI it is obvious that the player modeling area has the most immediate and direct impact on the player experience as it is the only area linked to the player-game interaction directly. From the remaining areas, PCG influences player experience the most as all games have some form of environment representation and mechanics. Finally, AI that plays as an NPC (either to win or for the experience of play) is constrained to games that include agents or non-player characters.

The areas not considered directly in this game AI perspective affect the player rather remotely. Research on AI tools that assist the generative process of content improves the game's quality as a whole and in retrospect the player experience since designers tend to maintain a *second-order player model* [378] while designing. Finally, AI that plays the game as a player character can be offered for testing both the content and the NPC behaviors of a game, but also the interaction between the player and the game (via e.g., player experience competitions), but is mainly directed to AI researchers (see Fig. 6.1).

6.2 How Game AI Areas Inform Each Other

In this section, we outline the core game AI areas and discuss how they inform or influence (the terms are used interchangeably) each other. All research areas could

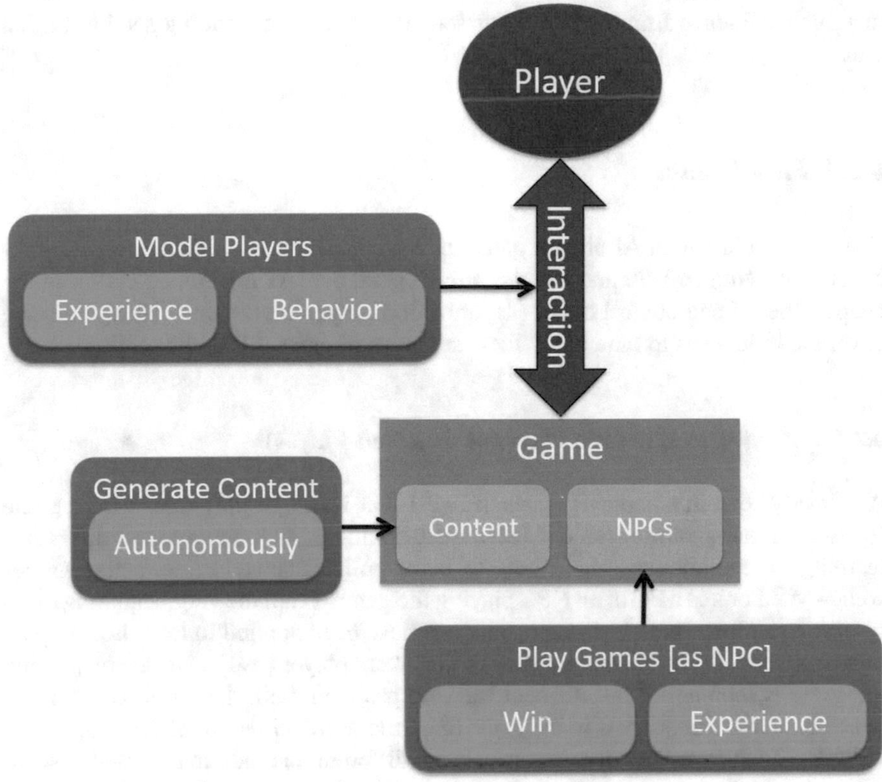

Fig. 6.2 Game AI areas and sub-areas viewed from a player-game interaction perspective.

be seen as potentially influencing each other to some degree; however, making a list of all such influences would be impractical and the result would be uninteresting. Therefore we only describe *direct* influences. Direct influences can be either **strong** (represented by a • as the bullet point style next to the corresponding influence in the following lists) or **weak** (represented by a ∘). We do not list influences we do not consider potentially important for the informed research area, or which only go through a third research area.

The sections below list **outgoing** influence. Therefore, to know how area A influences area B you should look in the section describing area A. Some influences are mutual, some not. The notation $A \rightarrow B$ in the headings of this section denotes that "A influences B". In addition to the text description each section provides a figure representing all outgoing influences of the area as arrows. Dark and light gray colored areas represent, respectively, *strong* and *weak* influence. Areas with white background are not influenced by the area under consideration. The figures also depict the incoming influences from other areas. Incoming *strong* and *weak* influences are represented, respectively, with a solid line and a dashed line around the game AI areas that influence the area under consideration. Note that the description of the incoming influence from an area is presented in the corresponding section of that area.

6.2.1 Play Games

The key area in which AI plays a game (as covered in Chapter 3) involves the subareas of **Playing to Win** and **Playing for Experience**. As mentioned earlier in the chapter the AI can control either player or non-player characters of the game. We cover the influences to (and from) these subareas of game AI in this section.

6.2.1.1 Playing to Win (as a Player or as a Non-Player)

As already seen in Chapter 3 research in AI that learns to play (and win) a game focuses on using **reinforcement learning** techniques such as temporal difference learning or evolutionary algorithms to learn policies/behaviors that play games well—whether it is a PC or an NPC playing the game. From the very beginning of AI research, reinforcement learning techniques have been applied to learn how to play board games (see for example Samuel's Checkers player [591]). Basically, playing the game is seen as a reinforcement learning problem, with the reinforcement tied to some measure of success in the game (e.g., the score, or length of time survived). As with all reinforcement learning problems, different methods can be used to solve the problem (find a good policy) [715] including TD learning [689], evolutionary computation [406], competitive co-evolution [24, 538, 589, 580], simulated annealing [42], other optimization algorithms and a large number of combinations between such algorithms [339]. In recent years a large number of papers that describe the application of various learning methods to different types of video games have appeared in the literature (including several overviews [470, 406, 632, 457]). Finally, using games to develop artificial general intelligence builds on the idea that games can be useful environments for algorithms to learn complex and useful behaviors; thus research in algorithms that learn to win is essential.

It is also worth noting that most existing game-based **benchmarks** measure how well an agent plays a game—see for example [322, 404, 504]. Methods for learning to play a game are vital for such benchmarks, as the benchmarks are only meaningful in the context of the algorithms. When algorithms are developed that "beat" existing benchmarks, new benchmarks need to be developed. For example, the success of an early planning agent in the first Mario AI competition necessitated that the software be augmented with a better level generator for the next competition [322], and for the Simulated Car Racing competition, the performance of the best agents on the original competition game spurred the change to a new more sophisticated racing game [710, 392].

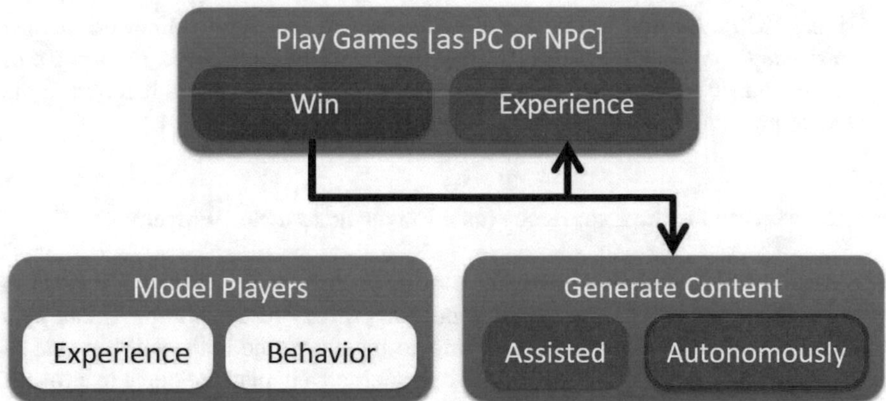

Fig. 6.3 Playing to Win: influence on (and from) other game AI research areas. **Outgoing** influence (represented by arrows): black and dark gray colored areas reached by arrows represent, respectively, **strong** and **weak** influence. **Incoming** influence is represented by red lines around the areas that influence the area under investigation (i.e., AI that plays to win in this figure): **strong** and **weak** influences are represented, respectively, by a solid and a dashed line.

Research in this area impacts game AI at large as three game AI subareas are directly affected; in turn, one subarea is directly affecting AI that plays to win (see Fig. 6.3).

- **Playing to Win → Playing for the Experience:** An agent cannot be believable or existent to augment the game's experience if it is not proficient. Being able to play a game well is in several ways a precondition for playing games in a believable manner though well playing agents can be developed without learning (e.g., via top-down approaches). In recent years, successful entries to competitions focused on believable agents, such as the 2K BotPrize and the Mario AI Championship Turing test track, have included a healthy dose of learning algorithms [719, 603].
- **Playing to Win → Generate Content (Autonomously):** Having an agent that is capable of playing a game proficiently is useful for **simulation-based testing** in procedural content generation, i.e., the testing of newly generated game content by playing through that content with an agent. For example, in a program generating levels for the platform game *Super Mario Bros* (Nintendo, 1985), the levels can be tested by allowing a trained agent to play them; those that the agent cannot complete can be discarded [335]. Browne's *Ludi* system, which generates complete board games, evaluates these games through simulated playthrough and uses learning algorithms to adapt the strategy to each game [74].
- **Playing to Win → Generate Content (Assisted):** Just as with autonomous procedural content generation, many tools for AI-assisted game design rely on being able to simulate playthroughs of some aspect of the game. For instance, the *Sentient Sketchbook* tool for level design uses simple simulations of game-playing agents to evaluate aspects of levels as they are being edited by a human de-

signer [379]. Another example is the automated playtesting framework named *Restricted Play* [295] which aims mostly at assisting designers on aspects of game balance during game design. A form of *Restricted Play* is featured in the *Ludocore* game engine [639].

6.2.1.2 Playing for the Experience (as a Player or as a Non-Player)

Research on AI that plays a game for a purpose other than winning is central to studies where playing the game well is not the primary research aim. AI can play the game as a *player* character attempting to maximize the believability value of play as, for instance, in [719, 619, 96]. It can alternatively play the game in a role of an NPC for the same purpose [268]. Work under this research subarea involves the study of believability, interestingness or playing experience in games and the investigations of mechanisms for the construction of **agent architectures** that appear to have e.g., believable or human-like characteristics. The approaches for developing such architectures can be either top-down behavior authoring (such as the FAtiMA model used in *My Dream Theatre* [100] and the Mind Module model [191] used in *The Pataphysic Institute*) or bottom-up attempting to imitate believable gameplay from human players such as the early work of Thurau et al. in *Quake II* (Activision, 1997) bots [696], the human imitation attempts in *Super Mario Bros* (Nintendo, 1985) [511], the *Unreal Tournament 2004* (Epic Games, 2004) believable bots of Schrum et al. [603] and the crowdsourcing studies of the *Restaurant game* [508]. Evidently, commercial games have for long benefited from agent believability research. Examples of this include popular games such as the *Sims* (Electronic Arts, 2000) series. The industry puts a strong emphasis on the design of believability in games as this contributes to more immersive game environments. The funding of believability research through game AI competitions such as the 2K BotPrize is one of the many clear indicators of the commercial value of agent believability.

Over the last few years there has been a growing academic (and commercial) interest in the establishment of **competitions** that can be used as assessment tools for agent believability [719]. Agent believability research has provided input and given substance to those game benchmarks. A number of game Turing competitions have been introduced to the benefit of agent believability research, including the 2K BotPrize on the *Unreal Tournament 2004* (Epic Games, 2004) [647, 264] game and the Mario AI Championship: Turing test track [619] on the *Super Mario Bros* (Nintendo, 1985) game. Recently, the community saw AI agents passing the Turing test in the 2K BotPrize [603].

The study of AI that plays games not for winning, but for other purposes, affects research on three other game AI areas as illustrated in Fig. 6.4, whereas it is affected by four other game AI areas.

- **Playing for the Experience $\rightarrow$ Model Players (Experience and Behavior):** There is a direct link between player modeling and believable agents as research carried out for the modeling of human, human-like, and supposedly believable playing behavior can inform the construction of more appropriate models for

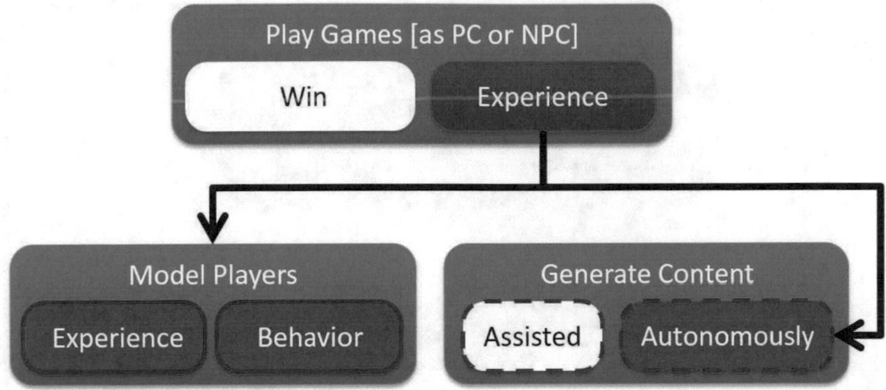

Fig. 6.4 Playing for the Experience: influence on (and from) other game AI research areas.

players. Examples include the imitation of human play styles in *Super Mario Bros* (Nintendo, 1985) [511] and *Quake II* (Activision, 1997) [696]. Though computational player modeling uses learning algorithms, it is only in some cases that it is the behavior of an NPC that is modeled. In particular, this is true when the in-game behavior of one or several players is modeled. This can be done using either reinforcement learning techniques, or supervised learning techniques such as backpropagation or decision trees. In either case, the intended outcome for the learning algorithm is not necessarily an NPC that plays as well as possible, but one that plays in the style of the modeled player [735, 511].

- **Playing for the Experience → Generate Content (Autonomously):** Believable characters may contribute to better levels [96], more believable stories [801, 401, 531] and, generally, better game representations [563]. A typical example of the integration of characters in the narrative and the drive of the latter based on the former includes the FAtiMa agents in *FearNot!* [516] and *My Dream Theater* [100]. Another example is the generation of *Super Mario Bros* (Nintendo, 1985) levels that maximize the believability of any Mario player [96].

6.2.2 Generate Content

As covered in detail in Chapter 4 AI can be used to design whole (or parts of) games in an autonomous or in an assisted fashion. This core game AI area includes the subareas of **autonomous (procedural) content generation** and **assisted** or **mixed-initiative (procedural) content generation**. The interactions of these subareas with the remaining areas of game AI are covered in this section.

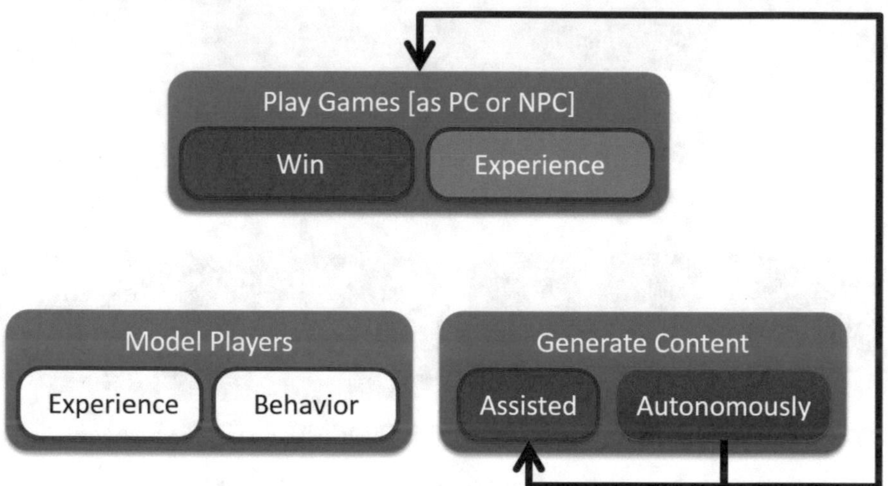

Fig. 6.5 Generate Content (Autonomously): influence on (and from) other game AI research areas.

6.2.2.1 Generate Content (Autonomously)

As stated in Chapter 4 procedural content generation has been included in limited roles in some commercial games since the early 1980s; however, recent years have seen an expansion of **research** on more controllable PCG for multiple types of game content [764], using techniques such as evolutionary search [720] and constraint solving [638]. The influence of PCG research beyond games is already evident in areas such as computational creativity [381] and interaction design (among others). There are several surveys of PCG available, including a recent book [616] and vision paper [702], as well as surveys of frameworks [783], sub-areas of PCG [554, 732] and methods [720, 638].

Autonomous content generation is one of the areas of recent academic research on AI in games which bears most promise for incorporation into **commercial games**. A number of recent games have been based heavily on PCG, including independent ("indie") game production successes such as *Spelunky* (Mossmouth, 2009) and *Minecraft* (Mojang, 2011), and mainstream AAA games such as *Diablo III* (Blizzard Entertainment, 2012) and *Civilization V* (2K Games, 2010). A notable example, as mentioned in Chapter 4 is *No Man's Sky* (Hello Games, 2016) with its quintillion different procedurally generated planets. Some games heavily based on PCG and developed by researchers have been released as commercial games on platforms such as Steam and Facebook; two good examples of this are *Petalz* [565, 566] and *Galactic Arms Race* [249].

Figure 6.5 depicts the three (and five) areas that are influenced by (and influence) autonomous PCG.

- **Generate Content (Autonomously) → Play to Win:** If an agent is trained to perform well in only a single game environment, it is easy to overspecialize the training and arrive at a policy/behavior that will not generalize to other levels. Therefore, it is important to have a large number of environments available for training. PCG can help with this, potentially providing an infinite supply of test environments. For example, when training players for the Mario AI Championship it is common practice to test each agent on a large set of freshly generated levels, to avoid overtraining [322]. There has also been research on adapting NPC behavior specifically to generated content [332]. Finally, one approach to **artificial general intelligence** is to train agents to be good at playing games in general, and test them on a large variety of games drawn from some genre or distribution. To avoid overfitting, this requires games to be generated automatically, a form of PCG [598]. The generation of new environments is very important for NPC behavior learning, and this extends to benchmarks that measure some aspect of NPC behavior. Apart from the Mario AI Championship, competitions such as the Simulated Car Racing Championship use freshly generated tracks, unseen by the participants, to test submitted controllers [102]. But there is also scope for benchmarks and competitions focused on measuring the capabilities of PCG systems themselves, such as the Level Generation track of the Mario AI Championship [620].

- ○ **Generate Content (Autonomously) → Play for the Experience:** Research on autonomous PCG naturally influences research on agent (PC or NPC) control for believability, interestingness or other aims aside from winning given that these agents are performing in a particular environment and under a specific game context. This influence is still in its infancy and the only study we can point the reader to is the one by Camilleri et al. [96] where the impact of level design on player character believability is examined in *Super Mario Bros* (Nintendo, 1985). Further, research on interactive narrative benefits from and influences the use of believable agents that interact with the player and are interwoven in the story plot. The narrative can yield more (or less) believability to agents and thus the relationship between the behavior of the agents and the emergent story is strong [801, 401, 531]. In that sense, the computational narrative of a game may define the arena for believable agent design.
- • **Generate Content (Autonomously) → Generate Content (Assisted):** As content design is a central part of game design, many AI-assisted design tools incorporate some form of assisted content design. Examples include *Tanagra*, which helps designers create complete platform game levels which ensure playability through the use of constraint solvers [641], and *SketchaWorld* [634]. Another example is *Sentient Sketchbook*, which assists humans in designing strategy game levels through giving immediate feedback on properties of levels and autonomously suggesting modifications [379].

6.2.2.2 Generate Content (Assisted)

Assisted content generation refers to the development of AI-powered tools that support the game design and development process. This is perhaps the AI research area

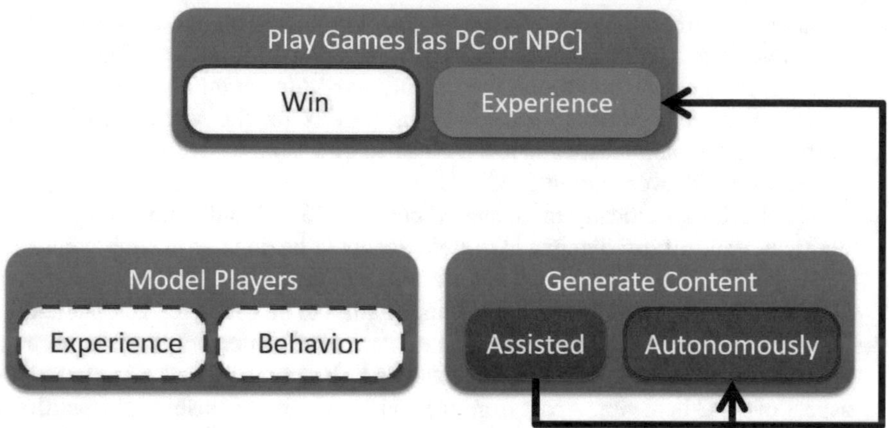

Fig. 6.6 Generate Content (Assisted): influence on (and from) other game AI research areas.

which is most promising for the development of better games [764]. In particular, AI can assist in the creation of game content varying from levels and maps to game mechanics and narratives. The impact of AI-enabled authoring tools on design and development influences the study of AI that plays games for believability, interestingness or player experience, and research in autonomous procedural content generation (see Fig. 6.6). **AI-assisted game design** tools range from those designed to assist with generation of complete game rulesets such as *MetaGame* [522] or *RuLearn* [699] to those focused on more specific domains such as strategy game levels [379], platform game levels [642], horror games [394] or physics-based puzzles [613].

It is worth noting that AI tools have been used extensively for supporting design and **commercial game development**. Examples such as the *SpeedTree* (Interactive Data Visualization Inc., 2013) generator for trees and other plants [287] have seen uses in several game productions. The mixed-initiative PCG tools mentioned above have a great potential in the near future as most of these are already tested on commercial games or developed with game industrial partners. Furthermore, there are tools designed for interactive modeling and analysis of game rules and mechanics, which are not focused on generating complete games but on prototyping and understanding aspects of complex games; such systems could be applied to existing commercial games [639].

- **Generate Content (Assisted) → Play for the Experience:** Authoring tools in forms of open-world sandboxes could potentially be used for the creation of more believable behaviors. While this is largely still an unexplored area of research and development, notable attempts include the NERO game AI platform where players can train game agents for efficient and believable first-person shooter bot behaviors [654]. An open version of this platform focused on crowdsourcing behaviors has been released recently [327]. A similar line of research is the generation of *Super Mario Bros* (Nintendo, 1985) players by means of interactive evolution [648].

- **Generate Content (Assisted) → Generate Content (Autonomously):** Research on methods of **mixed-initiative co-creation** [774] and design can feed input to and spur discussion on central topics in procedural content generation. Given the importance of content design in the development process as a whole, any form of mixed-initiative AI assistance in the generation process can support and augment procedural content generation. Notable examples of mixed-initiative PCG include the *Tanagra* platform game level design AI assistant [641], and the *SketchaWorld* [634], the *Sentient World* [380], the *Sentient Sketchbook* [379, 774] and the *Sonancia* [394] systems which generate game maps and worlds in a mixed-initiative design fashion following different approaches and levels of human computation. Further, tools can assist the authoring of narrative in games. In particular, **drama management** tools have long been investigated within the game AI community. An academic example is ABL which has allowed the authoring of narrative in *Façade* [441]. Among the few available and well-functional authoring tools the most notable is the *Versu* [197] storytelling system which was used in the game *Blood & Laurels* (Emily Short, 2014) and the *Inform 7* [480] software package that led to the design of *Mystery House Possessed* (Emily Short, 2005). More story generation tools as such can be found at the http://storygen.org/ repository, by Chris Martens and Rogelio E. Cardona-Rivera.

6.2.3 Model Players

As already explored in Chapter 5, modeling players involves the subtasks of modeling their **behavior** or their **experience**. Given the interwoven nature of these two tasks we present their influences to (and from) other game AI areas under one common section. In player modeling [782, 636], computational models are created for detecting how the player perceives and reacts to gameplay. As stated in Chapter 5 such models are often built using machine learning methods where data consisting of some aspect of the game or player-game interaction is associated with labels derived from some assessment of player *experience*, gathered for example from questionnaires [781]. However, the area of player modeling is also concerned with structuring observed player *behavior* even when no correlates to experience are available—e.g., for identifying player types or predicting player behavior.

Research and development in player modeling can inform attempts for player experience in **commercial-standard games**. Player experience detection methods and algorithms can advance the study of user experience in commercial games. In addition, the appropriateness of sensor technology, the technical plausibility of biofeedback sensors, and the suitability of various modalities of human input can inform industrial developments. Quantitative testing via game metrics—varying from behavioral data mining to in-depth low scale studies—is also improved [764, 178, 186].

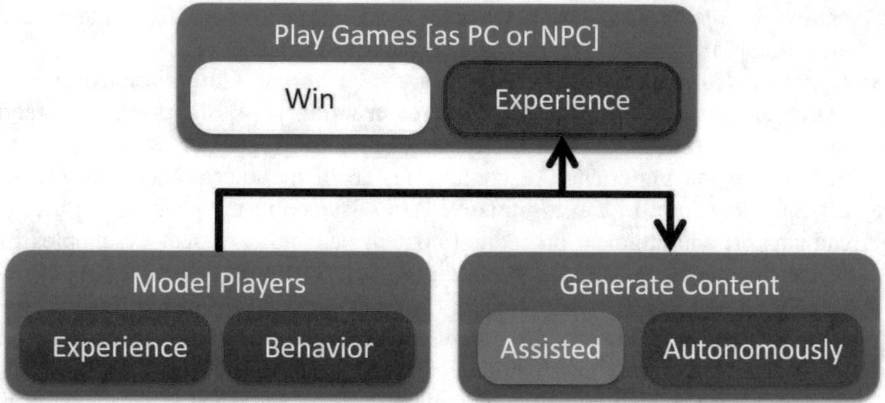

Fig. 6.7 Model Players: influence on (and from) other game AI research areas.

By now, a considerable number of academic studies use directly datasets from commercial games to induce models of players that could inform further development of the game. For example, we refer the reader to the experiments in clustering players of *Tomb Raider: Underworld* (Square Enix, 2008) into archetypes [176] and predicting their late-game performance based on early-game behavior [414]. Examples of player modeling components within high-profile commercial games include the arousal-driven appearance of NPCs in *Left 4 Dead 2* (Valve Corporation, 2009), the fearful combat skills of the opponent NPCs in *F.E.A.R.* (Monolith, 2005), and the avatars' emotion expression in the *Sims* series (Maxis, 2000) and *Black and White* (Lionhead Studios, 2001). A notable example of a game that is based on player experience modeling is *Nevermind* (Flying Mollusk, 2016); the game adapts its content based on the stress of the player, which is manifested via a number of physiological sensors.

Player modeling is considered to be one of the core non-traditional uses of AI in games [764] and affects research in AI-assisted game design, believable agents, computational narrative and procedural content generation (see Fig. 6.7).

- **Model Players → Play for the Experience:** Player models can inform and update believable agent architectures. Models of behavioral, affective and cognitive aspects of gameplay can improve the human-likeness and believability of any agent controller—whether it is ad-hoc designed or built on data derived from gameplay. While the link between player modeling and believable agent design is obvious and direct, research efforts towards this integration within games are still sparse. However, the few efforts made on the imitation of human game playing for the construction of believable architectures have resulted in successful outcomes. For example, human behavior imitation in platform [511] and racing games [735, 307] has provided human-like and believable agents while similar approaches for developing *Unreal Tournament 2004* (Epic Games, 2004) bots (e.g., in [328]) recently managed to pass the Turing test in the 2K BotPrize competition. Notably, one of the two agents that passed the Turing test in 2K BotPrize managed to do so by imitating (mirroring) aspects of human play [535]. A line of work that stands in between player modeling and playing games for the experience is the study on *procedural personas* [268, 267, 269]. As introduced in Chapter 5 procedural personas are NPCs that are trained to imitate realistically the decision making process of humans during play. Their study both influences our understanding about the internal (cognitive) processes of playing behavior and advances our knowledge on how to build believable characters in games.

- **Model Players → Generate Content (Autonomously):** There is an obvious link between computational models of players and PCG as player models can drive the generation of new personalized content for the player. The **experience-driven** role of PCG [783], as covered in Chapter 4, views game content as an indirect building block of a player's affective, cognitive and behavioral state and proposes adaptive mechanisms for synthesizing personalized game experiences. The "core loop" of an experience-driven PCG solution involves learning a model that can predict player experience, and then using this model as part of an evaluation function for evolving (or otherwise optimizing) game content; game content is evaluated based on how well it elicits a particular player experience, according to the model. Examples of PCG that are driven by player models include the generation of game rules [716], camera profiles [780, 85] and platform game levels [617]. Most work that goes under the label "game adaptation" can be said to implement the experience-driven architecture; this includes work on adapting the game content to the player using reinforcement learning [28] or semantic constraint solving [398] rather than evolution. Player models may also inform the generation of computational narrative. Predictive models of playing experience can drive the generation of individualized scenarios in a game. Examples of the coupling between player modeling and computational narrative include the affect-driven narrative systems met in *Façade* [441] and *FearNot!* [26], and the affect-centered game narratives such as the one of *Final Fantasy VII* (Square, 1997).
- ◦ **Model Players → Generate Content (Assisted):** User models can enhance authoring tools that, in turn, can assist the design process. The research area that bridges user modeling and AI-assisted design is in its infancy and only a few example studies can be identified. Indicatively, **designer models** [378] have been employed to personalize mixed-initiative design processes [774, 377, 379]. Such models may drive the procedural generation of designer-tailored content.

6.3 The Road Ahead

This chapter has initially identified the currently most active areas and subareas within game AI and placed them on three holistic frameworks: an AI method mapping, a game stakeholder (end user) taxonomy and the player-game interaction loop. This analysis revealed dominant AI algorithms within particular areas as well as room for exploration of new methods within areas. In addition, it revealed the dissimilar impact of different areas on different end users such as the AI researcher and the designer and, finally, outlined the influence of the different game AI areas on the player, the game and their interaction. From the high-level analysis of the game AI field we moved on to the detailed analysis of the game AI areas that compose it and thoroughly surveyed the meaningful interconnections between the different areas.

The total number of strong and weak influences is rather small compared to all possible interconnections between the areas, which clearly signals the research capacity of the game AI field for further explorations. We can distinguish a number of connections which are currently very active, meaning that much work currently goes on in one area that draws on work in another area. Here we see, for example, the connection between AI that plays to win in a general fashion in conjunction with the use of tree search algorithms: the MCTS algorithm was invented in the context of board game-playing, proved to be really useful in the general game playing competition, and is being investigated for use in games as different as *StarCraft* (Blizzard Entertainment, 1998) and *Super Mario Bros* (Nintendo, 1985). Improvements and modifications to the algorithm have been flowing back and forth between the various areas. Another indicative connection that is alive and strong is between player modeling and procedural content generation, where it is now common for newly devised PCG algorithms and experimental studies to include player behavioral or player experience models.

One can also study the currently strong areas by trying to cluster the trending topics in recent iterations of the IEEE CIG and AIIDE conferences. Such studies always include some form of selection bias, as papers can usually be counted into more than one area (e.g., depending on if you group by method or domain), but if you start from the session groupings made by the program chairs of each conference you achieve at least some inter-subjective validity. According to such a clustering, the most active topics over the last few years have been player (or emotion) modeling, game analytics, general game AI, real-time strategy game playing—especially *StarCraft* (Blizzard Entertainment, 1998)—and PCG (in general). Another perspective of the trend in game AI research is the varying percentage of studies on NPC (or game agent) behavior learning over other uses of AI in games at the two key conferences in the field (IEEE CIG and AIIDE). Our preliminary calculations suggest that while, initially, AI was mainly applied for NPC control and for playing board/card games well—more than 75% of CIG and AIIDE papers were linked to NPC behavior and agent game playing in 2005—that trend has drastically changed as entirely new (non-traditional) uses of AI became more common over the years—e.g., roughly 52% of the papers in CIG and AIIDE in 2011 did not involve game agents and NPC behavior. These facts indicate a shift in the use of AI in and for games towards multiple non-traditional applications—which tend to be traditional by now—for the development of better games [764].

But it is maybe even more interesting to look at all those connections that are unexploited or underexploited or potentially strong. For example, player modeling is potentially very important in the development of AI that controls believable, interesting or curious agents, but this has not been explored in enough depth yet; the same holds for the application of user (or else, designer) modeling principles towards the personalization of AI-assisted game design. Believable agents have, in turn, not been used enough in content generation (either autonomous or assisted). A grand vision of game AI for the years to come is to let it identify its own role within game design and development as it sees fit. In the last chapter of this book we discuss frontier research topics as such and identify unexplored roles of AI in games.

6.4 Summary

We hope that with this chapter of this book, we have been able to give our readers a sense of how this—by now rather large and multifaceted—research field hangs together, and what could be done to integrate it further. We realize that this is only our view of its dynamics and interconnections, and that there are (or could be) many competing views. We look forward to seeing those in upcoming studies in the field.

Finally, it is important to note that it would have been impossible to provide a complete survey of all the areas as, first, the game AI field is growing rapidly and, second, it is not the core objective of the book. This means that the bibliography is indicative rather than exhaustive and serves as a general guideline for the reader. The website of the book, instead of the book per se, will be kept up to date regarding important new readings for each area.

The next and final chapter of the book is dedicated to a few long-standing, yet rather unexplored, research frontiers of game AI. We believe that any advances made in these directions will lead to scientific breakthroughs not merely within game AI but largely in both games (their design, technology and analysis) and AI per se.

Chapter 7
Frontiers of Game AI Research

In this final chapter of the book we discuss a number of long-term visionary goals of game AI, putting an emphasis on the **generality** of AI and the **extensibility** of its roles within games. In particular, in Section 7.1 we discuss our vision for general behavior for each one of the three main uses of AI in games. Play needs to become general; generators are required to have general generative capacities across games, content types, designers and players; models of players also need to showcase general modeling abilities. In Section 7.2 we also discuss roles of AI that are still unexplored and certainly worth investigating in the future. The book ends with a discussion dedicated to general ethical considerations of game AI (Section 7.3).

7.1 General General Game AI

As evidenced from the large volume of studies the game AI research area has been supported by an active and healthy research community for more than a decade—at least since the start of the IEEE CIG and the AIIDE conference series in 2005. Before then, research had been conducted on AI in board games since the dawn of automatic computing. Initially, most of the work published at IEEE CIG or AIIDE was concerned with learning to play a particular game as well as possible, or using search/planning algorithms to play a game as well as possible without learning. Gradually, a number of new applications for AI in games and for games in AI have come to complement the original focus on AI for playing games [764]. Papers on procedural content generation, player modeling, game data mining, human-like playing behavior, automatic game testing and so on have become commonplace within the community. As we saw in the previous chapter there is also a recognition that all these research endeavors depend on each other [785]. However, almost all research projects in the game AI field are very *specific*. Most published papers describe a particular method—or a comparison of two or more methods—for performing a single task (playing, modeling, generating, etc.) in a single game. This is problematic in several ways, both for the scientific value and for the practical appli-

G. N. Yannakakis and J. Togelius, *Artificial Intelligence and Games*, https://doi.org/10.1007/978-3-319-63519-4_7

cability of the methods developed and studies made in the field. If an AI approach is only tested on a single task for a single game, how can we argue that is an advance in the scientific study of artificial intelligence? And how can we argue that it is a useful method for a game designer or developer, who is likely working on a completely different game than the one the method was tested on?

As discussed in several parts of this book general game playing is an area that has already been studied extensively and constitutes one of the key areas of game AI [785]. The focus of generality solely on play, however, is *very narrow* as the possible roles of AI and general intelligence in games are many, including game design, content design and player experience design. The richness of the cognitive skills and affective processes required to successfully complete these tasks has so far been largely ignored by game AI research. We thus argue, that while the focus on general AI needs to be retained, research on general game AI needs to expand beyond mere game playing. The new scope for ***general*** **general game AI** beyond game-playing broadens the applicability and capacity of AI algorithms and our understanding of intelligence as tested in a creative domain that interweaves problem solving, art, and engineering.

For general game AI to eventually be *truly* general, we argue that we need to extend the generality of general game playing to all other ways in which AI is (or can be) applied to games. More specifically we argue that the field should move towards methods, systems and studies that incorporate three different types of generality:

1. **Game generality**. We should develop AI methods that work with not just one game, but with any game (within a given range) that the method is applied to.
2. **Task generality**. We should develop methods that can do not only one task (playing, modeling, testing, etc) but a number of different, related tasks.
3. **User/designer/player generality**. We should develop methods that can model, respond to and/or reproduce the very large variability among humans in design style, playing style, preferences and abilities.

We further argue that all of this generality can be embodied into the concept of **general game design**, which can be thought of as a final frontier of AI research within games. Further details about the notion of *general* general game AI can be found in the vision paper we co-authored about this frontier research area [718]. It is important to note that we are not arguing that more focused investigations into methods for single tasks in single games are useless; these are often important as proofs-of-concept or industrial applications and they will continue to be important in the future, but there will be an increasing need to validate such case studies in a more general context. We are also not envisioning that everyone will suddenly start working on general methods. Rather, we are positing generalizations as a long-term goal for our entire research community. Finally, the general systems of game AI that we envision ought to have a real-world use. There is a risk that by making systems too general we might end up not finding applications of these general systems to any specific real-world problem. Thus, the system's applicability (or usefulness) sets our core constraint towards this vision of general game AI. More specifically, we envision *general* general game AI systems that are nevertheless integrated successfully within *specific* game platforms or game engines.

7.1.1 General Play

The problem of playing games is the one that has been most generalized so far. There already exist at least three serious benchmarks or competitions attempting to pose the problem of playing games *in general*, each in its own imperfect way. The General Game Playing Competition, often abbreviated GGP [223], the Arcade Learning Environment [40] and the General Video Game AI competition [528]; all three have been discussed in various places in this book. The results from these competitions so far indicate that general purpose search and learning algorithms by far outperform more domain-specific solutions and "clever hacks". Somewhat simplified, we can say that variations of Monte Carlo tree search perform best on GVGAI and GGP [202], and for ALE (where no forward model is available so learning a policy for each game is necessary) reinforcement learning with deep networks [464] and search-based iterative width [389, 301, 390] perform best. This is a very marked difference from the results of the game-specific competitions, indicating the lack of domain-independent solutions.

While these are each laudable initiatives and currently the focus of much research, in the future we will need to expand the scope of these competitions and benchmarks considerably, including expanding the range of games available to play and the conditions under which gameplay happens. We need game playing benchmarks and competitions capable of expressing any kind of game, including puzzle games, 2D arcade games, text adventures, 3D action-adventures and so on; this is the best way to test general AI capacities and reasoning skills. We also need a number of different ways of interfacing with these games—there is room for both benchmarks that give agents no information beyond the raw screen data but give them hours to learn how to play the game, and those that give agents access to a forward model and perhaps the game code itself, but expect them to play any game presented to them with no time to learn. These different modes test different AI capabilities and tend to privilege different types of algorithms. It is worth noting that the GVGAI competition is currently expanding to different types of playing modes, and has a long-term goal to include many more types of games [527].

We also need to differentiate away from just measuring how to play games optimally. In the past, several competitions have focused on agents that play games in a human-like manner; these competitions have been organized similarly to the classic Turing test [263, 619]. Playing games in a human-like manner is important for a number of reasons, such as being able to test levels and other game content as part of search-based generation, and to demonstrate new content to players. So far, the question of how to play games in a human-like manner *in general* is mostly unexplored; some preliminary work is reported in [337]. Making progress here will likely involve modeling how humans play games in general, including characteristics such as short-term memory, reaction time and perceptual capabilities, and then translating these characteristics to playing style in individual games.

7.1.2 General Game Generation and Orchestration

The study of PCG [616] for the design of game levels has reached a certain maturity and is, by far, the most popular domain for the application of PCG algorithms and approaches (e.g., see [720, 785, 783] among many). What is common in most of the content generation studies covered in this book, however, is their *specificity* and strong dependency of the representation chosen on the game genre examined. For the Mario AI Framework, for instance, the focus on a single level generation problem has been very much a mixed blessing: it has allowed for the proliferation and simple comparison of multiple approaches to solving the same problem, but has also led to a clear overfitting of methods. Even though some limited generalization is expected within game levels of the same genre, the level generators that have been explored so far clearly do not have the capacity of *general* level design. We argue that there needs to be a shift in how level generation is viewed. The obvious change of perspective is to create **general level generators**—level generators with general intelligence that can generate levels for any game (within a specified range). That would mean that levels are generated successfully across game genres and players and that the output of the generation process is meaningful and playable as well as entertaining for the player. Further, a general level generator should be able to coordinate the generative process with the other computational game designers who are responsible for the other parts of the game design.

To achieve general level design intelligence algorithms are required to capture as much of the level design space as possible at different representation resolutions. We can think of representation learning approaches such as deep autoencoders [739] capturing core elements of the level design space and fusing various game genres within a sole representation—as already showcased by a few methods, such as the Deep Learning Novelty Explorer [373]. The first attempt to create a benchmark for general level generation has recently been launched in the form of the Level Generation Track of the GVGAI competition. In this competition track, competitors submit level generators capable of generating levels for unseen games. The generators are then supplied with the description of several games, and produce levels which are judged by human judges [338]. Initial results suggest that constructing competent level generators that can produce levels for any game is much more challenging than constructing competent level generators for a single game. A related effort is the Video Game Level Corpus [669] which aims to provide a set of game levels across multiple games and genres which can be used for training level generators for data-driven procedural content generation.

While level generation, as discussed above, is one of the main examples of procedural content generation, there are many other aspects (or **facets**) of games that can be generated. These include visuals, such as textures and images; narrative, such as quests and backstories; audio, such as sound effects and music; and of course all kinds of things that go into game levels, such as items, weapons, enemies and personalities [381, 616]. However, an even greater challenge is the generation of complete games, including some or all of these facets together with the rules of the game. While, as covered in Chapter 4, there have been several attempts to generate games (including their rules) we are not aware of any approach to generating games that tries to generate more than two of the facets of games listed above. We are also not aware of any game generation system that even tries to generate games of more than one genre. Multi-faceted generation systems like *Sonancia* [394, 395] co-generate horror game levels with corresponding soundscapes but do not cater to the generation of rules. It is clear that the very domain-limited and facet-limited aspects of current game generation systems result from intentionally limiting design choices in order to make the very difficult problem of generating complete games tractable. Yet, in order to move beyond what could be argued to be toy domains and start to fulfill the promise of game generation, we need systems that can generate multiple facets of games at the same time, and that can generate games of different kinds.

This process has been defined as facet (domain) **orchestration** in games [371, 324]. Orchestration refers to the process of *harmonizing game generation*. Evidently, orchestration is a necessary process when we consider the output of *two or more* content type generators—such as visuals and audio—for the generation of a complete game. Drawing inspiration from music, orchestration may vary from a *top-down*, conductor-driven process to a *bottom-up*, free-from generation process [371]. A few years ago, something very much like general game generation and orchestration was outlined as the challenges of "multi-content, multi-domain PCG" and "generating complete games" [702]. It is interesting to note that there has not seemingly been any attempt to create more general game generators since then, perhaps due to the complexity of the task. A recent study by Karavolos et al. [324] moves towards the orchestration direction as it fuses level and game design parameters in first-person shooters via deep convolutional neural networks. The trained networks can be used to generate balanced games. Currently the only genre for which generators have been built that can generate high-quality (complete) games is abstract board games. Once more genres have been "conquered", we hope that the task of building more general level generators can begin.

Linked to the tasks of orchestration and general game generation there are important questions with respect to the **creative** capacity of the generation process that remain largely unanswered. For example, how creative can a generator be and how can we assess it? Is it, for instance, deemed to have appreciation, skill, and imagination [130]? When it comes to the evaluation of the creative capacity of current PCG algorithms a case can be made that most of them possess only skill. Does the creator manage to explore novel combinations within a constrained space, thereby resulting in **exploratory** game design creativity [53]; or, is on the other hand trying to break existing boundaries and constraints within game design to come up with entirely new designs, demonstrating **transformational** creativity [53]? If used in a mixed-initiative fashion, does it enhance the designer's creativity by boosting the possibility space for her? Arguably, the appropriateness of various evaluation methods for autonomous PCG creation or mixed-initiative co-creation [774] remains largely unexplored within both human and computational creativity research.

7.1.3 General Game Affective Loop

It stands to reason that general intelligence implies (and is tightly coupled with) general emotional intelligence [443]. The ability to recognize human behavior and emotion is a complex yet critical task for human communication that acts as a facilitator of general intelligence [157]. Throughout evolution, we have developed particular forms of advanced cognitive, emotive and social skills to address this challenge. Beyond these skills, we also have the capacity to detect affective patterns across people with different moods, cultural backgrounds and personalities. This generalization ability also extends, to a degree, across contexts and social settings. Despite their importance, the characteristics of social intelligence have not yet been transferred to AI in the form of general emotive, cognitive or behavioral models. While research in affective computing [530] has reached important milestones such as the capacity for real-time emotion recognition [794]—which can be faster than humans under particular conditions—all key findings suggest that any success of affective computing is heavily dependent on the domain, the task at hand, and the context in general. This *specificity* limitation is particularly evident in the domain of games [781] as most work in modeling player experience focuses on particular games, under well-controlled conditions with particular, small sets of players (see [783, 609, 610, 435] among many). In this section we identify and discuss two core unexplored and interwoven aspects of modeling players that are both important and necessary steps towards the long-term aim of game AI to realize truly adaptive games. The first aspect is the closure of the affective loop in games; the second aspect is the construction of general models capable of capturing experience across players and games.

As stated at the start of this book, affective computing is best realized within games in what we name the **game affective loop**. While the phases of emotion elicitation, affect modeling and affect expression have offered some robust solutions by now, the very loop of affective-based interaction has not been closed yet. Aside from a few studies demonstrating some affect-enabled adaptation of the game [772, 617] the area remains largely unexplored. It is not only the complexity of modeling players and their experience that is the main hurdle against any advancement. What is also far from trivial is the appropriate and meaningful integration of any of these models in a game. The questions of how often the system should adapt, what it should alter and by what degree are not easy to answer. As most of the questions are still open to the research community the only way to move forward is to do more research in adaptive games involving affective aspects of the experience. Existing commercial-standard games that already realize the affective loop such as *Nevermind* (Flying Mollusk, 2016) are the ambassadors for further work in this area.

Once the game affective loop is successfully realized within particular games the next goal for game AI is the **generality of affect-based interaction** across games. The game affective loop should not only be operational; it should ideally be general too. For AI in games to be general beyond game-playing it needs to be able to recognize general emotional and cognitive-behavioral patterns. This is essentially AI that can detect context-free emotive and cognitive reactions and expressions across contexts and builds general computational models of human behavior and experience which are grounded in a general gold standard of human behavior. So far we have only seen a few proof-of-concept studies in this direction. Early work within the game AI field focused on the ad-hoc design of general metrics of player interest that were tested across different prey-predator games [768, 767]. In other, more recent, studies predictors of player experience were tested for their ability to capture player experience across dissimilar games [431, 612, 97]. Another study on deep multimodal fusion can be seen as an embryo for further research in this direction [435], in which various modalities of player input such as player metrics, skin conductance and heart activity have been fused using stacked autoencoders. Discovering entirely new representations of player behavior and emotive manifestations across games, modalities of data, and player types is a first step towards achieving general player modeling. Such representations can, in turn, be used as the basis for approximating the *ground truth* of user experience in games.

7.2 AI in Other Roles in Games

The structure of this book reflects our belief that playing games, generating content and modeling players are the central applications of AI methods in games. However, there are many variants and use cases of game playing, player modeling or content generation that we have not had time to explore properly in the book, and which in some cases not have been explored in the literature at all. Further, there are some applications of AI in games that cannot be really classified as special cases of our "big three" AI applications in games, despite our best efforts. This section briefly sketches some of these applications, some of which may be important future research directions.

Playtesting: One of the many use cases for AI for playing games is to test the games. Testing games for bugs, balancing player experience and behavior, and other issues is important in game development, and one of the areas where game developers are looking for AI assistance. While playtesting is one of the AI capabilities within many of the mixed-initiative tools discussed in Chapter 4, there has also been work on AI-based playtesting outside of that context. For example, Denzinger et al. evolved action sequences to find exploits in sports games, with discouragingly good results [165]. For the particular case of finding bugs and exploits in games, one of the research challenges is to find a good and representative coverage of problems, so as to deliver an accurate picture to the development team of how many problems there are and how easy they are to run into, and allow prioritization of which problems to fix.

Critiquing Games: Can AI methods meaningfully judge and critique games? Game criticism is hard and generally depends on deep understanding of not only games but also the surrounding cultural context. Still, there might be automated metrics that are useful for game criticism, and can provide information to help reviewers, game curators and others in selecting which games to consider for reviewing for inclusion in app stores. The ANGELINA game generation system is one of the few examples towards this direction [136] in which AI generates the overview of the game to be played.

Hyper-formalist Game Studies: AI methods can be applied to corpora of games in order to understand distributions of game characteristics. For example, decision trees can be used to visualize patterns of resource systems in games [312]. There are likely many other ways of using game AI for game studies that are still to be discovered.

Game Directing: The outstanding feature of *Left 4 Dead* (Valve Corporation, 2008) was its AI director, which adjusted the onslaught of zombies to provide a dramatic curve of challenge for players. While simple and literally one-dimensional (only a single dimension of player experience was tracked), the AI director proved highly effective. There is much room for creating more sophisticated AI directors; the experience-driven PCG framework [783] is one potential way within which to work towards this.

Creative Inspiration: While designing a complete game that actually works likely requires a very complex generator, it can be simpler to generate an idea for new games, that are then designed by humans. Creative ideation tools range from simple word-recombination-based tools implemented as card games or Twitter bots, to elaborate computational creativity systems such as the *What If-Machine* [391].

Chat Monitoring: In-game chats are important in many online multi-player games, as they allow people to collaborate within games and socialize through them. Unfortunately, such chats can also be used to threaten or abuse other players. Given the very large volume of chat messages sent through a successful online game, it becomes impossible for the game developers to curate chats manually. In the efforts to combat toxic behavior, some game developers have therefore turned to machine learning. Notably, Riot Games have trained algorithms to recognize and remove toxic behavior in the MOBA *League of Legends* (Riot Games, 2009) [413]. Even worse, sexual predation can be seen in some games, where pedophiles use game chats to reach children; there have been attempts to use machine learning to detect sexual predators in game chats too [241].

AI-Based Game Design: Throughout most of the book, we have assumed the existence of a game or at least a game design, and discussed how AI can be used to play that game, generate content for it or model its players. However, one could also start from some AI method or capability and try to design a game that builds on that method or capability. This could be seen as an opportunity to showcase AI methods in the context of games, but it could also be seen as a way of advancing game design. Most classic game designs originate in an era where there were few effective AI algorithms, there was little knowledge among game designers about those AI algorithms that existed, and CPU and memory capacity of home computers was too limited to allow anything beyond simple heuristic AI and some best-first search to be used. One could even say that many classic video game designs are an attempt to design around the lack of AI—for example, the lack of good dialog AI for NPCs led to the use of dialog trees, the lack of AIs that could play FPS games believably and competently led to FPS game designs where most enemies are only on-screen for a few seconds so that you do not notice their lack of smarts, and the lack of level generation methods that guaranteed balance and playability led to game designs where levels did not need to be completable. The persistence of such design patterns may be responsible for the relatively low utilization of interesting AI methods within commercial game development. By starting with the AI and designing a game around it, new design patterns that actually exploit some of the recent AI advances can be found.

Several games have been developed within the game AI research community specifically to showcase AI capabilities, some of which have been discussed in this book. Three of the more prominent examples are based on Stanley et al.'s work on neuroevolution and the NEAT algorithm: *NERO*, which is an RTS-like game where the player trains an army through building a training environment rather than controlling it directly [654]; *Galactic Arms Race*, in which weapons controlled through neural networks are indirectly collectively evolved by thousands of players [250, 249]; and *Petalz*, which is a Facebook game about collecting flowers based on a similar idea of selection-based collective neuroevolution [565, 566]. Other games have been built to demonstrate various adaptation mechanisms, such as *Infinite Tower Defense* [25] and *Maze-Ball* [780]. Within interactive narrative it is relatively common to build games that showcase specific theories and methods; a famous example is *Façade* [441] and another prominent example is *Prom Week* [447]. Treanor et al. have attempted to identify AI-based game design patterns, and found a diverse array of roles in which AI can be or has been used in games, and a number of avenues for future AI-based game design [724].

7.3 Ethical Considerations

Like all technologies, artificial intelligence, including game AI, can be used for many purposes, some of them nefarious. Perhaps even more importantly, technology can have ethically negative or at least questionable effects even when there is no malicious intent. The ethical effects of using AI with and in games are not always obvious, and the topic is not receiving the attention it should. This short section looks at some of the ways in which game AI intersects with ethical questions. For general AI research issues, ethics and values we refer the interested reader to the Asilomar AI Principles[1] developed in conjunction with the 2017 Asilomar conference.

Player modeling is perhaps the part of game AI where the ethical questions are most direct, and perhaps most urgent. There is now a vigorous debate about the mass collection of data about us both by government entities (such as the US National Security Agency or the United Kingdom's GCHQ) and private entities (such as Google, Amazon, Facebook and Microsoft) [64, 502]. With methodological advances in data mining, it is becoming possible to learn more and more about individual people from their digital traces, including inferring sensitive information and predicting behavior. Given that player modeling involves large-scale data collection and mining, many of the same ethical challenges exist in player modeling as in the mining of data about humans in general. Mikkelsen et al. present an overview of ethical challenges for player modeling [458]. Below we give some examples of such challenges.

Privacy: It is becoming increasingly possible and even practicable to infer various real-life traits and properties of people from their in-game behavior. This can be done without the consent or even knowledge of the subject, and some of the information can be of a private and sensitive nature. For example, Yee and colleagues investigated how player choices in *World of Warcraft* (Blizzard Entertainment, 2004) correlated with the personalities of players. They used data about players' characters from the Armory database of *World of Warcraft* (Blizzard Entertainment, 2004) and correlated this information with personality tests administered to players; multiple strong correlations were found [788]. In a similar vein, a study investigated how players' life motives correlated with their *Minecraft* (Mojang, 2011) log files [101]. That research used the life motivation questionnaires of Steven Reiss, and found that players' self-reported life motives (independence, family, etc.) were expressed in a multitude of ways inside constructed *Minecraft* (Mojang, 2011) worlds. Using a very different type of game strong correlations have been found between playing style in the first-person shooter *Battlefield 3* (Electronic Arts, 2011) and player characteristics such as personality [687], age [686] and nationality [46]. It is entirely plausible that similar methods could be used to infer sexual preferences, political views, health status and religious beliefs. Such information could be used by advertising networks to serve targeted ads, by criminals looking to blackmail the player, by insurance companies looking to differentiate premiums, or by malevolent political regimes for various forms of suppression. We do not know yet what can be predicted and with what accuracy, but it is imperative that more research be done on this within the publicly available literature; it is clear that this kind of research will also be carried out behind locked doors.

[1] https://futureoflife.org/ai-principles/

Ownership of Data: Some player data can be used to recreate aspects of the player's behavior; this is the case for e.g., the *Drivatars* in Microsoft's *Forza Motorsport* series, and more generally for agents created according to the procedural persona concept [267]. It is currently not clear who owns this data, and if the game developer/publisher owns the data, what they can do with it. Will the game allow other people to play against a model of you, i.e., how you would have played the game? If so, can it identify you to other players as the origin of this data? Does it have to be faithful to the behavioral model of you, or can it add or distort aspects of your playing behavior?

Adaptation: Much of the research within game AI is concerned with adaptation of games, with the experience-driven PCG framework being the perhaps most complete account on how to combine player modeling with procedural content generation to create personalized game experiences [783]. However, it is not clear that it is always a good thing to adapt games to players. The "filter bubble" is a concept within discussion of social networks which refers to the phenomenon where collaborative filtering ensures that users are only provided with content that is already in line with their political, ethical, or aesthetic preferences, leading to a lack of healthy engagement with other perspectives. Excessive adaptation and personalization might have a similar effect, where players are funneled into a narrow set of game experiences.

Stereotypes: Anytime we train a model using some dataset, we run the risk of reproducing stereotypes within that dataset. For example, it has been shown that word embeddings trained on standard datasets of the English language reproduce gender-based stereotypes [93]. The same effects could be present when modeling player preferences and behavior, and the model might learn to reproduce prejudiced conceptions regarding gender, race, etc. Such problems can be exacerbated or ameliorated by the tools made available to players for expressing themselves in-game. For example, Lim and Harrell have developed quantitative methods for measuring and addressing bias in character creation tools [386].

Censorship: Of course, it is entirely possible, and advisable, to use AI methods to promote ethical behavior and uphold ethical values. Earlier, Section 7.2 discussed the examples of AI for filtering player chats in online multi-player games, and for detecting sexual predators. While such technologies are generally welcome, there are important ethical considerations in how they should be deployed. For example, a model that has been trained to recognize hate speech might also react to normal in-game jargon; setting the right decision threshold might involve a delicate tradeoff between ensuring a welcoming game environment and not restricting communications unduly.

AI Beyond Games: Finally, a somewhat more far-fetched concern, but one we believe still merits discussion is the following. Games are frequently used to train and test AI algorithms—this is the main aim of, for example, the General Video Game

AI Competition and the Arcade Learning Environment. However, given how many games are focused on violent competition, does this mean that we focus unduly on the development of violence in artificial intelligence? What effects could this have on AI that is trained on games but employed in other domains, such as transport or health care?

7.4 Summary

In this last chapter of this book we went through directions that we view as critical and important for the advancement of the game AI field. Initially we have argued that the general intelligence capacity of machines needs to be both explored and exploited to its full potential (1) across the different **tasks** that exist within the game design and development process, including but absolutely no longer limited to game playing; (2) across different **games** within the game design space; and (3) across different **users** (players or designers) of AI. We claim that, thus far, we have underestimated the potential for general AI within games. We also claim that the currently dominant practice of only designing AI for a specific task within a specific domain will eventually be detrimental to game AI research as algorithms, methods and epistemological procedures will remain specific to the task at hand. As a result, we will not manage to push the boundaries of AI and exploit its full capacity for game design. We are inspired by the general game-playing paradigm and the recent successes of AI algorithms in that domain and suggest that we become less specific about all subareas of the game AI field including player modeling and game generation. Doing so would allow us to detect and mimic different general cognitive and emotive skills of humans when designing games. It is worth noting, again, that we are not advocating that all research within the game AI field focuses on generality right now; studies on particular games and particular tasks are still valuable, given how little we still understand and can do. But over time, we predict that more and more research will focus on generality across tasks, games and users, because it is in the general problems that the interesting research questions of the future lie. It seems that we are not alone in seeing this need as other researchers have argued for the use of various game-related tasks (not just game playing) to be used in artificial general intelligence research [799].

The path towards achieving general game artificial intelligence is still largely unexplored. For AI to become less specific—yet remain relevant and useful for game design—we envision a number of immediate steps that could be taken: first and foremost, the game AI community needs to adopt an **open-source** accessible strategy so that methods and algorithms developed across the different tasks are shared among researchers for the advancement of this research area. Venues such as the current game AI research portal[2] could be expanded and used to host successful methods and algorithms. For the algorithms and methods to be of direct use particular tech-

[2] http://www.aigameresearch.org/

nical specifications need to be established—e.g., such as those established within game-based AI benchmarks—which will maximize the interoperability among the various tools and elements submitted. Examples of benchmarked specifications for the purpose of general game AI research include the general video game description language and the puzzle game engine PuzzleScript.[3] Finally, following the GVGAI competition paradigm, we envision a new set of competitions rewarding general player models, AI-assisted tools and game generation techniques. These competitions would further motivate researchers to work in this exciting research area and enrich the database of open access interoperable methods and algorithms, directly contributing to the state of the art in computational general game design.

Beyond generality we also put a focus on the extensibility of AI roles within games. In that regard, we outlined a number of AI roles that are underrepresented currently but nevertheless define very promising research frontiers for game AI. These include the roles of AI as playtester, game critic, game studies formalist, director, creative designer, and gameplay ethics judge. Further, we view the placement of AI at the very center of the design process (AI-based game design) as another critical research frontier.

This chapter, and the book itself, concluded with a discussion on the ethical implications of whatever we do in game AI research. In particular, we discussed aspects such as the privacy and ownership of data, the considerations about game adaptation, the emergence of stereotypes through computational models of players, the risks of AI acting as a censor, and finally the ethical constraints imposed on AI by "unethical" aspects of the very nature of games.

[3] http://www.puzzlescript.net/

References

1. Espen Aarseth. Genre trouble. *Electronic Book Review*, 3, 2004.
2. Martín Abadi, Ashish Agarwal, Paul Barham, Eugene Brevdo, Zhifeng Chen, Craig Citro, Greg S. Corrado, Andy Davis, Jeffrey Dean, Matthieu Devin, et al. TensorFlow: Large-scale machine learning on heterogeneous distributed systems. *arXiv preprint arXiv:1603.04467*, 2016.
3. Ryan Abela, Antonios Liapis, and Georgios N. Yannakakis. A constructive approach for the generation of underwater environments. In *Proceedings of the FDG workshop on Procedural Content Generation in Games*, 2015.
4. David H. Ackley, Geoffrey E. Hinton, and Terrence J. Sejnowski. A learning algorithm for Boltzmann machines. *Cognitive Science*, 9(1):147–169, 1985.
5. Alexandros Agapitos, Julian Togelius, Simon M. Lucas, Jürgen Schmidhuber, and Andreas Konstantinidis. Generating diverse opponents with multiobjective evolution. In *Computational Intelligence and Games, 2008. CIG'08. IEEE Symposium On*, pages 135–142. IEEE, 2008.
6. Rakesh Agrawal, Tomasz Imieliński, and Arun Swami. Mining association rules between sets of items in large databases. In *ACM SIGMOD Record*, pages 207–216. ACM, 1993.
7. Rakesh Agrawal and Ramakrishnan Srikant. Fast algorithms for mining association rules. In *Proceedings of the 20th International Conference on Very Large Data Bases, VLDB*, pages 487–499, 1994.
8. John B. Ahlquist and Jeannie Novak. *Game development essentials: Game artificial intelligence*. Delmar Pub, 2008.
9. Zach Aikman. Galak-Z: Forever: Building Space-Dungeons Organically. In *Game Developers Conference*, 2015.
10. Bob Alexander. The beauty of response curves. *AI Game Programming Wisdom*, page 78, 2002.
11. Krishna Aluru, Stefanie Tellex, John Oberlin, and James MacGlashan. Minecraft as an experimental world for AI in robotics. In *AAAI Fall Symposium*, 2015.
12. Samuel Alvernaz and Julian Togelius. Autoencoder-augmented neuroevolution for visual doom playing. In *IEEE Conference on Computational Intelligence and Games*. IEEE, 2017.
13. Omar Alzoubi, Rafael A. Calvo, and Ronald H. Stevens. Classification of EEG for Affect Recognition: An Adaptive Approach. In *AI 2009: Advances in Artificial Intelligence*, pages 52–61. Springer, 2009.
14. Mike Ambinder. Biofeedback in gameplay: How Valve measures physiology to enhance gaming experience. In *Game Developers Conference*, San Francisco, California, US, 2011.
15. Dan Amerson, Shaun Kime, and R. Michael Young. Real-time cinematic camera control for interactive narratives. In *Proceedings of the 2005 ACM SIGCHI International Conference on Advances in Computer Entertainment Technology*, pages 369–369. ACM, 2005.

G. N. Yannakakis and J. Togelius, *Artificial Intelligence and Games*, https://doi.org/10.1007/978-3-319-63519-4

16. Elisabeth André, Martin Klesen, Patrick Gebhard, Steve Allen, and Thomas Rist. Integrating models of personality and emotions into lifelike characters. In *Affective interactions*, pages 150–165. Springer, 2000.
17. John L. Andreassi. *Psychophysiology: Human Behavior and Physiological Response*. Psychology Press, 2000.
18. Rudolf Arnheim. *Art and visual perception: A psychology of the creative eye*. University of California Press, 1956.
19. Ivon Arroyo, David G. Cooper, Winslow Burleson, Beverly Park Woolf, Kasia Muldner, and Robert Christopherson. Emotion sensors go to school. In *Proceedings of Conference on Artificial Intelligence in Education (AIED)*, pages 17–24. IOS Press, 2009.
20. W. Ross Ashby. Principles of the self-organizing system. In *Facets of Systems Science*, pages 521–536. Springer, 1991.
21. Daniel Ashlock. *Evolutionary computation for modeling and optimization*. Springer, 2006.
22. Stylianos Asteriadis, Kostas Karpouzis, Noor Shaker, and Georgios N. Yannakakis. Does your profile say it all? Using demographics to predict expressive head movement during gameplay. In *Proceedings of UMAP Workshops*, 2012.
23. Stylianos Asteriadis, Paraskevi Tzouveli, Kostas Karpouzis, and Stefanos Kollias. Estimation of behavioral user state based on eye gaze and head pose—application in an e-learning environment. *Multimedia Tools and Applications*, 41(3):469–493, 2009.
24. Phillipa Avery, Sushil Louis, and Benjamin Avery. Evolving coordinated spatial tactics for autonomous entities using influence maps. In *Computational Intelligence and Games, 2009. CIG 2009. IEEE Symposium on*, pages 341–348. IEEE, 2009.
25. Phillipa Avery, Julian Togelius, Elvis Alistar, and Robert Pieter van Leeuwen. Computational intelligence and tower defence games. In *Evolutionary Computation (CEC), 2011 IEEE Congress on*, pages 1084–1091. IEEE, 2011.
26. Ruth Aylett, Sandy Louchart, Joao Dias, Ana Paiva, and Marco Vala. FearNot!–an experiment in emergent narrative. In *Intelligent Virtual Agents*, pages 305–316. Springer, 2005.
27. Simon E. Ortiz B., Koichi Moriyama, Ken-ichi Fukui, Satoshi Kurihara, and Masayuki Numao. Three-subagent adapting architecture for fighting videogames. In *Pacific Rim International Conference on Artificial Intelligence*, pages 649–654. Springer, 2010.
28. Sander Bakkes, Pieter Spronck, and Jaap van den Herik. Rapid and reliable adaptation of video game AI. *IEEE Transactions on Computational Intelligence and AI in Games*, 1(2):93–104, 2009.
29. Sander Bakkes, Shimon Whiteson, Guangliang Li, George Viorel Vişniuc, Efstathios Charitos, Norbert Heijne, and Arjen Swellengrebel. Challenge balancing for personalised game spaces. In *Games Media Entertainment (GEM), 2014 IEEE*, pages 1–8. IEEE, 2014.
30. Rainer Banse and Klaus R. Scherer. Acoustic profiles in vocal emotion expression. *Journal of Personality and Social Psychology*, 70(3):614, 1996.
31. Ray Barrera, Aung Sithu Kyaw, Clifford Peters, and Thet Naing Swe. *Unity AI Game Programming*. Packt Publishing Ltd, 2015.
32. Gabriella A. B. Barros, Antonios Liapis, and Julian Togelius. Data adventures. In *Proceedings of the FDG workshop on Procedural Content Generation in Games*, 2015.
33. Richard A. Bartle. *Designing virtual worlds*. New Riders, 2004.
34. Chris Bateman and Richard Boon. *21st Century Game Design (Game Development Series)*. Charles River Media, Inc., 2005.
35. Chris Bateman and Lennart E. Nacke. The neurobiology of play. In *Proceedings of the International Academic Conference on the Future of Game Design and Technology*, pages 1–8. ACM, 2010.
36. Christian Bauckhage, Anders Drachen, and Rafet Sifa. Clustering game behavior data. *IEEE Transactions on Computational Intelligence and AI in Games*, 7(3):266–278, 2015.
37. Yoann Baveye, Jean-Noël Bettinelli, Emmanuel Dellandrea, Liming Chen, and Christel Chamaret. A large video database for computational models of induced emotion. In *Proceedings of Affective Computing and Intelligent Interaction*, pages 13–18, 2013.
38. Jessica D. Bayliss. Teaching game AI through Minecraft mods. In *2012 IEEE International Games Innovation Conference (IGIC)*, pages 1–4. IEEE, 2012.

39. Farès Belhadj. Terrain modeling: a constrained fractal model. In *Proceedings of the 5th international conference on Computer graphics, virtual reality, visualisation and interaction in Africa*, pages 197–204. ACM, 2007.
40. Marc G. Bellemare, Yavar Naddaf, Joel Veness, and Michael Bowling. The arcade learning environment: An evaluation platform for general agents. *arXiv preprint arXiv:1207.4708*, 2012.
41. Yoshua Bengio. Learning deep architectures for AI. *Foundations and Trends in Machine Learning*, 2(1):1–127, 2009.
42. José Luis Bernier, C. Ilia Herráiz, J. J. Merelo, S. Olmeda, and Alberto Prieto. Solving Mastermind using GAs and simulated annealing: a case of dynamic constraint optimization. In *Parallel Problem Solving from Nature (PPSN) IV*, pages 553–563. Springer, 1996.
43. Kent C. Berridge. Pleasures of the brain. *Brain and Cognition*, 52(1):106–128, 2003.
44. Dimitri P. Bertsekas. *Dynamic programming and optimal control*. Athena Scientific Belmont, MA, 1995.
45. Nadav Bhonker, Shai Rozenberg, and Itay Hubara. Playing SNES in the Retro Learning Environment. *arXiv preprint arXiv:1611.02205*, 2016.
46. Mateusz Bialas, Shoshannah Tekofsky, and Pieter Spronck. Cultural influences on play style. In *Computational Intelligence and Games (CIG), 2014 IEEE Conference on*, pages 1–7. IEEE, 2014.
47. Nadia Bianchi-Berthouze and Christine L. Lisetti. Modeling multimodal expression of user's affective subjective experience. *User Modeling and User-Adapted Interaction*, 12(1):49–84, 2002.
48. Darse Billings, Denis Papp, Jonathan Schaeffer, and Duane Szafron. Opponent modeling in poker. In *AAAI/IAAI*, pages 493–499, 1998.
49. Christopher M. Bishop. *Pattern Recognition and Machine Learning*. 2006.
50. Staffan Björk and Jesper Juul. Zero-player games. In *Philosophy of Computer Games Conference, Madrid*, 2012.
51. Vikki Blake. Minecraft Has 55 Million Monthly Players, 122 Million Sales. *Imagine Games Network*, February 2017.
52. Paris Mavromoustakos Blom, Sander Bakkes, Chek Tien Tan, Shimon Whiteson, Diederik M. Roijers, Roberto Valenti, and Theo Gevers. Towards Personalised Gaming via Facial Expression Recognition. In *Proceedings of AIIDE*, 2014.
53. Margaret A. Boden. What is creativity. *Dimensions of creativity*, pages 75–117, 1994.
54. Margaret A. Boden. Creativity and artificial intelligence. *Artificial Intelligence*, 103(1):347–356, 1998.
55. Margaret A. Boden. *The creative mind: Myths and mechanisms*. Psychology Press, 2004.
56. Slawomir Bojarski and Clare Bates Congdon. REALM: A rule-based evolutionary computation agent that learns to play Mario. In *Computational Intelligence and Games (CIG), 2010 IEEE Symposium on*, pages 83–90. IEEE, 2010.
57. Luuk Bom, Ruud Henken, and Marco Wiering. Reinforcement learning to train Ms. Pac-Man using higher-order action-relative inputs. In *Adaptive Dynamic Programming and Reinforcement Learning (ADPRL), 2013 IEEE Symposium on*, pages 156–163. IEEE, 2013.
58. Blai Bonet and Héctor Geffner. Planning as heuristic search. *Artificial Intelligence*, 129(1-2):5–33, 2001.
59. Philip Bontrager, Ahmed Khalifa, Andre Mendes, and Julian Togelius. Matching games and algorithms for general video game playing. In *Twelfth Artificial Intelligence and Interactive Digital Entertainment Conference*, 2016.
60. Michael Booth. The AI systems of Left 4 Dead. In *Fifth Artificial Intelligence and Interactive Digital Entertainment Conference (Keynote)*, 2009.
61. Adi Botea, Martin Müller, and Jonathan Schaeffer. Near optimal hierarchical path-finding. *Journal of Game Development*, 1(1):7–28, 2004.
62. David M. Bourg and Glenn Seemann. *AI for game developers*. O'Reilly Media, Inc., 2004.
63. Michael Bowling, Neil Burch, Michael Johanson, and Oskari Tammelin. Heads-up limit holdem poker is solved. *Science*, 347(6218):145–149, 2015.

64. Danah Boyd and Kate Crawford. Six provocations for big data. In *A decade in internet time: Symposium on the dynamics of the internet and society*. Oxford Internet Institute, Oxford, 2011.
65. S. R. K. Branavan, David Silver, and Regina Barzilay. Learning to win by reading manuals in a Monte-Carlo framework. *Journal of Artificial Intelligence Research*, 43:661–704, 2012.
66. Michael E. Bratman, David J. Israel, and Martha E. Pollack. Plans and resource-bounded practical reasoning. *Computational Intelligence*, 4(3):349–355, 1988.
67. Leo Breiman, Jerome Friedman, Charles J. Stone, and Richard A. Olshen. *Classification and regression trees*. CRC Press, 1984.
68. Daniel Brewer. Tactical pathfinding on a navmesh. *Game AI Pro: Collected Wisdom of Game AI Professionals*, page 361, 2013.
69. Gerhard Brewka, Thomas Eiter, and Mirosław Truszczyński. Answer set programming at a glance. *Communications of the ACM*, 54(12):92–103, 2011.
70. Rodney Brooks. A robust layered control system for a mobile robot. *IEEE Journal on Robotics and Automation*, 2(1):14–23, 1986.
71. David S. Broomhead and David Lowe. Radial basis functions, multi-variable functional interpolation and adaptive networks. *Royals Signals & Radar Establishment*, 1988.
72. Anna Brown and Alberto Maydeu-Olivares. How IRT can solve problems of ipsative data in forced-choice questionnaires. *Psychological Methods*, 18(1):36, 2013.
73. Daniel Lankford Brown. Mezzo: An adaptive, real-time composition program for game soundtracks. In *Eighth Artificial Intelligence and Interactive Digital Entertainment Conference*, 2012.
74. Cameron Browne. *Automatic generation and evaluation of recombination games*. PhD thesis, Queensland University of Technology, 2008.
75. Cameron Browne. Yavalath. In *Evolutionary Game Design*, pages 75–85. Springer, 2011.
76. Cameron Browne and Frederic Maire. Evolutionary game design. *IEEE Transactions on Computational Intelligence and AI in Games*, 2(1):1–16, 2010.
77. Cameron B. Browne, Edward Powley, Daniel Whitehouse, Simon M. Lucas, Peter I. Cowling, Philipp Rohlfshagen, Stephen Tavener, Diego Perez, Spyridon Samothrakis, and Simon Colton. A survey of Monte Carlo tree search methods. *Computational Intelligence and AI in Games, IEEE Transactions on*, 4(1):1–43, 2012.
78. Nicholas J. Bryan, Gautham J. Mysore, and Ge Wang. ISSE: An Interactive Source Separation Editor. In *Proceedings of the SIGCHI Conference on Human Factors in Computing Systems*, pages 257–266, 2014.
79. Bobby D. Bryant and Risto Miikkulainen. Evolving stochastic controller networks for intelligent game agents. In *Evolutionary Computation, 2006. CEC 2006. IEEE Congress on*, pages 1007–1014. IEEE, 2006.
80. Mat Buckland. *Programming game AI by example*. Jones & Bartlett Learning, 2005.
81. Mat Buckland and Mark Collins. *AI techniques for game programming*. Premier Press, 2002.
82. Vadim Bulitko, Yngvi Björnsson, Nathan R. Sturtevant, and Ramon Lawrence. Real-time heuristic search for pathfinding in video games. In *Artificial Intelligence for Computer Games*, pages 1–30. Springer, 2011.
83. Vadim Bulitko, Greg Lee, Sergio Poo Hernandez, Alejandro Ramirez, and David Thue. Techniques for AI-Driven Experience Management in Interactive Narratives. In *Game AI Pro 2: Collected Wisdom of Game AI Professionals*, pages 523–534. AK Peters/CRC Press, 2015.
84. Paolo Burelli. Virtual cinematography in games: investigating the impact on player experience. *Foundations of Digital Games*, 2013.
85. Paolo Burelli and Georgios N. Yannakakis. Combining Local and Global Optimisation for Virtual Camera Control. In *Proceedings of the 2010 IEEE Conference on Computational Intelligence and Games*, Copenhagen, Denmark, August 2010. IEEE.
86. Christopher J. C. Burges. A tutorial on support vector machines for pattern recognition. *Data mining and Knowledge Discovery*, 2(2):121–167, 1998.
87. Michael Buro and David Churchill. Real-time strategy game competitions. *AI Magazine*, 33(3):106, 2012.

88. Carlos Busso, Zhigang Deng, Serdar Yildirim, Murtaza Bulut, Chul Min Lee, Abe Kazemzadeh, Sungbok Lee, Ulrich Neumann, and Shrikanth Narayanan. Analysis of emotion recognition using facial expressions, speech and multimodal information. In *Proceedings of the International Conference on Multimodal Interfaces (ICMI)*, pages 205–211. ACM, 2004.
89. Eric Butler, Adam M. Smith, Yun-En Liu, and Zoran Popovic. A mixed-initiative tool for designing level progressions in games. In *Proceedings of the 26th Annual ACM Symposium on User Interface Software and Technology*, pages 377–386. ACM, 2013.
90. Martin V. Butz and Thies D. Lonneker. Optimized sensory-motor couplings plus strategy extensions for the TORCS car racing challenge. In *IEEE Symposium on Computational Intelligence and Games*, pages 317–324. IEEE, 2009.
91. John T. Cacioppo, Gary G. Berntson, Jeff T. Larsen, Kirsten M. Poehlmann, and Tiffany A. Ito. The psychophysiology of emotion. *Handbook of emotions*, 2:173–191, 2000.
92. Francesco Calimeri, Michael Fink, Stefano Germano, Andreas Humenberger, Giovambattista Ianni, Christoph Redl, Daria Stepanova, Andrea Tucci, and Anton Wimmer. Angry-HEX: an artificial player for Angry Birds based on declarative knowledge bases. *IEEE Transactions on Computational Intelligence and AI in Games*, 8(2):128–139, 2016.
93. Aylin Caliskan, Joanna J. Bryson, and Arvind Narayanan. Semantics derived automatically from language corpora contain human-like biases. *Science*, 356(6334):183–186, 2017.
94. Gordon Calleja. *In-game: from immersion to incorporation.* MIT Press, 2011.
95. Rafael Calvo, Iain Brown, and Steve Scheding. Effect of experimental factors on the recognition of affective mental states through physiological measures. In *AI 2009: Advances in Artificial Intelligence*, pages 62–70. Springer, 2009.
96. Elizabeth Camilleri, Georgios N. Yannakakis, and Alexiei Dingli. Platformer Level Design for Player Believability. In *IEEE Computational Intelligence and Games Conference*. IEEE, 2016.
97. Elizabeth Camilleri, Georgios N. Yannakakis, and Antonios Liapis. Towards General Models of Player Affect. In *Affective Computing and Intelligent Interaction (ACII), 2017 International Conference on*, 2017.
98. Murray Campbell, A. Joseph Hoane, and Feng-hsiung Hsu. Deep blue. *Artificial intelligence*, 134(1-2):57–83, 2002.
99. Henrique Campos, Joana Campos, João Cabral, Carlos Martinho, Jeppe Herlev Nielsen, and Ana Paiva. My Dream Theatre. In *Proceedings of the 2013 International Conference on Autonomous Agents and Multi-Agent Systems*, pages 1357–1358. International Foundation for Autonomous Agents and Multiagent Systems, 2013.
100. Joana Campos, Carlos Martinho, Gordon Ingram, Asimina Vasalou, and Ana Paiva. My dream theatre: Putting conflict on center stage. In *FDG*, pages 283–290, 2013.
101. Alessandro Canossa, Josep B. Martinez, and Julian Togelius. Give me a reason to dig Minecraft and psychology of motivation. In *Computational Intelligence in Games (CIG), 2013 IEEE Conference on*. IEEE, 2013.
102. Luigi Cardamone, Daniele Loiacono, and Pier Luca Lanzi. Interactive evolution for the procedural generation of tracks in a high-end racing game. In *Proceedings of the 13th Annual Conference on Genetic and Evolutionary Computation*, pages 395–402. ACM, 2011.
103. Luigi Cardamone, Georgios N. Yannakakis, Julian Togelius, and Pier Luca Lanzi. Evolving interesting maps for a first person shooter. In *Applications of Evolutionary Computation*, pages 63–72. Springer, 2011.
104. Justine Cassell. *Embodied conversational agents.* MIT Press, 2000.
105. Justine Cassell, Timothy Bickmore, Mark Billinghurst, Lee Campbell, Kenny Chang, Hannes Vilhjálmsson, and Hao Yan. Embodiment in conversational interfaces: Rea. In *Proceedings of the SIGCHI conference on Human Factors in Computing Systems*, pages 520–527. ACM, 1999.
106. Marc Cavazza, Fred Charles, and Steven J. Mead. Character-based interactive storytelling. *IEEE Intelligent Systems*, 17(4):17–24, 2002.

107. Marc Cavazza, Fred Charles, and Steven J. Mead. Interacting with virtual characters in interactive storytelling. In *Proceedings of the First International Joint Conference on Autonomous Agents and Multiagent Systems: part 1*, pages 318–325. ACM, 2002.
108. Georgios Chalkiadakis, Edith Elkind, and Michael Wooldridge. Computational aspects of cooperative game theory. *Synthesis Lectures on Artificial Intelligence and Machine Learning*, 5(6):1–168, 2011.
109. Alex J. Champandard. *AI game development: Synthetic creatures with learning and reactive behaviors*. New Riders, 2003.
110. Alex J. Champandard. Behavior trees for next-gen game AI. In *Game Developers Conference, Audio Lecture*, 2007.
111. Alex J. Champandard. Understanding Behavior Trees. *AiGameDev. com*, 2007.
112. Alex J. Champandard. Getting started with decision making and control systems. *AI Game Programming Wisdom*, 4:257–264, 2008.
113. Jason C. Chan. Response-order effects in Likert-type scales. *Educational and Psychological Measurement*, 51(3):531–540, 1991.
114. Senthilkumar Chandramohan, Matthieu Geist, Fabrice Lefevre, and Olivier Pietquin. User simulation in dialogue systems using inverse reinforcement learning. In *Interspeech 2011*, pages 1025–1028, 2011.
115. Devendra Singh Chaplot and Guillaume Lample. Arnold: An autonomous agent to play FPS games. In *Thirty-First AAAI Conference on Artificial Intelligence*, 2017.
116. Darryl Charles and Michaela Black. Dynamic player modelling: A framework for player-centric digital games. In *Proceedings of the International Conference on Computer Games: Artificial Intelligence, Design and Education*, pages 29–35, 2004.
117. Fred Charles, Miguel Lozano, Steven J. Mead, Alicia Fornes Bisquerra, and Marc Cavazza. Planning formalisms and authoring in interactive storytelling. In *Proceedings of TIDSE*, 2003.
118. Guillaume M. J. B. Chaslot, Mark H. M. Winands, H. Jaap van Den Herik, Jos W. H. M. Uiterwijk, and Bruno Bouzy. Progressive strategies for Monte-Carlo tree search. *New Mathematics and Natural Computation*, 4(03):343–357, 2008.
119. Xiang 'Anthony' Chen, Tovi Grossman, Daniel J. Wigdor, and George Fitzmaurice. Duet: Exploring joint interactions on a smart phone and a smart watch. In *Proceedings of the SIGCHI Conference on Human Factors in Computing Systems*, pages 159–168, 2014.
120. Zhengxing Chen, Magy Seif El-Nasr, Alessandro Canossa, Jeremy Badler, Stefanie Tignor, and Randy Colvin. Modeling individual differences through frequent pattern mining on role-playing game actions. In *Eleventh Artificial Intelligence and Interactive Digital Entertainment Conference, AIIDE*, 2015.
121. Sonia Chernova, Jeff Orkin, and Cynthia Breazeal. Crowdsourcing HRI through online multiplayer games. In *AAAI Fall Symposium: Dialog with Robots*, pages 14–19, 2010.
122. Wei Chu and Zoubin Ghahramani. Preference learning with Gaussian processes. In *Proceedings of the International Conference on Machine learning (ICML)*, pages 137–144, 2005.
123. David Churchill and Michael Buro. Portfolio greedy search and simulation for large-scale combat in StarCraft. In *Computational Intelligence in Games (CIG), 2013 IEEE Conference on*. IEEE, 2013.
124. David Churchill, Mike Preuss, Florian Richoux, Gabriel Synnaeve, Alberto Uriarte, Santiago Ontañón, and Michal Certickỳ. StarCraft Bots and Competitions. In *Encyclopedia of Computer Graphics and Games*. Springer, 2016.
125. Andrea Clerico, Cindy Chamberland, Mark Parent, Pierre-Emmanuel Michon, Sebastien Tremblay, Tiago H. Falk, Jean-Christophe Gagnon, and Philip Jackson. Biometrics and classifier fusion to predict the fun-factor in video gaming. In *IEEE Computational Intelligence and Games Conference*. IEEE, 2016.
126. Carlos A. Coello Coello, Gary B. Lamont, and David A. van Veldhuizen. *Evolutionary algorithms for solving multi-objective problems*. Springer, 2007.
127. Nicholas Cole, Sushil J. Louis, and Chris Miles. Using a genetic algorithm to tune first-person shooter bots. In *Congress on Evolutionary Computation (CEC)*, pages 139–145. IEEE, 2004.

128. Karen Collins. An introduction to procedural music in video games. *Contemporary Music Review*, 28(1):5–15, 2009.
129. Karen Collins. *Playing with sound: a theory of interacting with sound and music in video games*. MIT Press, 2013.
130. Simon Colton. Creativity versus the perception of creativity in computational systems. In *AAAI Spring Symposium: Creative Intelligent Systems*, 2008.
131. Cristina Conati. Intelligent tutoring systems: New challenges and directions. In *IJCAI*, pages 2–7, 2009.
132. Cristina Conati, Abigail Gertner, and Kurt VanLehn. Using Bayesian networks to manage uncertainty in student modeling. *User Modeling and User-Adapted Interaction*, 12(4):371–417, 2002.
133. Cristina Conati and Heather Maclaren. Modeling user affect from causes and effects. *User Modeling, Adaptation, and Personalization*, pages 4–15, 2009.
134. John Conway. The game of life. *Scientific American*, 223(4):4, 1970.
135. Michael Cook and Simon Colton. Multi-faceted evolution of simple arcade games. In *IEEE Computational Intelligence and Games*, pages 289–296, 2011.
136. Michael Cook and Simon Colton. Ludus ex machina: Building a 3D game designer that competes alongside humans. In *Proceedings of the 5th International Conference on Computational Creativity*, 2014.
137. Michael Cook, Simon Colton, and Alison Pease. Aesthetic Considerations for Automated Platformer Design. In *AIIDE*, 2012.
138. Seth Cooper, Firas Khatib, Adrien Treuille, Janos Barbero, Jeehyung Lee, Michael Beenen, Andrew Leaver-Fay, David Baker, Zoran Popović, et al. Predicting protein structures with a multiplayer online game. *Nature*, 466(7307):756–760, 2010.
139. Corinna Cortes and Vladimir Vapnik. Support-vector networks. *Machine Learning*, 20(3):273–297, 1995.
140. Paul T. Costa and Robert R. MacCrae. *Revised NEO personality inventory (NEO PI-R) and NEO five-factor inventory (NEO-FFI): Professional manual*. Psychological Assessment Resources, Incorporated, 1992.
141. Rémi Coulom. Efficient selectivity and backup operators in Monte-Carlo tree search. In *International Conference on Computers and Games*, pages 72–83. Springer, 2006.
142. Rémi Coulom. Computing Elo ratings of move patterns in the game of Go. In *Computer Games Workshop*, 2007.
143. Roddy Cowie and Randolph R. Cornelius. Describing the emotional states that are expressed in speech. *Speech Communication*, 40(1):5–32, 2003.
144. Roddy Cowie, Ellen Douglas-Cowie, Susie Savvidou, Edelle McMahon, Martin Sawey, and Marc Schröder. 'FEELTRACE': An instrument for recording perceived emotion in real time. In *ISCA Tutorial and Research Workshop (ITRW) on Speech and Emotion*, 2000.
145. Roddy Cowie and Martin Sawey. GTrace-General trace program from Queen's University, Belfast, 2011.
146. Peter I. Cowling, Edward J. Powley, and Daniel Whitehouse. Information set Monte Carlo tree search. *IEEE Transactions on Computational Intelligence and AI in Games*, 4(2):120–143, 2012.
147. Koby Crammer and Yoram Singer. Pranking with ranking. *Advances in Neural Information Processing Systems*, 14:641–647, 2002.
148. Chris Crawford. *Chris Crawford on interactive storytelling*. New Riders, 2012.
149. Mihaly Csikszentmihalyi. *Creativity: Flow and the psychology of discovery and invention*. New York: Harper Collins, 1996.
150. Mihaly Csikszentmihalyi. *Beyond boredom and anxiety*. Jossey-Bass, 2000.
151. Mihaly Csikszentmihalyi. *Toward a psychology of optimal experience*. Springer, 2014.
152. George Cybenko. Approximation by superpositions of a sigmoidal function. *Mathematics of Control, Signals and Systems*, 2(4):303–314, 1989.
153. Ryan S. J. d. Baker, Gregory R. Moore, Angela Z. Wagner, Jessica Kalka, Aatish Salvi, Michael Karabinos, Colin A. Ashe, and David Yaron. The Dynamics between Student Affect and Behavior Occurring Outside of Educational Software. In *Affective Computing and Intelligent Interaction*, pages 14–24. Springer, 2011.

154. Anders Dahlbom and Lars Niklasson. Goal-Directed Hierarchical Dynamic Scripting for RTS Games. In *AIIDE*, pages 21–28, 2006.
155. Steve Dahlskog and Julian Togelius. Patterns as objectives for level generation. In *Proceedings of the International Conference on the Foundations of Digital Games*. ACM, 2013.
156. Steve Dahlskog, Julian Togelius, and Mark J. Nelson. Linear levels through n-grams. In *Proceedings of the 18th International Academic MindTrek Conference: Media Business, Management, Content & Services*, pages 200–206. ACM, 2014.
157. Antonio R. Damasio, Barry J. Everitt, and Dorothy Bishop. The somatic marker hypothesis and the possible functions of the prefrontal cortex [and discussion]. *Philosophical Transactions of the Royal Society B: Biological Sciences*, 351(1346):1413–1420, 1996.
158. Gustavo Danzi, Andrade Hugo Pimentel Santana, André Wilson Brotto Furtado, André Roberto Gouveia, Amaral Leitao, and Geber Lisboa Ramalho. Online adaptation of computer games agents: A reinforcement learning approach. In *II Workshop de Jogos e Entretenimento Digital*, pages 105–112, 2003.
159. Isaac M. Dart, Gabriele De Rossi, and Julian Togelius. SpeedRock: procedural rocks through grammars and evolution. In *Proceedings of the 2nd International Workshop on Procedural Content Generation in Games*. ACM, 2011.
160. Fernando de Mesentier Silva, Scott Lee, Julian Togelius, and Andy Nealen. AI-based Playtesting of Contemporary Board Games. In *Proceedings of Foundations of Digital Games (FDG)*, 2017.
161. Maarten de Waard, Diederik M. Roijers, and Sander Bakkes. Monte Carlo tree search with options for general video game playing. In *Computational Intelligence and Games (CIG), 2016 IEEE Conference on*. IEEE, 2016.
162. Edward L. Deci and Richard M. Ryan. *Intrinsic motivation*. Wiley Online Library, 1975.
163. Erik D. Demaine, Giovanni Viglietta, and Aaron Williams. Super Mario Bros. is Harder/Easier than We Thought. In *Proceedings of the 8th International Conference on Fun with Algorithms (FUN 2016)*, pages 13:1–13:14, La Maddalena, Italy, June 8–10 2016.
164. Jack Dennerlein, Theodore Becker, Peter Johnson, Carson Reynolds, and Rosalind W. Picard. Frustrating computer users increases exposure to physical factors. In *Proceedings of the International Ergonomics Association (IEA)*, 2003.
165. Jörg Denzinger, Kevin Loose, Darryl Gates, and John W. Buchanan. Dealing with Parameterized Actions in Behavior Testing of Commercial Computer Games. In *IEEE Symposium on Computational Intelligence and Games*, 2005.
166. L. Devillers, R. Cowie, J. C. Martin, E. Douglas-Cowie, S. Abrilian, and M. McRorie. Real life emotions in French and English TV video clips: an integrated annotation protocol combining continuous and discrete approaches. In *Proceedings of the 5th International Conference on Language Resources and Evaluation (LREC 2006), Genoa, Italy*, page 22, 2006.
167. Ravi Dhar and Itamar Simonson. The effect of forced choice on choice. *Journal of Marketing Research*, 40(2), 2003.
168. Joao Dias, Samuel Mascarenhas, and Ana Paiva. Fatima modular: Towards an agent architecture with a generic appraisal framework. In *Emotion Modeling*, pages 44–56. Springer, 2014.
169. Kevin Dill. A pattern-based approach to modular AI for Games. *Game Programming Gems*, 8:232–243, 2010.
170. Kevin Dill. Introducing GAIA: A Reusable, Extensible architecture for AI behavior. In *Proceedings of the 2012 Spring Simulation Interoperability Workshop*, 2012.
171. Kevin Dill and L. Martin. A game AI approach to autonomous control of virtual characters. In *Interservice/Industry Training, Simulation, and Education Conference (I/ITSEC)*, 2011.
172. Sidney D'Mello and Art Graesser. Automatic detection of learner's affect from gross body language. *Applied Artificial Intelligence*, 23(2):123–150, 2009.
173. Joris Dormans. Adventures in level design: generating missions and spaces for action adventure games. In *Proceedings of the 2010 Workshop on Procedural Content Generation in Games*. ACM, 2010.

174. Joris Dormans and Sander Bakkes. Generating missions and spaces for adaptable play experiences. *IEEE Transactions on Computational Intelligence and AI in Games*, 3(3):216–228, 2011.
175. Aanders Drachen, Lennart Nacke, Georgios N. Yannakakis, and Anja Lee Pedersen. Correlation between heart rate, electrodermal activity and player experience in first-person shooter games. In *Proceedings of the SIGGRAPH Symposium on Video Games*. ACM-SIGGRAPH Publishers, 2010.
176. Anders Drachen, Alessandro Canossa, and Georgios N. Yannakakis. Player modeling using self-organization in Tomb Raider: Underworld. In *Proceedings of the 2009 IEEE Symposium on Computational Intelligence and Games*, pages 1–8. IEEE, 2009.
177. Anders Drachen and Matthias Schubert. Spatial game analytics. In *Game Analytics*, pages 365–402. Springer, 2013.
178. Anders Drachen, Christian Thurau, Julian Togelius, Georgios N. Yannakakis, and Christian Bauckhage. Game Data Mining. In *Game Analytics*, pages 205–253. Springer, 2013.
179. H. Drucker, C.J. C. Burges, L. Kaufman, A. Smola, and V. Vapnik. Support vector regression machines. In *Advances in Neural Information Processing Systems (NIPS)*, pages 155–161. Morgan Kaufmann Publishers, 1997.
180. David S. Ebert. *Texturing & modeling: a procedural approach*. Morgan Kaufmann, 2003.
181. Marc Ebner, John Levine, Simon M. Lucas, Tom Schaul, Tommy Thompson, and Julian Togelius. Towards a video game description language. *Dagstuhl Follow-Ups*, 6, 2013.
182. Arthur S. Eddington. The Constants of Nature. In *The World of Mathematics 2*, pages 1074–1093. Simon & Schuster, 1956.
183. Arjan Egges, Sumedha Kshirsagar, and Nadia Magnenat-Thalmann. Generic personality and emotion simulation for conversational agents. *Computer animation and virtual worlds*, 15(1):1–13, 2004.
184. Agoston E. Eiben and James E. Smith. *Introduction to Evolutionary Computing*. Springer, 2003.
185. Magy Seif El-Nasr. Intelligent lighting for game environments. *Journal of Game Development*, 2005.
186. Magy Seif El-Nasr, Anders Drachen, and Alessandro Canossa. *Game analytics: Maximizing the value of player data*. Springer, 2013.
187. Magy Seif El-Nasr, Shree Durga, Mariya Shiyko, and Carmen Sceppa. Data-driven retrospective interviewing (DDRI): a proposed methodology for formative evaluation of pervasive games. *Entertainment Computing*, 11:1–19, 2015.
188. Magy Seif El-Nasr, Athanasios Vasilakos, Chinmay Rao, and Joseph Zupko. Dynamic intelligent lighting for directing visual attention in interactive 3-D scenes. *Computational Intelligence and AI in Games, IEEE Transactions on*, 1(2):145–153, 2009.
189. Magy Seif El-Nasr, John Yen, and Thomas R. Ioerger. Flame—fuzzy logic adaptive model of emotions. *Autonomous Agents and Multi-Agent Systems*, 3(3):219–257, 2000.
190. Mirjam Palosaari Eladhari and Michael Mateas. Semi-autonomous avatars in World of Minds: A case study of AI-based game design. In *Proceedings of the 2008 International Conference on Advances in Computer Entertainment Technology*, pages 201–208. ACM, 2008.
191. Mirjam Palosaari Eladhari and Michael Sellers. Good moods: outlook, affect and mood in dynemotion and the mind module. In *Proceedings of the 2008 Conference on Future Play: Research, Play, Share*, pages 1–8. ACM, 2008.
192. George Skaff Elias, Richard Garfield, K. Robert Gutschera, and Peter Whitley. *Characteristics of games*. MIT Press, 2012.
193. David K. Elson and Mark O. Riedl. A lightweight intelligent virtual cinematography system for machinima production. In *AIIDE*, pages 8–13, 2007.
194. Nathan Ensmenger. Is Chess the Drosophila of AI? A Social History of an Algorithm. *Social Studies of Science*, 42(1):5–30, 2012.
195. Ido Erev and Alvin E. Roth. Predicting how people play games: Reinforcement learning in experimental games with unique, mixed strategy equilibria. *American Economic Review*, pages 848–881, 1998.

196. Martin Ester, Hans-Peter Kriegel, Jörg Sander, and Xiaowei Xu. A density-based algorithm for discovering clusters in large spatial databases with noise. In *Proceedings of the International Conference on Knowledge Discovery and Data Mining (KDD)*, pages 226–231, 1996.
197. Richard Evans and Emily Short. Versu—a simulationist storytelling system. *IEEE Transactions on Computational Intelligence and AI in Games*, 6(2):113–130, 2014.
198. Vincent E. Farrugia, Héctor P. Martínez, and Georgios N. Yannakakis. The preference learning toolbox. *arXiv preprint arXiv:1506.01709*, 2015.
199. Bjarke Felbo, Alan Mislove, Anders Søgaard, Iyad Rahwan, and Sune Lehmann. Using millions of emoji occurrences to learn any-domain representations for detecting sentiment, emotion and sarcasm. *arXiv preprint arXiv:1708.00524*, 2017.
200. Lisa A. Feldman. Valence focus and arousal focus: Individual differences in the structure of affective experience. *Journal of personality and social psychology*, 69(1):153, 1995.
201. David Ferrucci, Eric Brown, Jennifer Chu-Carroll, James Fan, David Gondek, Aditya A. Kalyanpur, Adam Lally, J. William Murdock, Eric Nyberg, John Prager, Nico Schlaefer, and Chris Welty. Building Watson: An overview of the DeepQA project. *AI Magazine*, 31(3):59–79, 2010.
202. Hilmar Finnsson and Yngvi Björnsson. Learning simulation control in general game-playing agents. In *AAAI*, pages 954–959, 2010.
203. Jacob Fischer, Nikolaj Falsted, Mathias Vielwerth, Julian Togelius, and Sebastian Risi. Monte-Carlo Tree Search for Simulated Car Racing. In *Proceedings of FDG*, 2015.
204. John H. Flavell. *The developmental psychology of Jean Piaget*. Ardent Media, 1963.
205. Dario Floreano, Peter Dürr, and Claudio Mattiussi. Neuroevolution: from architectures to learning. *Evolutionary Intelligence*, 1(1):47–62, 2008.
206. Dario Floreano, Toshifumi Kato, Davide Marocco, and Eric Sauser. Coevolution of active vision and feature selection. *Biological Cybernetics*, 90(3):218–228, 2004.
207. David B. Fogel. *Blondie24: Playing at the Edge of AI*. Morgan Kaufmann, 2001.
208. David B. Fogel, Timothy J. Hays, Sarah L. Hahn, and James Quon. The Blondie25 chess program competes against Fritz 8.0 and a human chess master. In *Computational Intelligence and Games, 2006 IEEE Symposium on*, pages 230–235. IEEE, 2006.
209. Tom Forsyth. Cellular automata for physical modelling. *Game Programming Gems*, 3:200–214, 2002.
210. Alain Fournier, Don Fussell, and Loren Carpenter. Computer rendering of stochastic models. *Communications of the ACM*, 25(6):371–384, 1982.
211. Michael Freed, Travis Bear, Herrick Goldman, Geoffrey Hyatt, Paul Reber, A. Sylvan, and Joshua Tauber. Towards more human-like computer opponents. In *Working Notes of the AAAI Spring Symposium on Artificial Intelligence and Interactive Entertainment*, pages 22–26, 2000.
212. Nico Frijda. *The Emotions*. Cambridge University Press, Englewood Cliffs, NJ, 1986.
213. Frederik Frydenberg, Kasper R. Andersen, Sebastian Risi, and Julian Togelius. Investigating MCTS modifications in general video game playing. In *Computational Intelligence and Games (CIG), 2015 IEEE Conference on*, pages 107–113. IEEE, 2015.
214. Drew Fudenberg and David K. Levine. *The theory of learning in games*. MIT Press, 1998.
215. J. Fürnkranz and E. Hüllermeier. *Preference learning*. Springer, 2010.
216. Raluca D. Gaina, Jialin Liu, Simon M. Lucas, and Diego Pérez-Liébana. Analysis of Vanilla Rolling Horizon Evolution Parameters in General Video Game Playing. In *European Conference on the Applications of Evolutionary Computation*, pages 418–434. Springer, 2017.
217. Maurizio Garbarino, Simone Tognetti, Matteo Matteucci, and Andrea Bonarini. Learning general preference models from physiological responses in video games: How complex is it? In *Affective Computing and Intelligent Interaction*, pages 517–526. Springer, 2011.
218. Pablo García-Sánchez, Alberto Tonda, Giovanni Squillero, Antonio Mora, and Juan J. Merelo. Evolutionary deckbuilding in Hearthstone. In *Computational Intelligence and Games (CIG), 2016 IEEE Conference on*. IEEE, 2016.
219. Tom A. Garner. From Sinewaves to Physiologically-Adaptive Soundscapes: The Evolving Relationship Between Sound and Emotion in Video Games. In *Emotion in Games: Theory and Praxis*, pages 197–214. Springer, 2016.

220. Tom A. Garner and Mark Grimshaw. Sonic virtuality: Understanding audio in a virtual world. *The Oxford Handbook of Virtuality*, 2014.
221. H. P. Gasselseder. Re-scoring the games score: Dynamic music and immersion in the ludonarrative. In *Proceedings of the Intelligent Human Computer Interaction conference*, 2014.
222. Jakub Gemrot, Rudolf Kadlec, Michal Bída, Ondřej Burkert, Radek Píbil, Jan Havlíček, Lukáš Zemčák, Juraj Šimlovič, Radim Vansa, Michal Štolba, Tomáš Plch, and Cyril Brom. Pogamut 3 can assist developers in building AI (not only) for their videogame agents. In *Agents for games and simulations*, pages 1–15. Springer, 2009.
223. Michael Genesereth, Nathaniel Love, and Barney Pell. General game playing: Overview of the AAAI competition. *AI Magazine*, 26(2):62, 2005.
224. Michael Georgeff, Barney Pell, Martha Pollack, Milind Tambe, and Michael Wooldridge. The belief-desire-intention model of agency. In *International Workshop on Agent Theories, Architectures, and Languages*, pages 1–10. Springer, 1998.
225. Kallirroi Georgila, James Henderson, and Oliver Lemon. Learning user simulations for information state update dialogue systems. In *Interspeech*, pages 893–896, 2005.
226. Panayiotis G. Georgiou, Matthew P. Black, Adam C. Lammert, Brian R. Baucom, and Shrikanth S. Narayanan. "That's Aggravating, Very Aggravating": Is It Possible to Classify Behaviors in Couple Interactions Using Automatically Derived Lexical Features? In *Affective Computing and Intelligent Interaction*, pages 87–96. Springer, 2011.
227. Maryrose Gerardi, Barbara Olasov Rothbaum, Kerry Ressler, Mary Heekin, and Albert Rizzo. Virtual reality exposure therapy using a virtual Iraq: case report. *Journal of Traumatic Stress*, 21(2):209–213, 2008.
228. Malik Ghallab, Dana Nau, and Paolo Traverso. *Automated Planning: theory and practice*. Elsevier, 2004.
229. Spyridon Giannatos, Yun-Gyung Cheong, Mark J. Nelson, and Georgios N. Yannakakis. Generating narrative action schemas for suspense. In *Eighth Artificial Intelligence and Interactive Digital Entertainment Conference*, 2012.
230. Arthur Gill. *Introduction to the theory of Finite-State Machines*. McGraw-Hill, 1962.
231. Ian Goodfellow, Yoshua Bengio, and Aaron Courville. *Deep Learning*. MIT Press, 2016.
232. Ian Goodfellow, Jean Pouget-Abadie, Mehdi Mirza, Bing Xu, David Warde-Farley, Sherjil Ozair, Aaron Courville, and Yoshua Bengio. Generative adversarial nets. In *Advances in Neural Information Processing Systems*, pages 2672–2680, 2014.
233. Nitesh Goyal, Gilly Leshed, Dan Cosley, and Susan R. Fussell. Effects of implicit sharing in collaborative analysis. In *Proceedings of the SIGCHI Conference on Human Factors in Computing Systems*, pages 129–138, 2014.
234. Katja Grace, John Salvatier, Allan Dafoe, Baobao Zhang, and Owain Evans. When Will AI Exceed Human Performance? Evidence from AI Experts. *arXiv preprint arXiv:1705.08807*, 2017.
235. Thore Graepel, Ralf Herbrich, and Julian Gold. Learning to fight. In *Proceedings of the International Conference on Computer Games: Artificial Intelligence, Design and Education*, pages 193–200, 2004.
236. Joseph F. Grafsgaard, Kristy Elizabeth Boyer, and James C. Lester. Predicting facial indicators of confusion with hidden Markov models. In *Proceedings of International Conference on Affective Computing and Intelligent Interaction (ACII)*, pages 97–106. Springer, 2011.
237. Jonathan Gratch. Emile: Marshalling passions in training and education. In *Proceedings of the Fourth International Conference on Autonomous Agents*, pages 325–332. ACM, 2000.
238. Jonathan Gratch and Stacy Marsella. A domain-independent framework for modeling emotion. *Cognitive Systems Research*, 5(4):269–306, 2004.
239. Jonathan Gratch and Stacy Marsella. Evaluating a computational model of emotion. *Autonomous Agents and Multi-Agent Systems*, 11(1):23–43, 2005.
240. Daniele Gravina, Antonios Liapis, and Georgios N. Yannakakis. Constrained surprise search for content generation. In *Computational Intelligence and Games (CIG), 2016 IEEE Conference on*. IEEE, 2016.

241. Elin Rut Gudnadottir, Alaina K. Jensen, Yun-Gyung Cheong, Julian Togelius, Byung Chull Bae, and Christoffer Holmgård Pedersen. Detecting predatory behaviour in online game chats. In *The 2nd Workshop on Games and NLP*, 2014.
242. Johan Hagelbck. Potential-field based navigation in StarCraft. In *IEEE Conference on Computational Intelligence and Games (CIG)*. IEEE, 2012.
243. Mark Hall, Eibe Frank, Geoffrey Holmes, Bernhard Pfahringer, Peter Reutemann, and Ian H. Witten. The WEKA data mining software: an update. *ACM SIGKDD explorations newsletter*, 11(1):10–18, 2009.
244. Jiawei Han and Micheline Kamber. *Data mining: concepts and techniques*. Morgan Kaufmann, 2006.
245. Nikolaus Hansen and Andreas Ostermeier. Completely derandomized self-adaptation in evolution strategies. *Evolutionary Computation*, 9(2):159–195, 2001.
246. Daniel Damir Harabor and Alban Grastien. Online Graph Pruning for Pathfinding on Grid Maps. In *AAAI*, 2011.
247. Peter E. Hart, Nils J. Nilsson, and Bertram Raphael. Correction to a formal basis for the heuristic determination of minimum cost paths. *ACM SIGART Bulletin*, (37):28–29, 1972.
248. Ken Hartsook, Alexander Zook, Sauvik Das, and Mark O. Riedl. Toward supporting stories with procedurally generated game worlds. In *Computational Intelligence and Games (CIG), 2011 IEEE Conference on*, pages 297–304. IEEE, 2011.
249. Erin J. Hastings, Ratan K. Guha, and Kenneth O. Stanley. Automatic content generation in the Galactic Arms Race video game. *IEEE Transactions on Computational Intelligence and AI in Games*, 1(4):245–263, 2009.
250. Erin J. Hastings, Ratan K. Guha, and Kenneth O. Stanley. Evolving content in the Galactic Arms Race video game. In *IEEE Symposium on Computational Intelligence and Games*, pages 241–248. IEEE, 2009.
251. Matthew Hausknecht, Joel Lehman, Risto Miikkulainen, and Peter Stone. A neuroevolution approach to general Atari game playing. *IEEE Transactions on Computational Intelligence and AI in Games*, 6(4):355–366, 2014.
252. Brian Hawkins. *Real-Time Cinematography for Games (Game Development Series)*. Charles River Media, Inc., 2004.
253. Simon Haykin. *Neural Networks: A Comprehensive Foundation*. Macmillian College Publishing Company Inc., Upper Saddle River, NJ, USA, 1998.
254. Richard L. Hazlett. Measuring emotional valence during interactive experiences: boys at video game play. In *Proceedings of SIGCHI Conference on Human Factors in Computing Systems (CHI)*, pages 1023–1026. ACM, 2006.
255. Jennifer Healey. Recording affect in the field: Towards methods and metrics for improving ground truth labels. In *Affective Computing and Intelligent Interaction*, pages 107–116. Springer, 2011.
256. D. O. Hebb. *The Organization of Behavior*. Wiley, New York, 1949.
257. Norbert Heijne and Sander Bakkes. Procedural Zelda: A PCG Environment for Player Experience Research. In *Proceedings of the International Conference on the Foundations of Digital Games*. ACM, 2017.
258. Harry Helson. *Adaptation-level theory*. Harper & Row, 1964.
259. Ralf Herbrich, Michael E. Tipping, and Mark Hatton. Personalized behavior of computer controlled avatars in a virtual reality environment, August 15 2006. US Patent 7,090,576.
260. Javier Hernandez, Rob R. Morris, and Rosalind W. Picard. Call center stress recognition with person-specific models. In *Affective Computing and Intelligent Interaction*, pages 125–134. Springer, 2011.
261. David Hilbert. Über die stetige Abbildung einer Linie auf ein Flächenstück. *Mathematische Annalen*, 38(3):459–460, 1891.
262. Philip Hingston. A Turing test for computer game bots. *IEEE Transactions on Computational Intelligence and AI in Games*, 1(3):169–186, 2009.
263. Philip Hingston. A new design for a Turing test for bots. In *Computational Intelligence and Games (CIG), 2010 IEEE Symposium on*, pages 345–350. IEEE, 2010.

264. Philip Hingston. *Believable Bots: Can Computers Play Like People?* Springer, 2012.
265. Philip Hingston, Clare Bates Congdon, and Graham Kendall. Mobile games with intelligence: A killer application? In *Computational Intelligence in Games (CIG), 2013 IEEE Conference on*, pages 1–7. IEEE, 2013.
266. Sepp Hochreiter and Jürgen Schmidhuber. Long short-term memory. *Neural computation*, 9(8):1735–1780, 1997.
267. Christoffer Holmgård, Antonios Liapis, Julian Togelius, and Georgios N. Yannakakis. Evolving personas for player decision modeling. In *Computational Intelligence and Games (CIG), 2014 IEEE Conference on*. IEEE, 2014.
268. Christoffer Holmgård, Antonios Liapis, Julian Togelius, and Georgios N. Yannakakis. Generative agents for player decision modeling in games. In *FDG*, 2014.
269. Christoffer Holmgård, Antonios Liapis, Julian Togelius, and Georgios N. Yannakakis. Personas versus clones for player decision modeling. In *International Conference on Entertainment Computing*, pages 159–166. Springer, 2014.
270. Christoffer Holmgård, Georgios N. Yannakakis, Karen-Inge Karstoft, and Henrik Steen Andersen. Stress detection for PTSD via the Startlemart game. In *Affective Computing and Intelligent Interaction (ACII), 2013 Humaine Association Conference on*, pages 523–528. IEEE, 2013.
271. Christoffer Holmgård, Georgios N. Yannakakis, Héctor P. Martínez, and Karen-Inge Karstoft. To rank or to classify? Annotating stress for reliable PTSD profiling. In *Affective Computing and Intelligent Interaction (ACII), 2015 International Conference on*, pages 719–725. IEEE, 2015.
272. Christoffer Holmgård, Georgios N. Yannakakis, Héctor P. Martínez, Karen-Inge Karstoft, and Henrik Steen Andersen. Multimodal PTSD characterization via the Startlemart game. *Journal on Multimodal User Interfaces*, 9(1):3–15, 2015.
273. Nils Iver Holtar, Mark J. Nelson, and Julian Togelius. Audioverdrive: Exploring bidirectional communication between music and gameplay. In *Proceedings of the 2013 International Computer Music Conference*, pages 124–131, 2013.
274. Vincent Hom and Joe Marks. Automatic design of balanced board games. In *Proceedings of the AAAI Conference on Artificial Intelligence and Interactive Digital Entertainment (AIIDE)*, pages 25–30, 2007.
275. Amy K. Hoover, William Cachia, Antonios Liapis, and Georgios N. Yannakakis. AudioInSpace: Exploring the Creative Fusion of Generative Audio, Visuals and Gameplay. In *Evolutionary and Biologically Inspired Music, Sound, Art and Design*, pages 101–112. Springer, 2015.
276. Amy K. Hoover, Paul A. Szerlip, and Kenneth O. Stanley. Functional scaffolding for composing additional musical voices. *Computer Music Journal*, 2014.
277. Amy K. Hoover, Julian Togelius, and Georgios N. Yannakakis. Composing video game levels with music metaphors through functional scaffolding. In *First Computational Creativity and Games Workshop, ICCC*, 2015.
278. John J. Hopfield. Neural networks and physical systems with emergent collective computational abilities. *Proceedings of the National Academy of Sciences*, 79(8):2554–2558, 1982.
279. Kurt Hornik, Maxwell Stinchcombe, and Halbert White. Multilayer feedforward networks are universal approximators. *Neural Networks*, 2(5):359–366, 1989.
280. Ben Houge. Cell-based music organization in Tom Clancy's EndWar. In *Demo at the AIIDE 2012 Workshop on Musical Metacreation*, 2012.
281. Ryan Houlette. *Player Modeling for Adaptive Games. AI Game Programming Wisdom II*, pages 557–566. Charles River Media, Inc., 2004.
282. Andrew Howlett, Simon Colton, and Cameron Browne. Evolving pixel shaders for the prototype video game Subversion. In *The Thirty Sixth Annual Convention of the Society for the Study of Artificial Intelligence and Simulation of Behaviour (AISB10), De Montfort University, Leicester, UK, 30th March*, 2010.
283. Johanna Höysniemi, Perttu Hämäläinen, Laura Turkki, and Teppo Rouvi. Children's intuitive gestures in vision-based action games. *Communications of the ACM*, 48(1):44–50, 2005.

284. Chih-Wei Hsu and Chih-Jen Lin. A comparison of methods for multiclass support vector machines. *IEEE Transactions on Neural Networks*, 13(2):415–425, 2002.
285. Feng-Hsiung Hsu. *Behind Deep Blue: Building the computer that defeated the world chess champion*. Princeton University Press, 2002.
286. Wijnand IJsselsteijn, Karolien Poels, and Y. A. W. De Kort. The game experience questionnaire: Development of a self-report measure to assess player experiences of digital games. *TU Eindhoven, Eindhoven, The Netherlands*, 2008.
287. Interactive Data Visualization. SpeedTree, 2010. http://www.speedtree.com/.
288. Aaron Isaksen, Dan Gopstein, Julian Togelius, and Andy Nealen. Discovering unique game variants. In *Computational Creativity and Games Workshop at the 2015 International Conference on Computational Creativity*, 2015.
289. Aaron Isaksen, Daniel Gopstein, and Andrew Nealen. Exploring Game Space Using Survival Analysis. In *Proceedings of Foundations of Digital Games (FDG)*, 2015.
290. Katherine Isbister and Noah Schaffer. *Game usability: Advancing the player experience*. CRC Press, 2015.
291. Damian Isla. Handling complexity in the Halo 2 AI. In *Game Developers Conference*, 2005.
292. Damian Isla and Bruce Blumberg. New challenges for character-based AI for games. In *Proceedings of the AAAI Spring Symposium on AI and Interactive Entertainment*, pages 41–45. AAAI Press, 2002.
293. Susan A. Jackson and Robert C. Eklund. Assessing flow in physical activity: the flow state scale-2 and dispositional flow scale-2. *Journal of Sport & Exercise Psychology*, 24(2), 2002.
294. Emil Juul Jacobsen, Rasmus Greve, and Julian Togelius. Monte Mario: platforming with MCTS. In *Proceedings of the 2014 Annual Conference on Genetic and Evolutionary Computation*, pages 293–300. ACM, 2014.
295. Alexander Jaffe, Alex Miller, Erik Andersen, Yun-En Liu, Anna Karlin, and Zoran Popovic. Evaluating competitive game balance with restricted play. In *AIIDE*, 2012.
296. Rishabh Jain, Aaron Isaksen, Christoffer Holmgård, and Julian Togelius. Autoencoders for level generation, repair, and recognition. In *ICCC Workshop on Computational Creativity and Games*, 2016.
297. Daniel Jallov, Sebastian Risi, and Julian Togelius. EvoCommander: A Novel Game Based on Evolving and Switching Between Artificial Brains. *IEEE Transactions on Computational Intelligence and AI in Games*, 9(2):181–191, 2017.
298. Susan Jamieson. Likert scales: how to (ab) use them. *Medical Education*, 38(12):1217–1218, 2004.
299. Aki Järvinen. Gran stylissimo: The audiovisual elements and styles in computer and video games. In *Proceedings of Computer Games and Digital Cultures Conference*, 2002.
300. Arnav Jhala and R. Michael Young. Cinematic visual discourse: Representation, generation, and evaluation. *Computational Intelligence and AI in Games, IEEE Transactions on*, 2(2):69–81, 2010.
301. Yuu Jinnai and Alex S. Fukunaga. Learning to prune dominated action sequences in online black-box planning. In *AAAI*, pages 839–845, 2017.
302. Thorsten Joachims. Text categorization with support vector machines: Learning with many relevant features. *Machine Learning: ECML-98*, pages 137–142, 1998.
303. Thorsten Joachims. Optimizing search engines using clickthrough data. In *Proceedings of the ACM SIGKDD International Conference on Knowledge Discovery in Data Mining (KDD)*, pages 133–142. ACM, 2002.
304. Lawrence Johnson, Georgios N. Yannakakis, and Julian Togelius. Cellular automata for real-time generation of infinite cave levels. In *Proceedings of the 2010 Workshop on Procedural Content Generation in Games*. ACM, 2010.
305. Matthew Johnson, Katja Hofmann, Tim Hutton, and David Bignell. The Malmo Platform for Artificial Intelligence Experimentation. In *IJCAI*, pages 4246–4247, 2016.
306. Tom Johnstone and Klaus R. Scherer. Vocal communication of emotion. In *Handbook of emotions*, pages 220–235. Guilford Press, New York, 2000.

307. German Gutierrez Jorge Munoz and Araceli Sanchis. Towards imitation of human driving style in car racing games. In Philip Hingston, editor, *Believable Bots: Can Computers Play Like People?* Springer, 2012.
308. Patrik N. Juslin and Klaus R. Scherer. *Vocal expression of affect*. Oxford University Press, Oxford, UK, 2005.
309. Niels Justesen, Tobias Mahlmann, and Julian Togelius. Online evolution for multi-action adversarial games. In *European Conference on the Applications of Evolutionary Computation*, pages 590–603. Springer, 2016.
310. Niels Justesen and Sebastian Risi. Continual Online Evolutionary Planning for In-Game Build Order Adaptation in StarCraft. In *Proceedings of the Conference on Genetic and Evolutionary Computation (GECCO)*, 2017.
311. Niels Justesen, Bálint Tillman, Julian Togelius, and Sebastian Risi. Script-and cluster-based UCT for StarCraft. In *Computational Intelligence and Games (CIG), 2014 IEEE Conference on*. IEEE, 2014.
312. Tróndur Justinussen, Peter Hald Rasmussen, Alessandro Canossa, and Julian Togelius. Resource systems in games: An analytical approach. In *Computational Intelligence and Games (CIG), 2012 IEEE Conference on*, pages 171–178. IEEE, 2012.
313. Jesper Juul. Games telling stories. *Game Studies*, 1(1):45, 2001.
314. Jesper Juul. *A casual revolution: Reinventing video games and their players*. MIT Press, 2010.
315. Souhila Kaci. *Working with preferences: Less is more*. Springer, 2011.
316. Leslie Pack Kaelbling, Michael L. Littman, and Andrew W. Moore. Reinforcement learning: A survey. *Journal of Artificial Intelligence Research*, 4:237–285, 1996.
317. Daniel Kahneman. A perspective on judgment and choice: mapping bounded rationality. *American psychologist*, 58(9):697, 2003.
318. Daniel Kahneman and Jason Riis. Living, and thinking about it: Two perspectives on life. *The science of well-being*, pages 285–304, 2005.
319. Theofanis Kannetis and Alexandros Potamianos. Towards adapting fantasy, curiosity and challenge in multimodal dialogue systems for preschoolers. In *Proceedings of International Conference on Multimodal Interfaces (ICMI)*, pages 39–46. ACM, 2009.
320. Theofanis Kannetis, Alexandros Potamianos, and Georgios N. Yannakakis. Fantasy, curiosity and challenge as adaptation indicators in multimodal dialogue systems for preschoolers. In *Proceedings of the 2nd Workshop on Child, Computer and Interaction*. ACM, 2009.
321. Ashish Kapoor, Winslow Burleson, and Rosalind W. Picard. Automatic prediction of frustration. *International Journal of Human-Computer Studies*, 65(8):724–736, 2007.
322. Sergey Karakovskiy and Julian Togelius. The Mario AI benchmark and competitions. *IEEE Transactions on Computational Intelligence and AI in Games*, 4(1):55–67, 2012.
323. Daniël Karavolos, Anders Bouwer, and Rafael Bidarra. Mixed-initiative design of game levels: Integrating mission and space into level generation. In *Proceedings of the 10th International Conference on the Foundations of Digital Games*, 2015.
324. Daniel Karavolos, Antonios Liapis, and Georgios N. Yannakakis. Learning the patterns of balance in a multi-player shooter game. In *Proceedings of the FDG workshop on Procedural Content Generation in Games*, 2017.
325. Kostas Karpouzis and Georgios N. Yannakakis. *Emotion in Games: Theory and Praxis*. Springer, 2016.
326. Kostas Karpouzis, Georgios N. Yannakakis, Noor Shaker, and Stylianos Asteriadis. The Platformer Experience Dataset. In *Affective Computing and Intelligent Interaction (ACII), 2015 International Conference on*, pages 712–718. IEEE, 2015.
327. Igor V. Karpov, Leif Johnson, and Risto Miikkulainen. Evaluation methods for active human-guided neuroevolution in games. In *2012 AAAI Fall Symposium on Robots Learning Interactively from Human Teachers (RLIHT)*, 2012.
328. Igor V. Karpov, Jacob Schrum, and Risto Miikkulainen. Believable bot navigation via playback of human traces. In Philip Hingston, editor, *Believable Bots: Can Computers Play Like People?* Springer, 2012.

329. Leonard Kaufman and Peter J. Rousseeuw. *Clustering by means of medoids*. North-Holland, 1987.
330. Leonard Kaufman and Peter J. Rousseeuw. *Finding groups in data: an introduction to cluster analysis*. John Wiley & Sons, 2009.
331. Richard Kaye. Minesweeper is NP-complete. *The Mathematical Intelligencer*, 22(2):9–15, 2000.
332. Markus Kemmerling and Mike Preuss. Automatic adaptation to generated content via car setup optimization in TORCS. In *Computational Intelligence and Games (CIG), 2010 IEEE Symposium on*, pages 131–138. IEEE, 2010.
333. Michał Kempka, Marek Wydmuch, Grzegorz Runc, Jakub Toczek, and Wojciech Jaśkowski. Vizdoom: A doom-based AI research platform for visual reinforcement learning. *arXiv preprint arXiv:1605.02097*, 2016.
334. Graham Kendall, Andrew J. Parkes, and Kristian Spoerer. A Survey of NP-Complete Puzzles. *ICGA Journal*, 31(1):13–34, 2008.
335. Manuel Kerssemakers, Jeppe Tuxen, Julian Togelius, and Georgios N. Yannakakis. A procedural procedural level generator generator. In *Computational Intelligence and Games (CIG), 2012 IEEE Conference on*, pages 335–341. IEEE, 2012.
336. Rilla Khaled and Georgios N. Yannakakis. Village voices: An adaptive game for conflict resolution. In *Proceedings of FDG*, pages 425–426, 2013.
337. Ahmed Khalifa, Aaron Isaksen, Julian Togelius, and Andy Nealen. Modifying MCTS for Human-like General Video Game Playing. In *Proceedings of IJCAI*, 2016.
338. Ahmed Khalifa, Diego Perez-Liebana, Simon M. Lucas, and Julian Togelius. General video game level generation. In *Proceedings of IJCAI*, 2016.
339. K-J Kim, Heejin Choi, and Sung-Bae Cho. Hybrid of evolution and reinforcement learning for Othello players. In *Computational Intelligence and Games, 2007. CIG 2007. IEEE Symposium on*, pages 203–209. IEEE, 2007.
340. Kyung-Min Kim, Chang-Jun Nan, Jung-Woo Ha, Yu-Jung Heo, and Byoung-Tak Zhang. Pororobot: A deep learning robot that plays video Q&A games. In *AAAI 2015 Fall Symposium on AI for Human-Robot Interaction (AI-HRI 2015)*, 2015.
341. Steven Orla Kimbrough, Gary J. Koehler, Ming Lu, and David Harlan Wood. On a Feasible–Infeasible Two-Population (FI-2Pop) genetic algorithm for constrained optimization: Distance tracing and no free lunch. *European Journal of Operational Research*, 190(2):310–327, 2008.
342. Diederik P. Kingma and Max Welling. Auto-encoding variational Bayes. *arXiv preprint arXiv:1312.6114*, 2013.
343. A. Kleinsmith and N. Bianchi-Berthouze. Affective body expression perception and recognition: A survey. *IEEE Transactions on Affective Computing*, 2012.
344. Andrea Kleinsmith and Nadia Bianchi-Berthouze. Form as a cue in the automatic recognition of non-acted affective body expressions. In *Affective Computing and Intelligent Interaction*, pages 155–164. Springer, 2011.
345. Yana Knight, Héctor Perez Martínez, and Georgios N. Yannakakis. Space maze: Experience-driven game camera control. In *FDG*, pages 427–428, 2013.
346. Matthias J. Koepp, Roger N. Gunn, Andrew D. Lawrence, Vincent J. Cunningham, Alain Dagher, Tasmin Jones, David J. Brooks, C. J. Bench, and P. M. Grasby. Evidence for striatal dopamine release during a video game. *Nature*, 393(6682):266–268, 1998.
347. Teuvo Kohonen. *Self-Organizing Maps*. Springer, Secaucus, NJ, USA, 3rd edition, 2001.
348. Andrey N. Kolmogorov. On the representation of continuous functions of several variables by superposition of continuous functions of one variable and addition. *Russian, American Mathematical Society Translation 28 (1963) 55-59. Doklady Akademiia Nauk SSR*, 14(5):953–956, 1957.
349. Richard Konečný. Modeling of fighting game players. Master's thesis, Institute of Digital Games, University of Malta, 2016.
350. Michael Kosfeld, Markus Heinrichs, Paul J. Zak, Urs Fischbacher, and Ernst Fehr. Oxytocin increases trust in humans. *Nature*, 435(7042):673–676, 2005.

351. Raph Koster. *Theory of fun for game design*. O'Reilly Media, Inc., 2013.
352. Bartosz Kostka, Jaroslaw Kwiecien, Jakub Kowalski, and Pawel Rychlikowski. Text-based Adventures of the Golovin AI Agent. *arXiv preprint arXiv:1705.05637*, 2017.
353. Jan Koutník, Giuseppe Cuccu, Jürgen Schmidhuber, and Faustino Gomez. Evolving large-scale neural networks for vision-based reinforcement learning. In *Proceedings of the 15th Annual Conference on Genetic and Evolutionary Computation*, pages 1061–1068. ACM, 2013.
354. Jakub Kowalski and Andrzej Kisielewicz. Towards a Real-time Game Description Language. In *ICAART (2)*, pages 494–499, 2016.
355. Jakub Kowalski and Marek Szykuła. Evolving chess-like games using relative algorithm performance profiles. In *European Conference on the Applications of Evolutionary Computation*, pages 574–589. Springer, 2016.
356. John R. Koza. *Genetic programming: on the programming of computers by means of natural selection*. MIT Press, 1992.
357. Teofebano Kristo and Nur Ulfa Maulidevi. Deduction of fighting game countermeasures using Neuroevolution of Augmenting Topologies. In *Data and Software Engineering (ICoDSE), 2016 International Conference on*. IEEE, 2016.
358. Ben Kybartas and Rafael Bidarra. A semantic foundation for mixed-initiative computational storytelling. In *Interactive Storytelling*, pages 162–169. Springer, 2015.
359. Alexandros Labrinidis and Hosagrahar V. Jagadish. Challenges and opportunities with big data. *Proceedings of the VLDB Endowment*, 5(12):2032–2033, 2012.
360. John Laird and Michael van Lent. Human-level AI's killer application: Interactive computer games. *AI Magazine*, 22(2):15, 2001.
361. G. B. Langley and H. Sheppeard. The visual analogue scale: its use in pain measurement. *Rheumatology International*, 5(4):145–148, 1985.
362. Frank Lantz, Aaron Isaksen, Alexander Jaffe, Andy Nealen, and Julian Togelius. Depth in strategic games. In *Proceedings of the AAAI WNAIG Workshop*, 2017.
363. Pier Luca Lanzi, Wolfgang Stolzmann, and Stewart W. Wilson. *Learning classifier systems: from foundations to applications*. Springer, 2003.
364. Richard S. Lazarus. *Emotion and adaptation*. Oxford University Press, 1991.
365. Nicole Lazzaro. Why we play games: Four keys to more emotion without story. Technical report, XEO Design Inc., 2004.
366. Yann LeCun, Yoshua Bengio, and Geoffrey Hinton. Deep learning. *Nature*, 521(7553):436–444, 2015.
367. David Lee and Mihalis Yannakakis. Principles and methods of testing finite state machines—a survey. *Proceedings of the IEEE*, 84(8):1090–1123, 1996.
368. Alan Levinovitz. The mystery of Go, the ancient game that computers still can't win. *Wired Magazine*, 2014.
369. Mike Lewis and Kevin Dill. Game AI appreciation, revisited. In *Game AI Pro 2: Collected Wisdom of Game AI Professionals*, pages 3–18. AK Peters/CRC Press, 2015.
370. Boyang Li, Stephen Lee-Urban, Darren Scott Appling, and Mark O. Riedl. Crowdsourcing narrative intelligence. *Advances in Cognitive Systems*, 2(1), 2012.
371. Antonios Liapis. Creativity facet orchestration: the whys and the hows. *Artificial and Computational Intelligence in Games: Integration; Dagstuhl Follow-Ups*, 2015.
372. Antonios Liapis. Mixed-initiative Creative Drawing with webIconoscope. In *Proceedings of the 6th International Conference on Computational Intelligence in Music, Sound, Art and Design. (EvoMusArt)*. Springer, 2017.
373. Antonios Liapis, Héctor P. Martınez, Julian Togelius, and Georgios N. Yannakakis. Transforming exploratory creativity with DeLeNoX. In *Proceedings of the Fourth International Conference on Computational Creativity*, pages 56–63, 2013.
374. Antonios Liapis, Gillian Smith, and Noor Shaker. Mixed-initiative content creation. In *Procedural Content Generation in Games*, pages 195–214. Springer, 2016.
375. Antonios Liapis and Georgios N. Yannakakis. Boosting computational creativity with human interaction in mixed-initiative co-creation tasks. In *Proceedings of the ICCC Workshop on Computational Creativity and Games*, 2016.

376. Antonios Liapis, Georgios N. Yannakakis, and Julian Togelius. Neuroevolutionary constrained optimization for content creation. In *Computational Intelligence and Games (CIG), 2011 IEEE Conference on*, pages 71–78. IEEE, 2011.
377. Antonios Liapis, Georgios N. Yannakakis, and Julian Togelius. Adapting models of visual aesthetics for personalized content creation. *IEEE Transactions on Computational Intelligence and AI in Games*, 4(3):213–228, 2012.
378. Antonios Liapis, Georgios N. Yannakakis, and Julian Togelius. Designer modeling for personalized game content creation tools. In *Proceedings of the AIIDE Workshop on Artificial Intelligence & Game Aesthetics*, 2013.
379. Antonios Liapis, Georgios N. Yannakakis, and Julian Togelius. Sentient Sketchbook: Computer-aided game level authoring. In *Proceedings of ACM Conference on Foundations of Digital Games*, pages 213–220, 2013.
380. Antonios Liapis, Georgios N. Yannakakis, and Julian Togelius. Sentient World: Human-Based Procedural Cartography. In *Evolutionary and Biologically Inspired Music, Sound, Art and Design*, pages 180–191. Springer, 2013.
381. Antonios Liapis, Georgios N. Yannakakis, and Julian Togelius. Computational Game Creativity. In *Proceedings of the Fifth International Conference on Computational Creativity*, pages 285–292, 2014.
382. Antonios Liapis, Georgios N. Yannakakis, and Julian Togelius. Constrained novelty search: A study on game content generation. *Evolutionary Computation*, 23(1):101–129, 2015.
383. Vladimir Lifschitz. Answer set programming and plan generation. *Artificial Intelligence*, 138(1-2):39–54, 2002.
384. Rensis Likert. A technique for the measurement of attitudes. *Archives of Psychology*, 140:1–55, 1932.
385. Chong-U Lim, Robin Baumgarten, and Simon Colton. Evolving behaviour trees for the commercial game DEFCON. In *European Conference on the Applications of Evolutionary Computation*, pages 100–110. Springer, 2010.
386. Chong-U Lim and D. Fox Harrell. Revealing social identity phenomena in videogames with archetypal analysis. In *Proceedings of the 6th International AISB Symposium on AI and Games*, 2015.
387. Aristid Lindenmayer. Mathematical models for cellular interactions in development I. Filaments with one-sided inputs. *Journal of Theoretical Biology*, 18(3):280–299, 1968.
388. R. L. Linn and N. E. Gronlund. *Measurement and assessment in teaching*. Prentice-Hall, 2000.
389. Nir Lipovetzky and Hector Geffner. Width-based algorithms for classical planning: New results. In *Proceedings of the Twenty-first European Conference on Artificial Intelligence*, pages 1059–1060. IOS Press, 2014.
390. Nir Lipovetzky, Miquel Ramirez, and Hector Geffner. Classical Planning with Simulators: Results on the Atari Video Games. In *Proceedings of IJCAI*, pages 1610–1616, 2015.
391. Maria Teresa Llano, Michael Cook, Christian Guckelsberger, Simon Colton, and Rose Hepworth. Towards the automatic generation of fictional ideas for games. In *Experimental AI in Games (EXAG14), a Workshop collocated with the Tenth Annual AAAI Conference on Artificial Intelligence and Interactive Digital Entertainment (AIIDE14). AAAI Publications*, 2014.
392. Daniele Loiacono, Pier Luca Lanzi, Julian Togelius, Enrique Onieva, David A. Pelta, Martin V. Butz, Thies D. Lönneker, Luigi Cardamone, Diego Perez, Yago Sáez, Mike Preuss, and Jan Quadflieg. The 2009 simulated car racing championship. *Computational Intelligence and AI in Games, IEEE Transactions on*, 2(2):131–147, 2010.
393. Daniele Loiacono, Julian Togelius, Pier Luca Lanzi, Leonard Kinnaird-Heether, Simon M. Lucas, Matt Simmerson, Diego Perez, Robert G. Reynolds, and Yago Saez. The WCCI 2008 simulated car racing competition. In *IEEE Symposium on Computational Intelligence and Games*, pages 119–126. IEEE, 2008.
394. Phil Lopes, Antonios Liapis, and Georgios N. Yannakakis. Sonancia: Sonification of procedurally generated game levels. In *Proceedings of the ICCC workshop on Computational Creativity & Games*, 2015.

395. Phil Lopes, Antonios Liapis, and Georgios N. Yannakakis. Framing tension for game generation. In *Proceedings of the Seventh International Conference on Computational Creativity*, 2016.
396. Phil Lopes, Antonios Liapis, and Georgios N. Yannakakis. Modelling affect for horror soundscapes. *IEEE Transactions on Affective Computing*, 2017.
397. Phil Lopes, Georgios N. Yannakakis, and Antonios Liapis. RankTrace: Relative and Unbounded Affect Annotation. In *Affective Computing and Intelligent Interaction (ACII), 2017 International Conference on*, 2017.
398. Ricardo Lopes and Rafael Bidarra. Adaptivity challenges in games and simulations: a survey. *Computational Intelligence and AI in Games, IEEE Transactions on*, 3(2):85–99, 2011.
399. Sandy Louchart, Ruth Aylett, Joao Dias, and Ana Paiva. Unscripted narrative for affectively driven characters. In *AIIDE*, pages 81–86, 2005.
400. Nathaniel Love, Timothy Hinrichs, David Haley, Eric Schkufza, and Michael Genesereth. General game playing: Game description language specification. Technical Report LG-2006-01, Stanford Logic Group, Computer Science Department, Stanford University, 2008.
401. A. Bryan Loyall and Joseph Bates. Personality-rich believable agents that use language. In *Proceedings of the First International Conference on Autonomous Agents*, pages 106–113. ACM, 1997.
402. Feiyu Lu, Kaito Yamamoto, Luis H. Nomura, Syunsuke Mizuno, YoungMin Lee, and Ruck Thawonmas. Fighting game artificial intelligence competition platform. In *Consumer Electronics (GCCE), 2013 IEEE 2nd Global Conference on*, pages 320–323. IEEE, 2013.
403. Simon M. Lucas. Evolving a Neural Network Location Evaluator to Play Ms. Pac-Man. In *Proceedings of the IEEE Symposium on Computational Intelligence and Games*, pages 203–210, 2005.
404. Simon M. Lucas. Ms Pac-Man competition. *ACM SIGEVOlution*, 2(4):37–38, 2007.
405. Simon M. Lucas. Computational intelligence and games: Challenges and opportunities. *International Journal of Automation and Computing*, 5(1):45–57, 2008.
406. Simon M. Lucas and Graham Kendall. Evolutionary computation and games. *Computational Intelligence Magazine, IEEE*, 1(1):10–18, 2006.
407. Simon M. Lucas, Michael Mateas, Mike Preuss, Pieter Spronck, and Julian Togelius. Artificial and Computational Intelligence in Games (Dagstuhl Seminar 12191). *Dagstuhl Reports*, 2(5):43–70, 2012.
408. Simon M. Lucas and T. Jeff Reynolds. Learning finite-state transducers: Evolution versus heuristic state merging. *IEEE Transactions on Evolutionary Computation*, 11(3):308–325, 2007.
409. Jeremy Ludwig and Art Farley. A learning infrastructure for improving agent performance and game balance. In Georgios N. Yannakakis and John Hallam, editors, *Proceedings of the AIIDE'07 Workshop on Optimizing Player Satisfaction, Technical Report WS-07-01*, pages 7–12. AAAI Press, 2007.
410. Kevin Lynch. *The Image of the City*. MIT Press, 1960.
411. James MacQueen. Some methods for classification and analysis of multivariate observations. In *Proceedings of the Fifth Berkeley Symposium on Mathematical Statistics and Probability*, number 14, pages 281–297. Oakland, CA, USA, 1967.
412. Brian Magerko. Story representation and interactive drama. In *AIIDE*, pages 87–92, 2005.
413. Brendan Maher. Can a video game company tame toxic behaviour? *Nature*, 531(7596):568–571, 2016.
414. Tobias Mahlmann, Anders Drachen, Julian Togelius, Alessandro Canossa, and Georgios N. Yannakakis. Predicting player behavior in Tomb Raider: Underworld. In *Proceedings of the 2010 IEEE Conference on Computational Intelligence and Games*, pages 178–185. IEEE, 2010.
415. Tobias Mahlmann, Julian Togelius, and Georgios N. Yannakakis. Modelling and evaluation of complex scenarios with the strategy game description language. In *Computational Intelligence and Games (CIG), 2011 IEEE Conference on*, pages 174–181. IEEE, 2011.

416. Tobias Mahlmann, Julian Togelius, and Georgios N. Yannakakis. Evolving card sets towards balancing Dominion. In *Proceedings of the IEEE Congress on Evolutionary Computation (CEC)*. IEEE, 2012.
417. Kevin Majchrzak, Jan Quadflieg, and Günter Rudolph. Advanced dynamic scripting for fighting game AI. In *International Conference on Entertainment Computing*, pages 86–99. Springer, 2015.
418. Nikos Malandrakis, Alexandros Potamianos, Georgios Evangelopoulos, and Athanasia Zlatintsi. A supervised approach to movie emotion tracking. In *Acoustics, Speech and Signal Processing (ICASSP), 2011 IEEE International Conference on*, pages 2376–2379. IEEE, 2011.
419. Thomas W. Malone. What makes computer games fun? *Byte*, 6:258–277, 1981.
420. Regan L. Mandryk and M. Stella Atkins. A fuzzy physiological approach for continuously modeling emotion during interaction with play technologies. *International Journal of Human-Computer Studies*, 65(4):329–347, 2007.
421. Regan L. Mandryk, Kori M. Inkpen, and Thomas W. Calvert. Using psychophysiological techniques to measure user experience with entertainment technologies. *Behaviour & Information Technology*, 25(2):141–158, 2006.
422. Jacek Mandziuk. Computational intelligence in mind games. *Challenges for Computational Intelligence*, 63:407–442, 2007.
423. Jacek Mandziuk. *Knowledge-free and learning-based methods in intelligent game playing*. Springer, 2010.
424. Benjamin Mark, Tudor Berechet, Tobias Mahlmann, and Julian Togelius. Procedural Generation of 3D Caves for Games on the GPU. In *Proceedings of the Conference on the Foundations of Digital Games (FDG)*, 2015.
425. Dave Mark. *Behavioral Mathematics for game AI*. Charles River Media, 2009.
426. Dave Mark and Kevin Dill. Improving AI decision modeling through utility theory. In *Game Developers Conference*, 2010.
427. Gloria Mark, Yiran Wang, and Melissa Niiya. Stress and multitasking in everyday college life: An empirical study of online activity. In *Proceedings of the SIGCHI Conference on Human Factors in Computing Systems*, pages 41–50, 2014.
428. Stacy Marsella, Jonathan Gratch, and Paolo Petta. Computational models of emotion. *A Blueprint for Affective Computing—A sourcebook and manual*, 11(1):21–46, 2010.
429. Chris Martens. Ceptre: A language for modeling generative interactive systems. In *Eleventh Artificial Intelligence and Interactive Digital Entertainment Conference*, 2015.
430. Héctor P. Martínez, Yoshua Bengio, and Georgios N. Yannakakis. Learning deep physiological models of affect. *Computational Intelligence Magazine, IEEE*, 9(1):20–33, 2013.
431. Héctor P. Martínez, Maurizio Garbarino, and Georgios N. Yannakakis. Generic physiological features as predictors of player experience. In *Affective Computing and Intelligent Interaction*, pages 267–276. Springer, 2011.
432. Héctor P. Martínez, Kenneth Hullett, and Georgios N. Yannakakis. Extending neuroevolutionary preference learning through player modeling. In *Proceedings of the 2010 IEEE Conference on Computational Intelligence and Games*, pages 313–320. IEEE, 2010.
433. Héctor P. Martínez and Georgios N. Yannakakis. Genetic search feature selection for affective modeling: a case study on reported preferences. In *Proceedings of the 3rd International Workshop on Affective Interaction in Natural Environments*, pages 15–20. ACM, 2010.
434. Héctor P. Martínez and Georgios N. Yannakakis. Mining multimodal sequential patterns: a case study on affect detection. In *Proceedings of International Conference on Multimodal Interfaces (ICMI)*, pages 3–10. ACM, 2011.
435. Héctor P. Martínez and Georgios N. Yannakakis. Deep multimodal fusion: Combining discrete events and continuous signals. In *Proceedings of the 16th International Conference on Multimodal Interaction*, pages 34–41. ACM, 2014.
436. Héctor P. Martínez, Georgios N. Yannakakis, and John Hallam. Don't Classify Ratings of Affect; Rank them! *IEEE Transactions on Affective Computing*, 5(3):314–326, 2014.

437. Giovanna Martinez-Arellano, Richard Cant, and David Woods. Creating AI Characters for Fighting Games using Genetic Programming. *IEEE Transactions on Computational Intelligence and AI in Games*, 2016.
438. Michael Mateas. *Interactive Drama, Art and Artificial Intelligence*. PhD thesis, Carnegie Mellon University, Pittsburgh, PA, USA, 2002.
439. Michael Mateas. Expressive AI: Games and Artificial Intelligence. In *DIGRA Conference*, 2003.
440. Michael Mateas and Andrew Stern. A behavior language for story-based believable agents. *IEEE Intelligent Systems*, 17(4):39–47, 2002.
441. Michael Mateas and Andrew Stern. Façade: An experiment in building a fully-realized interactive drama. In *Game Developers Conference*, 2003.
442. Michael Mauderer, Simone Conte, Miguel A. Nacenta, and Dhanraj Vishwanath. Depth perception with gaze-contingent depth of field. In *Proceedings of the SIGCHI Conference on Human Factors in Computing Systems*, pages 217–226, 2014.
443. John D. Mayer and Peter Salovey. The intelligence of emotional intelligence. *Intelligence*, 17(4):433–442, 1993.
444. Allan Mazur, Elizabeth J. Susman, and Sandy Edelbrock. Sex difference in testosterone response to a video game contest. *Evolution and Human Behavior*, 18(5):317–326, 1997.
445. Andrew McAfee, Erik Brynjolfsson, Thomas H. Davenport, D. J. Patil, and Dominic Barton. Big data. *The management revolution. Harvard Bus Rev*, 90(10):61–67, 2012.
446. John McCarthy. Partial formalizations and the Lemmings game. Technical report, Stanford University, 1998.
447. Josh McCoy, Mike Treanor, Ben Samuel, Michael Mateas, and Noah Wardrip-Fruin. Prom week: social physics as gameplay. In *Proceedings of the 6th International Conference on Foundations of Digital Games*, pages 319–321. ACM, 2011.
448. Josh McCoy, Mike Treanor, Ben Samuel, Aaron A. Reed, Noah Wardrip-Fruin, and Michael Mateas. Prom week. In *Proceedings of the International Conference on the Foundations of Digital Games*, pages 235–237. ACM, 2012.
449. Robert R. McCrae and Paul T. Costa Jr. A five-factor theory of personality. *Handbook of personality: Theory and research*, 2:139–153, 1999.
450. Warren S. McCulloch and Walter Pitts. A logical calculus of the ideas immanent in nervous activity. *Bulletin of Mathematical Biophysics*, 5(4):115–133, 1943.
451. Scott W. McQuiggan, Sunyoung Lee, and James C. Lester. Early prediction of student frustration. In *Proceedings of International Conference on Affective Computing and Intelligent Interaction*, pages 698–709. Springer, 2007.
452. Scott W. McQuiggan, Bradford W. Mott, and James C. Lester. Modeling self-efficacy in intelligent tutoring systems: An inductive approach. *User Modeling and User-Adapted Interaction*, 18(1):81–123, 2008.
453. Andre Mendes, Julian Togelius, and Andy Nealen. Hyper-heuristic general video game playing. In *Computational Intelligence and Games (CIG), 2016 IEEE Conference on*. IEEE, 2016.
454. Daniel S. Messinger, Tricia D. Cassel, Susan I. Acosta, Zara Ambadar, and Jeffrey F. Cohn. Infant smiling dynamics and perceived positive emotion. *Journal of Nonverbal Behavior*, 32(3):133–155, 2008.
455. Angeliki Metallinou and Shrikanth Narayanan. Annotation and processing of continuous emotional attributes: Challenges and opportunities. In *10th IEEE International Conference and Workshops on Automatic Face and Gesture Recognition (FG)*. IEEE, 2013.
456. Zbigniew Michalewicz. Do not kill unfeasible individuals. In *Proceedings of the Fourth Intelligent Information Systems Workshop*, pages 110–123, 1995.
457. Risto Miikkulainen, Bobby D. Bryant, Ryan Cornelius, Igor V. Karpov, Kenneth O. Stanley, and Chern Han Yong. Computational intelligence in games. *Computational Intelligence: Principles and Practice*, pages 155–191, 2006.
458. Benedikte Mikkelsen, Christoffer Holmgård, and Julian Togelius. Ethical Considerations for Player Modeling. In *Proceedings of the AAAI WNAIG workshop*, 2017.

459. Tomas Mikolov, Kai Chen, Greg Corrado, and Jeffrey Dean. Efficient estimation of word representations in vector space. *arXiv preprint arXiv:1301.3781*, 2013.
460. George A. Miller. The magical number seven, plus or minus two: some limits on our capacity for processing information. *Psychological Review*, 63(2):81, 1956.
461. Ian Millington and John Funge. *Artificial intelligence for games*. CRC Press, 2009.
462. Talya Miron-Shatz, Arthur Stone, and Daniel Kahneman. Memories of yesterday's emotions: Does the valence of experience affect the memory-experience gap? *Emotion*, 9(6):885, 2009.
463. Volodymyr Mnih, Adria Puigdomenech Badia, Mehdi Mirza, Alex Graves, Timothy Lillicrap, Tim Harley, David Silver, and Koray Kavukcuoglu. Asynchronous methods for deep reinforcement learning. In *International Conference on Machine Learning*, pages 1928–1937, 2016.
464. Volodymyr Mnih, Koray Kavukcuoglu, David Silver, Andrei A. Rusu, Joel Veness, Marc G. Bellemare, Alex Graves, Martin Riedmiller, Andreas K. Fidjeland, Georg Ostrovski, Stig Petersen, Charles Beattie, Amir Sadik, Ioannis Antonoglou, Helen King, Dharshan Kumaran, Daan Wierstra, Shane Legg, and Demis Hassabis. Human-level control through deep reinforcement learning. *Nature*, 518(7540):529–533, 2015.
465. Mathew Monfort, Matthew Johnson, Aude Oliva, and Katja Hofmann. Asynchronous data aggregation for training end to end visual control networks. In *Proceedings of the 16th Conference on Autonomous Agents and Multi-Agent Systems*, pages 530–537. International Foundation for Autonomous Agents and Multiagent Systems, May 2017.
466. Nick Montfort and Ian Bogost. *Racing the beam: The Atari video computer system*. MIT Press, 2009.
467. Matej Moravčík, Martin Schmid, Neil Burch, Viliam Lisý, Dustin Morrill, Nolan Bard, Trevor Davis, Kevin Waugh, Michael Johanson, and Michael Bowling. Deepstack: Expert-level artificial intelligence in no-limit poker. *arXiv preprint arXiv:1701.01724*, 2017.
468. Jon D. Morris. Observations: SAM: The self-assessment manikin—An efficient cross-cultural measurement of emotional response. *Journal of Advertising Research*, 35(6):63–68, 1995.
469. Jorge Munoz, Georgios N. Yannakakis, Fiona Mulvey, Dan Witzner Hansen, German Gutierrez, and Araceli Sanchis. Towards gaze-controlled platform games. In *Computational Intelligence and Games (CIG), 2011 IEEE Conference on*, pages 47–54. IEEE, 2011.
470. Hector Munoz-Avila, Christian Bauckhage, Michal Bida, Clare Bates Congdon, and Graham Kendall. Learning and Game AI. *Dagstuhl Follow-Ups*, 6, 2013.
471. Roger B. Myerson. Game theory: analysis of conflict. 1991. *Cambridge: Mass, Harvard University*.
472. Roger B. Myerson. *Game theory*. Harvard University Press, 2013.
473. Lennart Nacke and Craig A. Lindley. Flow and immersion in first-person shooters: measuring the player's gameplay experience. In *Proceedings of the 2008 Conference on Future Play: Research, Play, Share*, pages 81–88. ACM, 2008.
474. Frederik Nagel, Reinhard Kopiez, Oliver Grewe, and Eckart Altenmüller. Emujoy: Software for continuous measurement of perceived emotions in music. *Behavior Research Methods*, 39(2):283–290, 2007.
475. Karthik Narasimhan, Tejas Kulkarni, and Regina Barzilay. Language understanding for text-based games using deep reinforcement learning. *arXiv preprint arXiv:1506.08941*, 2015.
476. Alexander Nareyek. Intelligent agents for computer games. In T.A. Marsland and I. Frank, editors, *Computers and Games, Second International Conference, CG 2002*, pages 414–422, 2002.
477. Alexander Nareyek. Game AI is dead. Long live game AI! *IEEE Intelligent Systems*, (1):9–11, 2007.
478. John F. Nash. Equilibrium points in n-person games. In *Proceedings of the National Academy of Sciences*, number 1, pages 48–49, 1950.
479. Steve Nebel, Sascha Schneider, and Günter Daniel Rey. Mining learning and crafting scientific experiments: a literature review on the use of Minecraft in education and research. *Journal of Educational Technology & Society*, 19(2):355, 2016.

480. Graham Nelson. Natural language, semantic analysis, and interactive fiction. *IF Theory Reader*, 141, 2006.
481. Mark J. Nelson. Game Metrics Without Players: Strategies for Understanding Game Artifacts. In *AIIDE Workshop on Artificial Intelligence in the Game Design Process*, 2011.
482. Mark J. Nelson, Simon Colton, Edward J. Powley, Swen E. Gaudl, Peter Ivey, Rob Saunders, Blanca Pérez Ferrer, and Michael Cook. Mixed-initiative approaches to on-device mobile game design. In *Proceedings of the CHI Workshop on Mixed-Initiative Creative Interfaces*, 2017.
483. Mark J. Nelson and Michael Mateas. Search-Based Drama Management in the Interactive Fiction Anchorhead. In *Proceedings of the First Artificial Intelligence and Interactive Digital Entertainment Conference*, pages 99–104, 2005.
484. Mark J. Nelson and Michael Mateas. An interactive game-design assistant. In *Proceedings of the 13th International Conference on Intelligent User Interfaces*, pages 90–98, 2008.
485. Mark J. Nelson and Adam M. Smith. ASP with applications to mazes and levels. In *Procedural Content Generation in Games*, pages 143–157. Springer, 2016.
486. Mark J. Nelson, Julian Togelius, Cameron Browne, and Michael Cook. Rules and mechanics. In *Procedural Content Generation in Games*, pages 99–121. Springer, 2016.
487. John Von Neumann. *Theory of Self-Reproducing Automata*. University of Illinois Press, Champaign, IL, USA, 1966.
488. Truong-Huy D. Nguyen, Shree Subramanian, Magy Seif El-Nasr, and Alessandro Canossa. Strategy Detection in Wuzzit: A Decision Theoretic Approach. In *International Conference on Learning Science—Workshop on Learning Analytics for Learning and Becoming a Practice*, 2014.
489. Jakob Nielsen. Usability 101: Introduction to usability, 2003. Available at http://www.useit.com/alertbox/20030825.html.
490. Jon Lau Nielsen, Benjamin Fedder Jensen, Tobias Mahlmann, Julian Togelius, and Georgios N. Yannakakis. AI for General Strategy Game Playing. *Handbook of Digital Games*, pages 274–304, 2014.
491. Thorbjørn S. Nielsen, Gabriella A. B. Barros, Julian Togelius, and Mark J. Nelson. General Video Game Evaluation Using Relative Algorithm Performance Profiles. In *Applications of Evolutionary Computation*, pages 369–380. Springer, 2015.
492. Thorbjørn S. Nielsen, Gabriella A. B. Barros, Julian Togelius, and Mark J. Nelson. Towards generating arcade game rules with VGDL. In *Proceedings of the 2015 IEEE Conference on Computational Intelligence and Games*, 2015.
493. Anton Nijholt. BCI for games: A state of the art survey. In *Entertainment Computing-ICEC 2008*, pages 225–228. Springer, 2009.
494. Nils J. Nilsson. Shakey the robot. Technical report, DTIC Document, 1984.
495. Kai Ninomiya, Mubbasir Kapadia, Alexander Shoulson, Francisco Garcia, and Norman Badler. Planning approaches to constraint-aware navigation in dynamic environments. *Computer Animation and Virtual Worlds*, 26(2):119–139, 2015.
496. Stefano Nolfi and Dario Floreano. *Evolutionary robotics: The biology, intelligence, and technology of self-organizing machines*. MIT Press, 2000.
497. David G. Novick and Stephen Sutton. What is mixed-initiative interaction. In *Proceedings of the AAAI Spring Symposium on Computational Models for Mixed Initiative Interaction*, pages 114–116, 1997.
498. Gabriela Ochoa. On genetic algorithms and Lindenmayer systems. In *Parallel Problem Solving from Nature—PPSN V*, pages 335–344. Springer, 1998.
499. Jacob Kaae Olesen, Georgios N. Yannakakis, and John Hallam. Real-time challenge balance in an RTS game using rtNEAT. In *Computational Intelligence and Games, 2008. CIG'08. IEEE Symposium On*, pages 87–94. IEEE, 2008.
500. Jacob Olsen. Realtime procedural terrain generation. 2004.
501. Peter Thorup Ølsted, Benjamin Ma, and Sebastian Risi. Interactive evolution of levels for a competitive multiplayer FPS. In *Evolutionary Computation (CEC), 2015 IEEE Congress on*, pages 1527–1534. IEEE, 2015.

502. Cathy O'Neil. *Weapons of math destruction: How big data increases inequality and threatens democracy*. Crown Publishing Group (NY), 2016.
503. Santiago Ontañón. The combinatorial multi-armed bandit problem and its application to real-time strategy games. In *Ninth Artificial Intelligence and Interactive Digital Entertainment Conference*, 2013.
504. Santiago Ontañón, Gabriel Synnaeve, Alberto Uriarte, Florian Richoux, David Churchill, and Mike Preuss. A survey of real-time strategy game AI research and competition in StarCraft. *IEEE Transactions on Computational Intelligence and AI in Games*, 5(4):293–311, 2013.
505. Santiago Ontañón, Gabriel Synnaeve, Alberto Uriarte, Florian Richoux, David Churchill, and Mike Preuss. RTS AI: Problems and Techniques. In *Encyclopedia of Computer Graphics and Games*. Springer, 2015.
506. Jeff Orkin. Applying goal-oriented action planning to games. *AI game programming wisdom*, 2:217–228, 2003.
507. Jeff Orkin. Three states and a plan: the AI of F.E.A.R. In *Game Developers Conference*, 2006.
508. Jeff Orkin and Deb Roy. The restaurant game: Learning social behavior and language from thousands of players online. *Journal of Game Development*, 3(1):39–60, 2007.
509. Mauricio Orozco, Juan Silva, Abdulmotaleb El Saddik, and Emil Petriu. The role of haptics in games. In *Haptics Rendering and Applications*. InTech, 2012.
510. Brian O'Neill and Mark Riedl. Emotion-driven narrative generation. In *Emotion in Games: Theory and Praxis*, pages 167–180. Springer, 2016.
511. Juan Ortega, Noor Shaker, Julian Togelius, and Georgios N. Yannakakis. Imitating human playing styles in Super Mario Bros. *Entertainment Computing*, 4(2):93–104, 2013.
512. Andrew Ortony, Gerald L. Clore, and Allan Collins. *The cognitive structure of emotions*. Cambridge University Press, 1990.
513. Martin J. Osborne. *An introduction to game theory*. Oxford University Press, 2004.
514. Alexander Osherenko. *Opinion Mining and Lexical Affect Sensing. Computer-aided analysis of opinions and emotions in texts*. PhD thesis, University of Augsburg, 2010.
515. Seth Ovadia. Ratings and rankings: Reconsidering the structure of values and their measurement. *International Journal of Social Research Methodology*, 7(5):403–414, 2004.
516. Ana Paiva, Joao Dias, Daniel Sobral, Ruth Aylett, Polly Sobreperez, Sarah Woods, Carsten Zoll, and Lynne Hall. Caring for agents and agents that care: Building empathic relations with synthetic agents. In *Proceedings of the Third International Joint Conference on Autonomous Agents and Multiagent Systems*, pages 194–201. IEEE Computer Society, 2004.
517. Bo Pang and Lillian Lee. Opinion mining and sentiment analysis. *Foundations and Trends in Information Retrieval*, 2(1–2):1–135, 2008.
518. Matt Parker and Bobby D. Bryant. Visual control in Quake II with a cyclic controller. In *Computational Intelligence and Games, 2008. CIG'08. IEEE Symposium On*, pages 151–158. IEEE, 2008.
519. Matt Parker and Bobby D. Bryant. Neurovisual control in the Quake II environment. *IEEE Transactions on Computational Intelligence and AI in Games*, 4(1):44–54, 2012.
520. Chris Pedersen, Julian Togelius, and Georgios N. Yannakakis. Modeling Player Experience in Super Mario Bros. In *Proceedings of the IEEE Symposium on Computational Intelligence and Games*, pages 132–139. IEEE, 2009.
521. Chris Pedersen, Julian Togelius, and Georgios N. Yannakakis. Modeling Player Experience for Content Creation. *IEEE Transactions on Computational Intelligence and AI in Games*, 2(1):54–67, 2010.
522. Barney Pell. *Strategy generation and evaluation for meta-game playing*. PhD thesis, University of Cambridge, 1993.
523. Peng Peng, Quan Yuan, Ying Wen, Yaodong Yang, Zhenkun Tang, Haitao Long, and Jun Wang. Multiagent Bidirectionally-Coordinated Nets for Learning to Play StarCraft Combat Games. *arXiv preprint arXiv:1703.10069*, 2017.
524. Tom Pepels, Mark H. M. Winands, and Marc Lanctot. Real-time Monte Carlo tree search in Ms Pac-Man. *IEEE Transactions on Computational Intelligence and AI in Games*, 6(3):245–257, 2014.

525. Diego Perez, Edward J. Powley, Daniel Whitehouse, Philipp Rohlfshagen, Spyridon Samothrakis, Peter I. Cowling, and Simon M. Lucas. Solving the physical traveling salesman problem: Tree search and macro actions. *IEEE Transactions on Computational Intelligence and AI in Games*, 6(1):31–45, 2014.
526. Diego Perez, Spyridon Samothrakis, Simon Lucas, and Philipp Rohlfshagen. Rolling horizon evolution versus tree search for navigation in single-player real-time games. In *Proceedings of the 15th Annual Conference on Genetic and Evolutionary Computation*, pages 351–358. ACM, 2013.
527. Diego Perez-Liebana, Spyridon Samothrakis, Julian Togelius, Tom Schaul, and Simon M. Lucas. General video game AI: Competition, challenges and opportunities. In *Proceedings of the Thirtieth AAAI Conference on Artificial Intelligence*, 2016.
528. Diego Perez-Liebana, Spyridon Samothrakis, Julian Togelius, Tom Schaul, Simon M. Lucas, Adrien Couëtoux, Jerry Lee, Chong-U Lim, and Tommy Thompson. The 2014 general video game playing competition. *IEEE Transactions on Computational Intelligence and AI in Games*, 8(3):229–243, 2016.
529. Ken Perlin. An image synthesizer. *ACM SIGGRAPH Computer Graphics*, 19(3):287–296, 1985.
530. Rosalind W. Picard. *Affective Computing*. MIT Press, Cambridge, MA, 1997.
531. Grant Pickett, Foaad Khosmood, and Allan Fowler. Automated generation of conversational non player characters. In *Eleventh Artificial Intelligence and Interactive Digital Entertainment Conference*, 2015.
532. Michele Pirovano. The use of Fuzzy Logic for Artificial Intelligence in Games. Technical report, University of Milano, Milano, 2012.
533. Jacques Pitrat. Realization of a general game-playing program. In *IFIP congress (2)*, pages 1570–1574, 1968.
534. Isabella Poggi, Catherine Pelachaud, Fiorella de Rosis, Valeria Carofiglio, and Berardina De Carolis. GRETA. A believable embodied conversational agent. In *Multimodal intelligent information presentation*, pages 3–25. Springer, 2005.
535. Mihai Polceanu. Mirrorbot: Using human-inspired mirroring behavior to pass a Turing test. In *Computational Intelligence in Games (CIG), 2013 IEEE Conference on*. IEEE, 2013.
536. Riccardo Poli, William B. Langdon, and Nicholas F. McPhee. *A field guide to genetic programming*. 2008. Published via http://lulu.com and freely available at http://www.gp-field-guide.org.uk (With contributions by J. R. Koza).
537. Jordan B. Pollack and Alan D. Blair. Co-evolution in the successful learning of backgammon strategy. *Machine learning*, 32(3):225–240, 1998.
538. Jordan B. Pollack, Alan D. Blair, and Mark Land. Coevolution of a backgammon player. In *Artificial Life V: Proceedings of the Fifth International Workshop on the Synthesis and Simulation of Living Systems*, pages 92–98. Cambridge, MA: The MIT Press, 1997.
539. Jonathan Posner, James A. Russell, and Bradley S. Peterson. The circumplex model of affect: An integrative approach to affective neuroscience, cognitive development, and psychopathology. *Development and psychopathology*, 17(03):715–734, 2005.
540. David Premack and Guy Woodruff. Does the chimpanzee have a theory of mind? *Behavioral and brain sciences*, 1(04):515–526, 1978.
541. Mike Preuss, Daniel Kozakowski, Johan Hagelbäck, and Heike Trautmann. Reactive strategy choice in StarCraft by means of Fuzzy Control. In *Computational Intelligence in Games (CIG), 2013 IEEE Conference on*. IEEE, 2013.
542. Przemyslaw Prusinkiewicz and Aristid Lindenmayer. *The algorithmic beauty of plants*. Springer, 1990.
543. Jan Quadflieg, Mike Preuss, and Günter Rudolph. Driving as a human: a track learning based adaptable architecture for a car racing controller. *Genetic Programming and Evolvable Machines*, 15(4):433–476, 2014.
544. J. Ross Quinlan. Induction of decision trees. *Machine Learning*, 1(1):81–106, 1986.
545. J. Ross Quinlan. *C4. 5: programs for machine learning*. Elsevier, 2014.
546. Steve Rabin. *AI Game Programming Wisdom*. Charles River Media, Inc., 2002.

547. Steve Rabin. *AI Game Programming Wisdom 2*. Charles River Media, Inc., 2003.
548. Steve Rabin. *AI Game Programming Wisdom 3*. Charles River Media, Inc., 2006.
549. Steve Rabin. *AI Game Programming Wisdom 4*. Nelson Education, 2014.
550. Steve Rabin and Nathan Sturtevant. Pathfinding Architecture Optimizations. In *Game AI Pro: Collected Wisdom of Game AI Professionals*. CRC Press, 2013.
551. Steve Rabin and Nathan Sturtevant. Combining Bounding Boxes and JPS to Prune Grid Pathfinding. In *AAAI Conference on Artificial Intelligence*, 2016.
552. Steven Rabin. *Game AI Pro: Collected Wisdom of Game AI Professionals*. CRC Press, 2013.
553. Steven Rabin. *Game AI Pro 2: Collected Wisdom of Game AI Professionals*. CRC Press, 2015.
554. William L. Raffe, Fabio Zambetta, and Xiaodong Li. A survey of procedural terrain generation techniques using evolutionary algorithms. In *IEEE Congress on Evolutionary Computation (CEC)*. IEEE, 2012.
555. Judith Ramey, Ted Boren, Elisabeth Cuddihy, Joe Dumas, Zhiwei Guan, Maaike J. van den Haak, and Menno D. T. De Jong. Does think aloud work? How do we know? In *CHI'06 Extended Abstracts on Human Factors in Computing Systems*, pages 45–48. ACM, 2006.
556. Pramila Rani, Nilanjan Sarkar, and Changchun Liu. Maintaining optimal challenge in computer games through real-time physiological feedback. In *Proceedings of the 11th International Conference on Human Computer Interaction*, pages 184–192, 2005.
557. Jakob Rasmussen. *Are Behavior Trees a Thing of the Past?* Gamasutra, 2016.
558. Niklas Ravaja, Timo Saari, Mikko Salminen, Jari Laarni, and Kari Kallinen. Phasic emotional reactions to video game events: A psychophysiological investigation. *Media Psychology*, 8(4):343–367, 2006.
559. Genaro Rebolledo-Mendez, Ian Dunwell, Erika Martínez-Mirón, Maria Dolores Vargas-Cerdán, Sara De Freitas, Fotis Liarokapis, and Alma R. García-Gaona. Assessing Neurosky's usability to detect attention levels in an assessment exercise. *Human-Computer Interaction. New Trends*, pages 149–158, 2009.
560. Jochen Renz, Xiaoyu Ge, Stephen Gould, and Peng Zhang. The Angry Birds AI Competition. *AI Magazine*, 36(2):85–87, 2015.
561. Antonio Ricciardi and Patrick Thill. Adaptive AI for Fighting Games. Technical report, Stanford University, 2008.
562. Mark O. Riedl and Vadim Bulitko. Interactive narrative: An intelligent systems approach. *AI Magazine*, 34(1):67, 2012.
563. Mark O. Riedl and Andrew Stern. Believable agents and intelligent story adaptation for interactive storytelling. *Technologies for Interactive Digital Storytelling and Entertainment*, pages 1–12, 2006.
564. Mark O. Riedl and Alexander Zook. AI for game production. In *IEEE Conference on Computational Intelligence in Games (CIG)*. IEEE, 2013.
565. Sebastian Risi, Joel Lehman, David B. D'Ambrosio, Ryan Hall, and Kenneth O. Stanley. Combining Search-Based Procedural Content Generation and Social Gaming in the Petalz Video Game. In *Proceedings of AIIDE*, 2012.
566. Sebastian Risi, Joel Lehman, David B. D'Ambrosio, Ryan Hall, and Kenneth O. Stanley. Petalz: Search-based procedural content generation for the casual gamer. *IEEE Transactions on Computational Intelligence and AI in Games*, 8(3):244–255, 2016.
567. Sebastian Risi and Julian Togelius. Neuroevolution in games: State of the art and open challenges. *IEEE Transactions on Computational Intelligence and AI in Games*, 9(1):25–41, 2017.
568. David L. Roberts, Harikrishna Narayanan, and Charles L. Isbell. Learning to influence emotional responses for interactive storytelling. In *Proceedings of the 2009 AAAI Symposium on Intelligent Narrative Technologies II*, 2009.
569. Glen Robertson and Ian D. Watson. A review of real-time strategy game AI. *AI Magazine*, 35(4):75–104, 2014.
570. Glen Robertson and Ian D. Watson. An Improved Dataset and Extraction Process for StarCraft AI. In *FLAIRS Conference*, 2014.

571. Michael D. Robinson and Gerald L. Clore. Belief and feeling: evidence for an accessibility model of emotional self-report. *Psychological Bulletin*, 128(6):934, 2002.
572. Jennifer Robison, Scott McQuiggan, and James Lester. Evaluating the consequences of affective feedback in intelligent tutoring systems. In *Proceedings of International Conference on Affective Computing and Intelligent Interaction (ACII)*. IEEE, 2009.
573. Philipp Rohlfshagen, Jialin Liu, Diego Perez-Liebana, and Simon M. Lucas. Pac-Man Conquers Academia: Two Decades of Research Using a Classic Arcade Game. *IEEE Transactions on Computational Intelligence and AI in Games*, 2017.
574. Philipp Rohlfshagen and Simon M. Lucas. Ms Pac-Man versus Ghost team CEC 2011 competition. In *IEEE Congress on Evolutionary Computation (CEC)*, pages 70–77. IEEE, 2011.
575. Edmund T. Rolls. The orbitofrontal cortex and reward. *Cerebral Cortex*, 10(3):284–294, 2000.
576. Frank Rosenblatt. The perceptron: a probabilistic model for information storage and organization in the brain. *Psychological Review*, 65(6):386, 1958.
577. Jonathan Rowe, Bradford Mott, Scott McQuiggan, Jennifer Robison, Sunyoung Lee, and James Lester. Crystal Island: A narrative-centered learning environment for eighth grade microbiology. In *Workshop on Intelligent Educational Games at the 14th International Conference on Artificial Intelligence in Education, Brighton, UK*, pages 11–20, 2009.
578. Jonathan P. Rowe, Lucy R. Shores, Bradford W. Mott, and James C. Lester. Integrating learning, problem solving, and engagement in narrative-centered learning environments. *International Journal of Artificial Intelligence in Education*, 21(1-2):115–133, 2011.
579. David E. Rumelhart, Geoffrey E. Hinton, and Ronald J. Williams. Learning representations by back-propagating errors. *Nature*, 323(6088):533–536, 1986.
580. Thomas Philip Runarsson and Simon M. Lucas. Coevolution versus self-play temporal difference learning for acquiring position evaluation in small-board Go. *IEEE Transactions on Evolutionary Computation*, 9(6):628–640, 2005.
581. James A. Russell. A circumplex model of affect. *Journal of Personality and Social Psychology*, 39(6):1161, 1980.
582. Stuart Russell and Peter Norvig. *Artificial Intelligence: A Modern Approach*. Prentice-Hall, Englewood Cliffs, 1995.
583. Richard M. Ryan, C. Scott Rigby, and Andrew Przybylski. The motivational pull of video games: A self-determination theory approach. *Motivation and emotion*, 30(4):344–360, 2006.
584. Jennifer L. Sabourin and James C. Lester. Affect and engagement in Game-Based Learning environments. *IEEE Transactions on Affective Computing*, 5(1):45–56, 2014.
585. Owen Sacco, Antonios Liapis, and Georgios N. Yannakakis. A holistic approach for semantic-based game generation. In *Computational Intelligence and Games (CIG), 2016 IEEE Conference on*. IEEE, 2016.
586. Frantisek Sailer, Michael Buro, and Marc Lanctot. Adversarial planning through strategy simulation. In *Computational Intelligence and Games, 2007. CIG 2007. IEEE Symposium on*, pages 80–87. IEEE, 2007.
587. Katie Salen and Eric Zimmerman. *Rules of play: Game design fundamentals*. MIT Press, 2004.
588. Christoph Salge, Christian Lipski, Tobias Mahlmann, and Brigitte Mathiak. Using genetically optimized artificial intelligence to improve gameplaying fun for strategical games. In *Sandbox '08: Proceedings of the 2008 ACM SIGGRAPH symposium on Video games*, pages 7–14, New York, NY, USA, 2008. ACM.
589. Spyridon Samothrakis, Simon M. Lucas, Thomas Philip Runarsson, and David Robles. Coevolving game-playing agents: Measuring performance and intransitivities. *Evolutionary Computation, IEEE Transactions on*, 17(2):213–226, 2013.
590. Spyridon Samothrakis, David Robles, and Simon M. Lucas. Fast approximate max-n Monte Carlo tree search for Ms Pac-Man. *IEEE Transactions on Computational Intelligence and AI in Games*, 3(2):142–154, 2011.
591. Arthur L. Samuel. Some studies in machine learning using the game of Checkers. *IBM Journal of research and development*, 3(3):210–229, 1959.

592. Frederik Schadd, Sander Bakkes, and Pieter Spronck. Opponent modeling in real-time strategy games. In *GAMEON*, pages 61–70, 2007.
593. Jonathan Schaeffer, Neil Burch, Yngvi Björnsson, Akihiro Kishimoto, Martin Müller, Robert Lake, Paul Lu, and Steve Sutphen. Checkers is solved. *Science*, 317(5844):1518–1522, 2007.
594. Jonathan Schaeffer, Robert Lake, Paul Lu, and Martin Bryant. Chinook: the world man-machine Checkers champion. *AI Magazine*, 17(1):21, 1996.
595. Jost Schatzmann, Karl Weilhammer, Matt Stuttle, and Steve Young. A survey of statistical user simulation techniques for reinforcement-learning of dialogue management strategies. *Knowledge Engineering Review*, 21(2):97–126, 2006.
596. Tom Schaul. A video game description language for model-based or interactive learning. In *Computational Intelligence in Games (CIG), 2013 IEEE Conference on*. IEEE, 2013.
597. Tom Schaul. An extensible description language for video games. *IEEE Transactions on Computational Intelligence and AI in Games*, 6(4):325–331, 2014.
598. Tom Schaul, Julian Togelius, and Jürgen Schmidhuber. Measuring intelligence through games. *arXiv preprint arXiv:1109.1314*, 2011.
599. Jesse Schell. *The Art of Game Design: A book of lenses*. CRC Press, 2014.
600. Klaus R. Scherer. What are emotions? and how can they be measured? *Social Science Information*, 44(4):695–729, 2005.
601. Klaus R. Scherer, Angela Schorr, and Tom Johnstone. *Appraisal processes in emotion: Theory, methods, research*. Oxford University Press, 2001.
602. Jürgen Schmidhuber. Developmental robotics, optimal artificial curiosity, creativity, music, and the fine arts. *Connection Science*, 18(2):173–187, 2006.
603. Jacob Schrum, Igor V. Karpov, and Risto Miikkulainen. UT^2: Human-like behavior via neuroevolution of combat behavior and replay of human traces. In *Computational Intelligence and Games (CIG), 2011 IEEE Conference on*, pages 329–336. IEEE, 2011.
604. Brian Schwab. *AI game engine programming*. Nelson Education, 2009.
605. Brian Schwab, Dave Mark, Kevin Dill, Mike Lewis, and Richard Evans. GDC: Turing tantrums: AI developers rant, 2011.
606. Marco Scirea, Yun-Gyung Cheong, Mark J. Nelson, and Byung-Chull Bae. Evaluating musical foreshadowing of videogame narrative experiences. In *Proceedings of the 9th Audio Mostly: A Conference on Interaction With Sound*. ACM, 2014.
607. Ben Seymour and Samuel M. McClure. Anchors, scales and the relative coding of value in the brain. *Current Opinion in Neurobiology*, 18(2):173–178, 2008.
608. Mohammad Shaker, Mhd Hasan Sarhan, Ola Al Naameh, Noor Shaker, and Julian Togelius. Automatic generation and analysis of physics-based puzzle games. In *Computational Intelligence in Games (CIG), 2013 IEEE Conference on*. IEEE, 2013.
609. Noor Shaker, Stylianos Asteriadis, Georgios N. Yannakakis, and Kostas Karpouzis. A game-based corpus for analysing the interplay between game context and player experience. In *Affective Computing and Intelligent Interaction*, pages 547–556. Springer, 2011.
610. Noor Shaker, Stylianos Asteriadis, Georgios N. Yannakakis, and Kostas Karpouzis. Fusing visual and behavioral cues for modeling user experience in games. *Cybernetics, IEEE Transactions on*, 43(6):1519–1531, 2013.
611. Noor Shaker, Miguel Nicolau, Georgios N. Yannakakis, Julian Togelius, and Michael O'Neil. Evolving levels for Super Mario Bros using grammatical evolution. In *IEEE Conference on Computational Intelligence and Games*, pages 304–311. IEEE, 2012.
612. Noor Shaker, Mohammad Shaker, and Mohamed Abou-Zleikha. Towards generic models of player experience. In *Proceedings, the Eleventh AAAI Conference on Artificial Intelligence and Interactive Digital Entertainment*. AAAI Press, 2015.
613. Noor Shaker, Mohammad Shaker, and Julian Togelius. Evolving Playable Content for Cut the Rope through a Simulation-Based Approach. In *AIIDE*, 2013.
614. Noor Shaker, Mohammad Shaker, and Julian Togelius. Ropossum: An Authoring Tool for Designing, Optimizing and Solving Cut the Rope Levels. In *AIIDE*, 2013.
615. Noor Shaker, Gillian Smith, and Georgios N. Yannakakis. Evaluating content generators. In *Procedural Content Generation in Games*, pages 215–224. Springer, 2016.

616. Noor Shaker, Julian Togelius, and Mark J. Nelson, editors. *Procedural Content Generation in Games*. Springer, 2016.
617. Noor Shaker, Julian Togelius, and Georgios N. Yannakakis. Towards Automatic Personalized Content Generation for Platform Games. In *Proceedings of the AAAI Conference on Artificial Intelligence and Interactive Digital Entertainment (AIIDE)*. AAAI Press, October 2010.
618. Noor Shaker, Julian Togelius, and Georgios N. Yannakakis. The experience-driven perspective. In *Procedural Content Generation in Games*, pages 181–194. Springer, 2016.
619. Noor Shaker, Julian Togelius, Georgios N. Yannakakis, Likith Poovanna, Vinay S. Ethiraj, Stefan J. Johansson, Robert G. Reynolds, Leonard K. Heether, Tom Schumann, and Marcus Gallagher. The Turing test track of the 2012 Mario AI championship: entries and evaluation. In *Computational Intelligence in Games (CIG), 2013 IEEE Conference on*. IEEE, 2013.
620. Noor Shaker, Julian Togelius, Georgios N. Yannakakis, Ben Weber, Tomoyuki Shimizu, Tomonori Hashiyama, Nathan Sorenson, Philippe Pasquier, Peter Mawhorter, Glen Takahashi, Gillian Smith, and Robin Baumgarten. The 2010 Mario AI championship: Level generation track. *Computational Intelligence and AI in Games, IEEE Transactions on*, 3(4):332–347, 2011.
621. Noor Shaker, Georgios N. Yannakakis, and Julian Togelius. Crowdsourcing the aesthetics of platform games. *Computational Intelligence and AI in Games, IEEE Transactions on*, 5(3):276–290, 2013.
622. Amirhosein Shantia, Eric Begue, and Marco Wiering. Connectionist reinforcement learning for intelligent unit micro management in StarCraft. In *Neural Networks (IJCNN), The 2011 International Joint Conference on*, pages 1794–1801. IEEE, 2011.
623. Manu Sharma, Manish Mehta, Santiago Ontañón, and Ashwin Ram. Player modeling evaluation for interactive fiction. In *Proceedings of the AIIDE 2007 Workshop on Optimizing Player Satisfaction*, pages 19–24, 2007.
624. Nandita Sharma and Tom Gedeon. Objective measures, sensors and computational techniques for stress recognition and classification: A survey. *Computer methods and programs in biomedicine*, 108(3):1287–1301, 2012.
625. Peter Shizgal and Andreas Arvanitogiannis. Gambling on dopamine. *Science*, 299(5614):1856–1858, 2003.
626. Yoav Shoham and Kevin Leyton-Brown. *Multiagent systems: Algorithmic, game-theoretic, and logical foundations*. Cambridge University Press, 2008.
627. Alexander Shoulson, Francisco M. Garcia, Matthew Jones, Robert Mead, and Norman I. Badler. Parameterizing behavior trees. In *International Conference on Motion in Games*, pages 144–155. Springer, 2011.
628. Nikolaos Sidorakis, George Alex Koulieris, and Katerina Mania. Binocular eye-tracking for the control of a 3D immersive multimedia user interface. In *Everyday Virtual Reality (WEVR), 2015 IEEE 1st Workshop on*, pages 15–18. IEEE, 2015.
629. David Silver, Aja Huang, Chris J. Maddison, Arthur Guez, Laurent Sifre, George van Den Driessche, Julian Schrittwieser, Ioannis Antonoglou, Veda Panneershelvam, Marc Lanctot, et al. Mastering the game of Go with deep neural networks and tree search. *Nature*, 529(7587):484–489, 2016.
630. Herbert A. Simon. A behavioral model of rational choice. *The quarterly journal of economics*, 69(1):99–118, 1955.
631. Shawn Singh, Mubbasir Kapadia, Glenn Reinman, and Petros Faloutsos. Footstep navigation for dynamic crowds. *Computer Animation and Virtual Worlds*, 22(2-3):151–158, 2011.
632. Moshe Sipper. *Evolved to Win*. Lulu.com, 2011.
633. Burrhus Frederic Skinner. *The behavior of organisms: An experimental analysis*. BF Skinner Foundation, 1990.
634. Ruben M. Smelik, Tim Tutenel, Klaas Jan de Kraker, and Rafael Bidarra. Interactive creation of virtual worlds using procedural sketching. In *Proceedings of Eurographics*, 2010.
635. Adam M. Smith, Erik Andersen, Michael Mateas, and Zoran Popović. A case study of expressively constrainable level design automation tools for a puzzle game. In *Proceedings of the International Conference on the Foundations of Digital Games*, pages 156–163. ACM, 2012.

636. Adam M. Smith, Chris Lewis, Kenneth Hullett, Gillian Smith, and Anne Sullivan. An inclusive taxonomy of player modeling. Technical Report UCSC-SOE-11-13, University of California, Santa Cruz, 2011.
637. Adam M. Smith and Michael Mateas. Variations forever: Flexibly generating rulesets from a sculptable design space of mini-games. In *Computational Intelligence and Games (CIG), 2010 IEEE Symposium on*, pages 273–280. IEEE, 2010.
638. Adam M. Smith and Michael Mateas. Answer set programming for procedural content generation: A design space approach. *Computational Intelligence and AI in Games, IEEE Transactions on*, 3(3):187–200, 2011.
639. Adam M. Smith, Mark J. Nelson, and Michael Mateas. Ludocore: A logical game engine for modeling videogames. In *Computational Intelligence and Games (CIG), 2010 IEEE Symposium on*, pages 91–98. IEEE, 2010.
640. Gillian Smith and Jim Whitehead. Analyzing the expressive range of a level generator. In *Proceedings of the 2010 Workshop on Procedural Content Generation in Games*. ACM, 2010.
641. Gillian Smith, Jim Whitehead, and Michael Mateas. Tanagra: A mixed-initiative level design tool. In *Proceedings of the Fifth International Conference on the Foundations of Digital Games*, pages 209–216. ACM, 2010.
642. Gillian Smith, Jim Whitehead, and Michael Mateas. Tanagra: Reactive planning and constraint solving for mixed-initiative level design. *Computational Intelligence and AI in Games, IEEE Transactions on*, 3(3):201–215, 2011.
643. Ian Sneddon, Gary McKeown, Margaret McRorie, and Tijana Vukicevic. Cross-cultural patterns in dynamic ratings of positive and negative natural emotional behaviour. *PloS ONE*, 6(2), 2011.
644. Sam Snodgrass and Santiago Ontañón. A Hierarchical MdMC Approach to 2D Video Game Map Generation. In *Eleventh Artificial Intelligence and Interactive Digital Entertainment Conference*, 2015.
645. Dennis Soemers. Tactical planning using MCTS in the game of StarCraft, 2014. Bachelor Thesis, Department of Knowledge Engineering, Maastricht University.
646. Andreas Sonderegger, Andreas Uebelbacher, Manuela Pugliese, and Juergen Sauer. The influence of aesthetics in usability testing: the case of dual-domain products. In *Proceedings of the Conference on Human Factors in Computing Systems*, pages 21–30, 2014.
647. Bhuman Soni and Philip Hingston. Bots trained to play like a human are more fun. In *IEEE International Joint Conference on Neural Networks (IJCNN); IEEE World Congress on Computational Intelligence*, pages 363–369. IEEE, 2008.
648. Patrikk D. Sørensen, Jeppeh M. Olsen, and Sebastian Risi. Interactive Super Mario Bros Evolution. In *Proceedings of the 2016 Genetic and Evolutionary Computation Conference*, pages 41–42. ACM, 2016.
649. Nathan Sorenson and Philippe Pasquier. Towards a generic framework for automated video game level creation. *Applications of Evolutionary Computation*, pages 131–140, 2010.
650. Pieter Spronck, Marc Ponsen, Ida Sprinkhuizen-Kuyper, and Eric Postma. Adaptive game AI with dynamic scripting. *Machine Learning*, 63(3):217–248, 2006.
651. Pieter Spronck, Ida Sprinkhuizen-Kuyper, and Eric Postma. Difficulty scaling of game AI. In *Proceedings of the 5th International Conference on Intelligent Games and Simulation (GAME-ON 2004)*, pages 33–37, 2004.
652. Ramakrishnan Srikant and Rakesh Agrawal. Mining sequential patterns: Generalizations and performance improvements. In *International Conference on Extending Database Technology*, pages 1–17. Springer, 1996.
653. Kenneth O. Stanley. Compositional Pattern Producing Networks: A novel abstraction of development. *Genetic Programming and Evolvable Machines*, 8(2):131–162, 2007.
654. Kenneth O. Stanley, Bobby D. Bryant, and Risto Miikkulainen. Real-time neuroevolution in the NERO video game. *Evolutionary Computation, IEEE Transactions on*, 9(6):653–668, 2005.
655. Kenneth O. Stanley and Risto Miikkulainen. Evolving neural networks through augmenting topologies. *Evolutionary Computation*, 10(2):99–127, 2002.

656. Kenneth O. Stanley and Risto Miikkulainen. Evolving a roving eye for Go. In *Genetic and Evolutionary Computation Conference*, pages 1226–1238. Springer, 2004.
657. Stanley Smith Stevens. On the Theory of Scales of Measurement. *Science*, 103(2684):677–680, 1946.
658. Neil Stewart, Gordon D. A. Brown, and Nick Chater. Absolute identification by relative judgment. *Psychological Review*, 112(4):881, 2005.
659. Andreas Stiegler, Keshav Dahal, Johannes Maucher, and Daniel Livingstone. Symbolic Reasoning for Hearthstone. *IEEE Transactions on Computational Intelligence and AI in Games*, 2017.
660. Jeff Stuckman and Guo-Qiang Zhang. Mastermind is NP-complete. *arXiv preprint cs/0512049*, 2005.
661. Nathan Sturtevant. Memory-Efficient Pathfinding Abstractions. In *AI Programming Wisdom 4*. Charles River Media, 2008.
662. Nathan Sturtevant and Steve Rabin. Canonical orderings on grids. In *Proceedings of the International Joint Conference on Artificial Intelligence*, pages 683–689, 2016.
663. Nathan R. Sturtevant. Benchmarks for grid-based pathfinding. *IEEE Transactions on Computational Intelligence and AI in Games*, 4(2):144–148, 2012.
664. Nathan R. Sturtevant and Richard E. Korf. On pruning techniques for multi-player games. *Proceedings of The National Conference on Artificial Intelligence (AAAI)*, pages 201–208, 2000.
665. Nathan R. Sturtevant, Jason Traish, James Tulip, Tansel Uras, Sven Koenig, Ben Strasser, Adi Botea, Daniel Harabor, and Steve Rabin. The Grid-Based Path Planning Competition: 2014 Entries and Results. In *Eighth Annual Symposium on Combinatorial Search*, pages 241–251, 2015.
666. Adam James Summerville and Michael Mateas. Mystical Tutor: A Magic: The Gathering Design Assistant via Denoising Sequence-to-Sequence Learning. In *Twelfth Artificial Intelligence and Interactive Digital Entertainment Conference*, 2016.
667. Adam James Summerville, Shweta Philip, and Michael Mateas. MCMCTS PCG 4 SMB: Monte Carlo Tree Search to Guide Platformer Level Generation. In *Eleventh Artificial Intelligence and Interactive Digital Entertainment Conference*, 2015.
668. Adam James Summerville, Sam Snodgrass, Matthew Guzdial, Christoffer Holmgård, Amy K. Hoover, Aaron Isaksen, Andy Nealen, and Julian Togelius. Procedural Content Generation via Machine Learning (PCGML). *arXiv preprint arXiv:1702.00539*, 2017.
669. Adam James Summerville, Sam Snodgrass, Michael Mateas, and Santiago Ontañón Villar. The VGLC: The Video Game Level Corpus. *arXiv preprint arXiv:1606.07487*, 2016.
670. Petra Sundström. *Exploring the affective loop*. PhD thesis, Stockholm University, 2005.
671. Ben Sunshine-Hill, Michael Robbins, and Chris Jurney. Off the Beaten Path: Non-Traditional Uses of AI. In *Game Developers Conference, AI Summit*, 2012.
672. Richard S. Sutton and Andrew G. Barto. *Reinforcement learning: An introduction.* MIT Press, 1998.
673. Reid Swanson and Andrew S. Gordon. Say anything: Using textual case-based reasoning to enable open-domain interactive storytelling. *ACM Transactions on Interactive Intelligent Systems (TiiS)*, 2(3):16, 2012.
674. William R. Swartout, Jonathan Gratch, Randall W. Hill Jr, Eduard Hovy, Stacy Marsella, Jeff Rickel, and David Traum. Toward virtual humans. *AI Magazine*, 27(2):96, 2006.
675. Penelope Sweetser, Daniel M. Johnson, and Peta Wyeth. Revisiting the GameFlow model with detailed heuristics. *Journal: Creative Technologies*, 2012(3), 2012.
676. Penelope Sweetser and Janet Wiles. Scripting versus emergence: issues for game developers and players in game environment design. *International Journal of Intelligent Games and Simulations*, 4(1):1–9, 2005.
677. Penelope Sweetser and Janet Wiles. Using cellular automata to facilitate emergence in game environments. In *Proceedings of the 4th International Conference on Entertainment Computing (ICEC05)*, 2005.
678. Penelope Sweetser and Peta Wyeth. GameFlow: a model for evaluating player enjoyment in games. *Computers in Entertainment (CIE)*, 3(3):3–3, 2005.

679. Maciej Świechowski and Jacek Mańdziuk. Self-adaptation of playing strategies in general game playing. *IEEE Transactions on Computational Intelligence and AI in Games*, 6(4):367–381, 2014.
680. Gabriel Synnaeve and Pierre Bessière. Multiscale Bayesian Modeling for RTS Games: An Application to StarCraft AI. *IEEE Transactions on Computational intelligence and AI in Games*, 8(4):338–350, 2016.
681. Gabriel Synnaeve, Nantas Nardelli, Alex Auvolat, Soumith Chintala, Timothée Lacroix, Zeming Lin, Florian Richoux, and Nicolas Usunier. TorchCraft: a Library for Machine Learning Research on Real-Time Strategy Games. *arXiv preprint arXiv:1611.00625*, 2016.
682. Nicolas Szilas. IDtension: a narrative engine for Interactive Drama. In *Proceedings of the Technologies for Interactive Digital Storytelling and Entertainment (TIDSE) Conference*, pages 1–11, 2003.
683. Niels A. Taatgen, Marcia van Oploo, Jos Braaksma, and Jelle Niemantsverdriet. How to construct a believable opponent using cognitive modeling in the game of set. In *Proceedings of the Fifth International Conference on Cognitive Modeling*, pages 201–206, 2003.
684. Nima Taghipour, Ahmad Kardan, and Saeed Shiry Ghidary. Usage-based web recommendations: a reinforcement learning approach. In *Proceedings of the 2007 ACM Conference on Recommender Systems*, pages 113–120. ACM, 2007.
685. Bulent Tastan and Gita Reese Sukthankar. Learning policies for first person shooter games using inverse reinforcement learning. In *Seventh Artificial Intelligence and Interactive Digital Entertainment Conference*, 2011.
686. Shoshannah Tekofsky, Pieter Spronck, Aske Plaat, Jaap van Den Herik, and Jan Broersen. Play style: Showing your age. In *Computational Intelligence in Games (CIG), 2013 IEEE Conference on*. IEEE, 2013.
687. Shoshannah Tekofsky, Pieter Spronck, Aske Plaat, Jaap van den Herik, and Jan Broersen. Psyops: Personality assessment through gaming behavior. In *BNAIC 2013: Proceedings of the 25th Benelux Conference on Artificial Intelligence, Delft, The Netherlands, November 7-8, 2013*, 2013.
688. Gerald Tesauro. Practical issues in temporal difference learning. *Machine learning*, 8(3-4):257–277, 1992.
689. Gerald Tesauro. Temporal difference learning and TD-Gammon. *Communications of the ACM*, 38(3):58–68, 1995.
690. Ruck Thawonmas, Yoshitaka Kashifuji, and Kuan-Ta Chen. Detection of MMORPG bots based on behavior analysis. In *Proceedings of the 2008 International Conference on Advances in Computer Entertainment Technology*, pages 91–94. ACM, 2008.
691. Michael Thielscher. A General Game Description Language for Incomplete Information Games. In *AAAI*, pages 994–999, 2010.
692. William R. Thompson. On the likelihood that one unknown probability exceeds another in view of the evidence of two samples. *Biometrika*, 25(3/4):285–294, 1933.
693. David Thue, Vadim Bulitko, Marcia Spetch, and Eric Wasylishen. Interactive Storytelling: A Player Modelling Approach. In *AIIDE*, pages 43–48, 2007.
694. Christian Thurau, Christian Bauckhage, and Gerhard Sagerer. Learning human-like opponent behavior for interactive computer games. *Pattern Recognition, Lecture Notes in Computer Science 2781*, pages 148–155, 2003.
695. Christian Thurau, Christian Bauckhage, and Gerhard Sagerer. Imitation learning at all levels of game AI. In *Proceedings of the International Conference on Computer Games, Artificial Intelligence, Design and Education*, 2004.
696. Christian Thurau, Christian Bauckhage, and Gerhard Sagerer. Learning human-like Movement Behavior for Computer Games. In S. Schaal, A. Ijspeert, A. Billard, S. Vijayakumar, J. Hallam, and J.-A. Meyer, editors, *From Animals to Animats 8: Proceedings of the Eighth International Conference on Simulation of Adaptive Behavior (SAB-04)*, pages 315–323, Santa Monica, CA, July 2004. The MIT Press.
697. Tim J. W. Tijs, Dirk Brokken, and Wijnand A. Ijsselsteijn. Dynamic game balancing by recognizing affect. In *Proceedings of International Conference on Fun and Games*, pages 88–93. Springer, 2008.

698. Julian Togelius. Evolution of a subsumption architecture neurocontroller. *Journal of Intelligent & Fuzzy Systems*, 15(1):15–20, 2004.
699. Julian Togelius. A procedural critique of deontological reasoning. In *Proceedings of DiGRA*, 2011.
700. Julian Togelius. AI researchers, Video Games are your friends! In *Computational Intelligence*, pages 3–18. Springer, 2015.
701. Julian Togelius. How to run a successful game-based AI competition. *IEEE Transactions on Computational Intelligence and AI in Games*, 8(1):95–100, 2016.
702. Julian Togelius, Alex J. Champandard, Pier Luca Lanzi, Michael Mateas, Ana Paiva, Mike Preuss, and Kenneth O. Stanley. Procedural Content Generation in Games: Goals, Challenges and Actionable Steps. *Dagstuhl Follow-Ups*, 6, 2013.
703. Julian Togelius, Renzo De Nardi, and Simon M. Lucas. Making racing fun through player modeling and track evolution. In *Proceedings of the SAB'06 Workshop on Adaptive Approaches for Optimizing Player Satisfaction in Computer and Physical Games*, 2006.
704. Julian Togelius, Renzo De Nardi, and Simon M. Lucas. Towards automatic personalised content creation for racing games. In *Computational Intelligence and Games, 2007. CIG 2007. IEEE Symposium on*, pages 252–259. IEEE, 2007.
705. Julian Togelius, Sergey Karakovskiy, and Robin Baumgarten. The 2009 Mario AI competition. In *Evolutionary Computation (CEC), 2010 IEEE Congress on*. IEEE, 2010.
706. Julian Togelius, Sergey Karakovskiy, Jan Koutník, and Jürgen Schmidhuber. Super Mario evolution. In *Computational Intelligence and Games, 2009. CIG 2009. IEEE Symposium on*, pages 156–161. IEEE, 2009.
707. Julian Togelius and Simon M. Lucas. Evolving controllers for simulated car racing. In *IEEE Congress on Evolutionary Computation*, pages 1906–1913. IEEE, 2005.
708. Julian Togelius and Simon M. Lucas. Arms races and car races. In *Parallel Problem Solving from Nature-PPSN IX*, pages 613–622. Springer, 2006.
709. Julian Togelius and Simon M. Lucas. Evolving robust and specialized car racing skills. In *IEEE Congress on Evolutionary Computation (CEC)*, pages 1187–1194. IEEE, 2006.
710. Julian Togelius, Simon M. Lucas, Ho Duc Thang, Jonathan M. Garibaldi, Tomoharu Nakashima, Chin Hiong Tan, Itamar Elhanany, Shay Berant, Philip Hingston, Robert M. MacCallum, Thomas Haferlach, Aravind Gowrisankar, and Pete Burrow. The 2007 IEEE CEC Simulated Car Racing Competition. *Genetic Programming and Evolvable Machines*, 9(4):295–329, 2008.
711. Julian Togelius, Mark J. Nelson, and Antonios Liapis. Characteristics of generatable games. In *Proceedings of the Fifth Workshop on Procedural Content Generation in Games*, 2014.
712. Julian Togelius, Mike Preuss, Nicola Beume, Simon Wessing, Johan Hagelbäck, and Georgios N. Yannakakis. Multiobjective exploration of the StarCraft map space. In *Computational Intelligence and Games (CIG), 2010 IEEE Symposium on*, pages 265–272. IEEE, 2010.
713. Julian Togelius, Mike Preuss, and Georgios N. Yannakakis. Towards multiobjective procedural map generation. In *Proceedings of the 2010 Workshop on Procedural Content Generation in Games*. ACM, 2010.
714. Julian Togelius, Tom Schaul, Jürgen Schmidhuber, and Faustino Gomez. Countering poisonous inputs with memetic neuroevolution. In *International Conference on Parallel Problem Solving from Nature*, pages 610–619. Springer, 2008.
715. Julian Togelius, Tom Schaul, Daan Wierstra, Christian Igel, Faustino Gomez, and Jürgen Schmidhuber. Ontogenetic and phylogenetic reinforcement learning. *Künstliche Intelligenz*, 23(3):30–33, 2009.
716. Julian Togelius and Jürgen Schmidhuber. An experiment in automatic game design. In *Computational Intelligence and Games, 2008. CIG'08. IEEE Symposium On*, pages 111–118. IEEE, 2008.
717. Julian Togelius, Noor Shaker, Sergey Karakovskiy, and Georgios N. Yannakakis. The Mario AI championship 2009-2012. *AI Magazine*, 34(3):89–92, 2013.
718. Julian Togelius and Georgios N. Yannakakis. General General Game AI. In *2016 IEEE Conference on Computational Intelligence and Games (CIG)*. IEEE, 2016.

719. Julian Togelius, Georgios N. Yannakakis, Sergey Karakovskiy, and Noor Shaker. Assessing believability. In Philip Hingston, editor, *Believable bots*, pages 215–230. Springer, 2012.
720. Julian Togelius, Georgios N. Yannakakis, Kenneth O. Stanley, and Cameron Browne. Search-based procedural content generation: A taxonomy and survey. *Computational Intelligence and AI in Games, IEEE Transactions on*, 3(3):172–186, 2011.
721. Simone Tognetti, Maurizio Garbarino, Andrea Bonarini, and Matteo Matteucci. Modeling enjoyment preference from physiological responses in a car racing game. In *Computational Intelligence and Games (CIG), 2010 IEEE Symposium on*, pages 321–328. IEEE, 2010.
722. Paul Tozour and I. S. Austin. Building a near-optimal navigation mesh. *AI Game Programming Wisdom*, 1:298–304, 2002.
723. Mike Treanor, Bryan Blackford, Michael Mateas, and Ian Bogost. Game-O-Matic: Generating Videogames that Represent Ideas. In *Procedural Content Generation Workshop at the Foundations of Digital Games Conference*. ACM, 2012.
724. Mike Treanor, Alexander Zook, Mirjam P. Eladhari, Julian Togelius, Gillian Smith, Michael Cook, Tommy Thompson, Brian Magerko, John Levine, and Adam Smith. AI-based game design patterns. 2015.
725. Alan M. Turing. Digital computers applied to games. *Faster than thought*, 101, 1953.
726. Hiroto Udagawa, Tarun Narasimhan, and Shim-Young Lee. Fighting Zombies in Minecraft With Deep Reinforcement Learning. Technical report, Stanford University, 2016.
727. Alfred Ultsch. Data mining and knowledge discovery with emergent self-organizing feature maps for multivariate time series. *Kohonen Maps*, 46:33–46, 1999.
728. Alberto Uriarte and Santiago Ontañón. Automatic learning of combat models for RTS games. In *Eleventh Artificial Intelligence and Interactive Digital Entertainment Conference*, 2015.
729. Nicolas Usunier, Gabriel Synnaeve, Zeming Lin, and Soumith Chintala. Episodic Exploration for Deep Deterministic Policies: An Application to StarCraft Micromanagement Tasks. *arXiv preprint arXiv:1609.02993*, 2016.
730. Josep Valls-Vargas, Santiago Ontañón, and Jichen Zhu. Towards story-based content generation: From plot-points to maps. In *Computational Intelligence in Games (CIG), 2013 IEEE Conference on*. IEEE, 2013.
731. Wouter van den Hoogen, Wijnand A. IJsselsteijn, and Yvonne de Kort. Exploring behavioral expressions of player experience in digital games. In *Proceedings of the Workshop on Facial and Bodily Expression for Control and Adaptation of Games (ECAG)*, pages 11–19, 2008.
732. Roland van der Linden, Ricardo Lopes, and Rafael Bidarra. Procedural generation of dungeons. *Computational Intelligence and AI in Games, IEEE Transactions on*, 6(1):78–89, 2014.
733. Pascal van Hentenryck. *Constraint satisfaction in logic programming*. MIT Press, Cambridge, 1989.
734. Niels van Hoorn, Julian Togelius, and Jürgen Schmidhuber. Hierarchical controller learning in a first-person shooter. In *Computational Intelligence and Games, 2009. CIG 2009. IEEE Symposium on*, pages 294–301. IEEE, 2009.
735. Niels van Hoorn, Julian Togelius, Daan Wierstra, and Jürgen Schmidhuber. Robust player imitation using multiobjective evolution. In *IEEE Congress on Evolutionary Computation (CEC)*, pages 652–659. IEEE, 2009.
736. Giel van Lankveld, Sonny Schreurs, Pieter Spronck, and Jaap van Den Herik. Extraversion in games. In *International Conference on Computers and Games*, pages 263–275. Springer, 2010.
737. Giel van Lankveld, Pieter Spronck, Jaap van den Herik, and Arnoud Arntz. Games as personality profiling tools. In *Computational Intelligence and Games (CIG), 2011 IEEE Conference on*, pages 197–202. IEEE, 2011.
738. Harm van Seijen, Mehdi Fatemi, Joshua Romoff, Romain Laroche, Tavian Barnes, and Jeffrey Tsang. Hybrid Reward Architecture for Reinforcement Learning. *arXiv preprint arXiv:1706.04208*, 2017.
739. Pascal Vincent, Hugo Larochelle, Yoshua Bengio, and Pierre-Antoine Manzagol. Extracting and composing robust features with denoising autoencoders. In *Proceedings of the 25th International Conference on Machine Learning (ICML)*, pages 1096–1103. ACM, 2008.

740. Madhubalan Viswanathan. Measurement of individual differences in preference for numerical information. *Journal of Applied Psychology*, 78(5):741–752, 1993.
741. Thurid Vogt and Elisabeth André. Comparing feature sets for acted and spontaneous speech in view of automatic emotion recognition. In *Proceedings of IEEE International Conference on Multimedia and Expo (ICME)*, pages 474–477. IEEE, 2005.
742. John Von Neumann. The general and logical theory of automata. *Cerebral Mechanisms in Behavior*, 1(41):1–2, 1951.
743. John Von Neumann and Oskar Morgenstern. *Theory of games and economic behavior*. Princeton University Press, 1944.
744. Karol Walędzik and Jacek Mańdziuk. An automatically generated evaluation function in general game playing. *IEEE Transactions on Computational Intelligence and AI in Games*, 6(3):258–270, 2014.
745. Che Wang, Pan Chen, Yuanda Li, Christoffer Holmgård, and Julian Togelius. Portfolio Online Evolution in StarCraft. In *Twelfth Artificial Intelligence and Interactive Digital Entertainment Conference*, 2016.
746. Colin D. Ward and Peter I. Cowling. Monte Carlo search applied to card selection in Magic: The Gathering. In *IEEE Symposium on Computational Intelligence and Games (CIG)*, pages 9–16. IEEE, 2009.
747. Joe H. Ward Jr. Hierarchical grouping to optimize an objective function. *Journal of the American Statistical Association*, 58(301):236–244, 1963.
748. Christopher J. C. H. Watkins and Peter Dayan. Q-learning. *Machine Learning*, 8(3-4):279–292, 1992.
749. Ben G. Weber. ABL versus Behavior Trees. *Gamasutra*, 2012.
750. Ben G. Weber and Michael Mateas. A data mining approach to strategy prediction. In *2009 IEEE Symposium on Computational Intelligence and Games*, pages 140–147. IEEE, 2009.
751. Joseph Weizenbaum. ELIZA—a computer program for the study of natural language communication between man and machine. *Communications of the ACM*, 9(1):36–45, 1966.
752. Paul John Werbos. *Beyond regression: new tools for prediction and analysis in the behavioral sciences*. PhD thesis, Harvard University, 1974.
753. Daan Wierstra, Tom Schaul, Jan Peters, and Juergen Schmidhuber. Natural evolution strategies. In *IEEE Congress on Evolutionary Computation (CEC) 2008. (IEEE World Congress on Computational Intelligence).*, pages 3381–3387. IEEE, 2008.
754. Geraint A. Wiggins. A preliminary framework for description, analysis and comparison of creative systems. *Knowledge-Based Systems*, 19(7):449–458, 2006.
755. Minecraft Wiki. Minecraft. *Mojang AB, Stockholm, Sweden*, 2013.
756. David H. Wolpert and William G. Macready. No free lunch theorems for optimization. *IEEE Transactions on Evolutionary Computation*, 1(1):67–82, 1997.
757. Robert F. Woodbury. Searching for designs: Paradigm and practice. *Building and Environment*, 26(1):61–73, 1991.
758. Steven Woodcock. Game AI: The State of the Industry 2000-2001: It's not Just Art, It's Engineering. *Game Developer Magazine*, 2001.
759. Xindong Wu, Vipin Kumar, J. Ross Quinlan, Joydeep Ghosh, Qiang Yang, Hiroshi Motoda, Geoffrey J. McLachlan, Angus Ng, Bing Liu, S. Yu Philip, Zhi-Hua Zhou, Michael Steinbach, David J. Hand, and Dan Steinberg. Top 10 algorithms in data mining. *Knowledge and Information Systems*, 14(1):1–37, 2008.
760. Kaito Yamamoto, Syunsuke Mizuno, Chun Yin Chu, and Ruck Thawonmas. Deduction of fighting-game countermeasures using the k-nearest neighbor algorithm and a game simulator. In *Computational Intelligence and Games (CIG), 2014 IEEE Conference on*. IEEE, 2014.
761. Yi-Hsuan Yang and Homer H. Chen. Ranking-based emotion recognition for music organization and retrieval. *Audio, Speech, and Language Processing, IEEE Transactions on*, 19(4):762–774, 2011.
762. Georgios N. Yannakakis. *AI in Computer Games: Generating Interesting Interactive Opponents by the use of Evolutionary Computation*. PhD thesis, University of Edinburgh, November 2005.

763. Georgios N. Yannakakis. Preference learning for affective modeling. In *Affective Computing and Intelligent Interaction and Workshops, 2009. ACII 2009. 3rd International Conference on*, pages 1–6. IEEE, 2009.
764. Georgios N. Yannakakis. Game AI revisited. In *Proceedings of the 9th conference on Computing Frontiers*, pages 285–292. ACM, 2012.
765. Georgios N. Yannakakis, Roddy Cowie, and Carlos Busso. The Ordinal Nature of Emotions. In *Affective Computing and Intelligent Interaction (ACII), 2017 International Conference on*, 2017.
766. Georgios N. Yannakakis and John Hallam. Evolving Opponents for Interesting Interactive Computer Games. In S. Schaal, A. Ijspeert, A. Billard, S. Vijayakumar, J. Hallam, and J.-A. Meyer, editors, *From Animals to Animats 8: Proceedings of the* 8th *International Conference on Simulation of Adaptive Behavior (SAB-04)*, pages 499–508, Santa Monica, CA, July 2004. The MIT Press.
767. Georgios N. Yannakakis and John Hallam. A Generic Approach for Generating Interesting Interactive Pac-Man Opponents. In *Proceedings of the IEEE Symposium on Computational Intelligence and Games*, 2005.
768. Georgios N. Yannakakis and John Hallam. A generic approach for obtaining higher entertainment in predator/prey computer games. *Journal of Game Development*, 1(3):23–50, December 2005.
769. Georgios N. Yannakakis and John Hallam. Modeling and augmenting game entertainment through challenge and curiosity. *International Journal on Artificial Intelligence Tools*, 16(06):981–999, 2007.
770. Georgios N. Yannakakis and John Hallam. Towards optimizing entertainment in computer games. *Applied Artificial Intelligence*, 21(10):933–971, 2007.
771. Georgios N. Yannakakis and John Hallam. Entertainment modeling through physiology in physical play. *International Journal of Human-Computer Studies*, 66(10):741–755, 2008.
772. Georgios N. Yannakakis and John Hallam. Real-time game adaptation for optimizing player satisfaction. *IEEE Transactions on Computational Intelligence and AI in Games*, 1(2):121–133, 2009.
773. Georgios N. Yannakakis and John Hallam. Rating vs. preference: A comparative study of self-reporting. In *Affective Computing and Intelligent Interaction*, pages 437–446. Springer, 2011.
774. Georgios N. Yannakakis, Antonios Liapis, and Constantine Alexopoulos. Mixed-initiative co-creativity. In *Proceedings of the 9th Conference on the Foundations of Digital Games*, 2014.
775. Georgios N. Yannakakis, Henrik Hautop Lund, and John Hallam. Modeling children's entertainment in the playware playground. In *2006 IEEE Symposium on Computational Intelligence and Games*, pages 134–141. IEEE, 2006.
776. Georgios N. Yannakakis and Manolis Maragoudakis. Player modeling impact on player's entertainment in computer games. In *Proceedings of International Conference on User Modeling (UM)*. Springer, 2005.
777. Georgios N. Yannakakis and Héctor P. Martínez. Grounding truth via ordinal annotation. In *Affective Computing and Intelligent Interaction (ACII), 2015 International Conference on*, pages 574–580. IEEE, 2015.
778. Georgios N. Yannakakis and Héctor P. Martínez. Ratings are Overrated! *Frontiers in ICT*, 2:13, 2015.
779. Georgios N. Yannakakis, Héctor P. Martínez, and Maurizio Garbarino. Psychophysiology in games. In *Emotion in Games: Theory and Praxis*, pages 119–137. Springer, 2016.
780. Georgios N. Yannakakis, Héctor P. Martínez, and Arnav Jhala. Towards affective camera control in games. *User Modeling and User-Adapted Interaction*, 20(4):313–340, 2010.
781. Georgios N. Yannakakis and Ana Paiva. Emotion in games. *Handbook on Affective Computing*, pages 459–471, 2014.
782. Georgios N. Yannakakis, Pieter Spronck, Daniele Loiacono, and Elisabeth André. Player modeling. *Dagstuhl Follow-Ups*, 6, 2013.

783. Georgios N. Yannakakis and Julian Togelius. Experience-driven procedural content generation. *Affective Computing, IEEE Transactions on*, 2(3):147–161, 2011.
784. Georgios N. Yannakakis and Julian Togelius. Experience-driven procedural content generation. In *Affective Computing and Intelligent Interaction (ACII), 2015 International Conference on*, pages 519–525. IEEE, 2015.
785. Georgios N. Yannakakis and Julian Togelius. A panorama of artificial and computational intelligence in games. *IEEE Transactions on Computational Intelligence and AI in Games*, 7(4):317–335, 2015.
786. Xin Yao. Evolving artificial neural networks. *Proceedings of the IEEE*, 87(9):1423–1447, 1999.
787. Nick Yee. The demographics, motivations, and derived experiences of users of massively multi-user online graphical environments. *Presence: Teleoperators and virtual environments*, 15(3):309–329, 2006.
788. Nick Yee, Nicolas Ducheneaut, Les Nelson, and Peter Likarish. Introverted elves & conscientious gnomes: the expression of personality in World of WarCraft. In *Proceedings of the SIGCHI Conference on Human Factors in Computing Systems*, pages 753–762. ACM, 2011.
789. Serdar Yildirim, Shrikanth Narayanan, and Alexandros Potamianos. Detecting emotional state of a child in a conversational computer game. *Computer Speech & Language*, 25(1):29–44, 2011.
790. Shubu Yoshida, Makoto Ishihara, Taichi Miyazaki, Yuto Nakagawa, Tomohiro Harada, and Ruck Thawonmas. Application of Monte-Carlo tree search in a fighting game AI. In *Consumer Electronics, 2016 IEEE 5th Global Conference on*. IEEE, 2016.
791. David Young. *Learning game AI programming with Lua*. Packt Publishing Ltd, 2014.
792. R. Michael Young, Mark O. Riedl, Mark Branly, Arnav Jhala, R. J. Martin, and C. J. Saretto. An architecture for integrating plan-based behavior generation with interactive game environments. *Journal of Game Development*, 1(1):51–70, 2004.
793. Mohammed J. Zaki. SPADE: An efficient algorithm for mining frequent sequences. *Machine Learning*, 42(1-2):31–60, 2001.
794. Zhihong Zeng, Maja Pantic, Glenn I. Roisman, and Thomas S. Huang. A survey of affect recognition methods: Audio, visual, and spontaneous expressions. *Pattern Analysis and Machine Intelligence, IEEE Transactions on*, 31(1):39–58, 2009.
795. Jiakai Zhang and Kyunghyun Cho. Query-efficient imitation learning for end-to-end autonomous driving. *arXiv preprint arXiv:1605.06450*, 2016.
796. Peng Zhang and Jochen Renz. Qualitative Spatial Representation and Reasoning in Angry Birds: The Extended Rectangle Algebra. In *Proceedings of the Fourteenth International Conference on Principles of Knowledge Representation and Reasoning*, 2014.
797. Martin Zinkevich, Michael Johanson, Michael Bowling, and Carmelo Piccione. Regret minimization in games with incomplete information. In *Advances in Neural Information Processing Systems*, pages 1729–1736, 2008.
798. Albert L. Zobrist. *Feature extraction and representation for pattern recognition and the game of Go*. PhD thesis, The University of Wisconsin, Madison, 1970.
799. Alexander Zook. Game AGI beyond Characters. In *Integrating Cognitive Architectures into Virtual Character Design*, pages 266–293. IGI Global, 2016.
800. Alexander Zook and Mark O. Riedl. A Temporal Data-Driven Player Model for Dynamic Difficulty Adjustment. In *8th AAAI Conference on Artificial Intelligence and Interactive Digital Entertainment*. AAAI, 2012.
801. Robert Zubek and Ian Horswill. Hierarchical Parallel Markov Models of Interaction. In *AIIDE*, pages 141–146, 2005.

Index

G. N. Yannakakis and J. Togelius, *Artificial Intelligence and Games*, https://doi.org/10.1007/978-3-319-63519-4

Printed in the United States
By Bookmasters